T0179802

Efficiency and Productivity Growth

STATISTICS IN PRACTICE

Series Advisory Editors

Marian Scott
University of Glasgow, UK

Stephen Senn
CRP-Santé, Luxembourg

Wolfgang Jank
University of Maryland, USA

Founding Editor

Vic Barnett
Nottingham Trent University, UK

Statistics in Practice is an important international series of texts which provide detailed coverage of statistical concepts, methods and worked case studies in specific fields of investigation and study.

With sound motivation and many worked practical examples, the books show in down-to-earth terms how to select and use an appropriate range of statistical techniques in a particular practical field within each title's special topic area.

The books provide statistical support for professionals and research workers across a range of employment fields and research environments. Subject areas covered include medicine and pharmaceutics; industry, finance and commerce; public services; the earth and environmental sciences, and so on.

The books also provide support to students studying statistical courses applied to the above areas. The demand for graduates to be equipped for the work environment has led to such courses becoming increasingly prevalent at universities and colleges.

It is our aim to present judiciously chosen and well-written workbooks to meet everyday practical needs. Feedback of views from readers will be most valuable to monitor the success of this aim.

A complete list of titles in this series appears at the end of the volume.

Efficiency and Productivity Growth
Modelling in the Financial Services Industry

Edited by

Fotios Pasiouras

University of Surrey, UK and
Technical University of Crete, Greece

A John Wiley & Sons, Ltd., Publication

Library of Congress Cataloging-in-Publication Data

Pasiouras, Fotios.
Efficiency and productivity growth : modelling in the financial services industry / edited by Fotios Pasiouras.
    pages   cm
  Includes bibliographical references and index.
  ISBN 978-1-119-96752-1 (cloth)
1. Banks and banking–Econometric models.   2. Financial services industry–Econometric models.   I. Title.
  HG1601.P29 2013
  332.101′5195–dc23

                                    2012045881

A catalogue record for this book is available from the British Library.

ISBN: 978-1-119-96752-1

Set in 10/12pt Times by SPi Publisher Services, Pondicherry, India
Printed and bound in Singapore by Markono Print Media Pte Ltd

In memory of my best friend, Manos Kerpinis.

He was a remarkable engineer and operations manager, always trying to operate at the frontier.

Contents

Jaap W.B. Bos and James W. Kolari

Claudia Curi and Ana Lozano-Vivas

Preface

In recent years, several studies have used frontier techniques to examine the efficiency of banks and other financial institutions. One could broadly classify the literature in two strands. The first strand consists of studies that use the efficiency frontier techniques to provide answers to important questions in relation to the operation of banks and other financial institutions (e.g. the impact of corporate governance, ownership and environmental conditions on efficiency). The second strand consists of more technical studies that introduce new innovative approaches or compare existing ones in search of the most efficient way to estimate the production frontier.

In both cases, a number of questions remain unanswered, and this book attempts to tackle some of them. Following up from the aforementioned classification, the first seven chapters of this book link efficiency with a variety of topics like Latin American banking, market discipline and governance, economics of scale, off-balance-sheet (OBS) activities, productivity of foreign banks, mergers and acquisitions (M&As), and mutual fund ratings. The next five chapters compare traditional techniques or present applications of techniques that were only recently employed in the bank efficiency literature, including, among others, network data envelopment analysis, multicriteria decision aid and quantile regression.

Despite the plethora of cross-country studies for the European banking market, less is known about the efficiency of the Latin American banking sector. Chapter 1 by Philip Molyneux and Jonathan Williams focuses on the efficiency of banks operating in Argentina, Brazil, Chile and Mexico. This study is of particular interest, not only because it examines these four very important economies but also because it does so by using a unique panel dataset which covers a quarter of a century, ranging from 1985 to 2010. Additionally, the authors use a relatively new approach that deals with the problems associated with firm heterogeneity over time. Molyneux and Williams draw some very interesting conclusions with regards to the trend of efficiency over different sub-periods, the impact of ownership on efficiency and the use of the proposed technique.

Market discipline and whether private agents can be successful in monitoring and disciplining banks has been subject to debate in literature. The interest in the topic was enhanced due to the introduction of the third pillar in Basel II, and several questions were raised during the financial crisis. Market discipline may interact with another mechanism that can be useful in monitoring managerial actions and mitigating agency problems, that of corporate governance. As a number of scholars mention, banks' governance differs from the one of non-financial firms for a number of reasons, and during the recent crisis, several banks were criticized for inadequate governance frameworks. Chapter 2 by Joseph Hughes and Loretta Mester examines the relationship between market discipline, corporate governance and market-value inefficiency. The latter is derived from a market-value frontier and is the difference between the best-practice potential (frontier) value and the noise-adjusted, observed market value of assets as a proportion of the potential value. As such, they deviate from the vast majority of existing studies which examine technical, cost and, more recently, profit efficiency.

Questions about the optimal bank size, relating to both efficiency and stability, have been central in the discussion of both academics and policy makers. Chapter 3 by Robert DeYoung provides a review and critique of the findings of studies on economies of scales written over 50 years by scholars and policy makers. Then, using a basic framework, DeYoung attempts to answer the questions 'how large is too large for banks?' and 'how small is too small for banks?' He concludes that there is bad news and good news. As he explains, the bad news is that there is currently no consensus about whether large banks truly exhibit economies of scale. The good news is that whereas existing studies provide no clear answers about scale economies at the largest banks, they provide very consistent results at the other end of the size spectrum, and they seem to agree that banks with less than about $500 million in assets have access to important economies of scale.

In recent years, OBS activities have become a major part of the banking services. Therefore, it is not surprising that various studies have examined the impact of these activities on risk-taking, productivity and efficiency. However, existing literature has ignored the potential existence of product-specific economies of scale and scope. Chapter 4 by Jaap Bos and James Kolari aims to close this gap in the literature. Using a sample of European banks over the period 1989–2005, they reach two very interesting conclusions. First, OBS activities have a downward sloping average cost curve rather than the U-shaped cost curves as in the cases of bank loans and investments. Second, scope economies between OBS activities and both loans and investments exist but not between loans and investments.

The operation of foreign banks and whether they perform better or worse than domestic banks is another topic that has received considerable attention in the literature. Actually, this question is related to the 'liability of foreigners' discussed in the literature on international business and management, whereas in the case of banking it has resulted in the development of the so-called *home field advantage* and *global advantage* hypotheses. Chapter 5 by Claudia Curi and Ana Lozano-Vivas focuses on the productivity of foreign banks operating in Luxembourg over the period 1995–2010. The authors report various results illustrating among other things the importance of nationality, size and organizational form. However, one of the most interesting findings is that the financial crisis exercised a positive effect on the performance banks, which responded with important improvements in technology, managing to reverse a negative productivity pattern to a positive one.

M&As is a very important phenomenon not only because it is associated with deals that bear a significant monetary value, but also because it has to do with the change of corporate control and the shaping of the structure of the market. Yet, one could easily argue that despite their popularity, many of the M&As fail to deliver the expected outcomes. This could simply reflect the complexity of these deals or that many of them are related to managerial motives (e.g. empire building) rather than to the maximization of shareholders' wealth. Chapter 6 by Franco Fiordelisi and Francesco Saverio Stentella Lopes examines the impact of bank M&As on efficiency and market power. In the first step of their analysis, the authors estimate the Lerner index, making use of the marginal cost that is obtained through an application of stochastic frontier approach. They then run various linear multiple regression models, aiming to explain the change in the Lerner index for banks involved in an M&A deal.

While the banking sector is the largest and probably the most important one in the financial services industry, the market could not operate efficiently without the existence of other important players like insurance firms and mutual funds. Thus, Chapter 7 by

Olivier Brandouy, Kristiaan Kerstens and Ignace Van de Woestyne deviates from the rest of the studies that are included in this book by providing a detailed discussion along with empirical evidence on the backtesting of super-fund portfolio strategies based on frontier-based mutual fund ratings. The authors reach a number of very interesting conclusions. Probably one of the most interesting results is that without portfolio optimization strategies and ignoring transaction costs, frontier mutual fund ratings perform poorly compared to traditional financial performance measures; however, with transaction costs, this finding is reversed.

Assuming that one wants to estimate the efficiency of banks, one must take some decisions relating to, among other things, the sample of the banks, the inputs and outputs, and the frontier technique that will be used. Chapter 8 by Necmi Avkiran starts with a theoretical comparison of the two most commonly used techniques, namely data envelopment analysis and stochastic frontier analysis. Then, Avkiran provides an introduction to network DEA (NDEA), an approach that was only recently applied in banking. To illustrate the concept of NDEA, he uses a numerical example using retail bank branches and simulated data. Thus, in addition to using NDEA, this chapter also departs from other applications in this book by focusing on branches rather than banking institutions as a whole. The chapter continues with a comparison between jackknifing and bootstrapping, and it closes with a discussion of some key issues related to the future of DEA.

Chapter 9 by Hirofumi Fukuyama and William Weber is an interesting read for at least three reasons. First, it is the only application in the book that focuses on the Japanese banking sector, one of the most important sectors in the world, as documented by the frequent inclusion of Japan's big three banks among the world's 30 largest banking institutions. Second, instead of examining commercial banks, it focuses on Shinkin banks, which are cooperative financial institutions. Third, the authors develop a dynamic network production technology to measure productivity change. This network consists of two stages. In the first stage, labour, physical capital and equity capital are used to produce deposits. In the second stage, deposits are used to produce loans and securities investments and an undesirable output of non-performing loans. Those non-performing loans become an undesirable input in the first stage of production during the next period, and as Fukuyama and Weber explain, any increase in non-performing loans requires an increase in first stage inputs, such as equity capital, to offset their effects. Apparently, this setting allows managers to maximize production possibilities over multiple periods.

As discussed in Chapter 8, both DEA and SFA have various advantages and disadvantages, which has resulted in a debate as to which approach should be preferred. While Chapter 8 offered a first insight into the differences between these two approaches, this was not accompanied by empirical evidence. Thus, Chapter 10 by Michael Koetter and Aljar Meesters aims to close this gap by comparing the two approaches on the basis of some predetermined consistency conditions criteria (e.g. efficiency distributions, rank correlations). Among other things, their analysis considers the impact of outliers, the use of different output and input prices, and the construction of common or group-specific frontiers. The authors conclude that cost efficiency measures differ depending on specification choices and the employed technique, and they suggest the use of multiple benchmarking methods.

One of the most attractive features of efficiency analysis is that it summarizes the performance of banks in a single indicator that considers simultaneously various inputs and outputs. Therefore, one can assess how efficient a bank is while taking into account the quality of its loan portfolio (e.g. by using non-performing loan as illustrated in Chapter 9), the portfolio mix (e.g. loans, securities), and the relationship between short-term liabilities (e.g. deposits) and illiquid assets (e.g. loans). Apparently, this offers an advantage over the use of simple financial ratios analysis that may lead to conflicting results. Yet, despite the popularity of frontier techniques in the banking literature, other approaches can also be used in estimating an overall indicator of performance while taking into account opposing managerial objectives. Chapter 11 by Michael Doumpos and Constantin Zopounidis describes the use of such an approach, namely the simulation-based stochastic multicriteria acceptability analysis, while using a large sample of European cooperative banks. The proposed method is based on the aggregation of financial ratios through an additive value function model that provides an overall evaluation and rank of the banks. The authors also discuss the results obtained with the use of DEA, and provide an interesting comparison between the two techniques.

In recent years, quantile regression analysis has been frequently employed in the econometrics literature; however, there are only a few studies in the context of efficiency estimation. Chapter 12 by Anastasia Koutsomanoli-Filippaki, Emmanuel Mamatzakis and Fotios Pasiouras aims to provide an overview of this promising alternative approach, along with an empirical application in a large international dataset. As the authors mention, quantile regression can be particularly useful in the context of efficiency analysis because it is well-suited for efficiency estimations when there is considerable heterogeneity in the firm-level data, making it robust against outliers and the associated problems discussed in Chapter 10. Additionally, quantile regression avoids some criticism against DEA and SFA related to the random error or the imposition of a particular form on the distribution of the inefficient term.

Fotios Pasiouras

Contributors

Necmi K. Avkiran, The University of Queensland, UQ Business School, Australia.

Jaap W.B. Bos, Finance Department, Maastricht University, The Netherlands.

Olivier Brandouy, IAE – Sorbonne Graduate Business School, Université Paris 1, France.

Claudia Curi, School of Economics and Management, Free University of Bolzano, Italy.

Robert DeYoung, KU School of Business, University of Kansas, USA.

Michael Doumpos, Financial Engineering Laboratory, Department of Production Engineering and Management, Technical University of Crete, Greece.

Franco Fiordelisi, Department of Business Studies, University of Rome III, Italy and Bangor Business School, Bangor University, UK.

Hirofumi Fukuyama, Department of Business Management, Faculty of Commerce, Fukuoka University, Japan.

Joseph P. Hughes, Department of Economics, Rutgers University, USA.

Kristiaan Kerstens, CNRS-LEM (UMR 8179), IESEG School of Management, France.

Michael Koetter, Finance Department, Frankfurt School of Finance and Management, Germany.

James W. Kolari, Finance Department, Texas A&M University, USA.

Anastasia Koutsomanoli-Filippaki, Bank of Greece, Greece.

Francesco Saverio Stentella Lopes, Department of Business Studies, University of Rome III, Italy and Finance Department, Tilburg University, The Netherlands.

Ana Lozano-Vivas, Department of Economic Theory and Economic History, University of Malaga, Spain.

Emmanuel Mamatzakis, Department of Business and Management, University of Sussex, UK.

Aljar Meesters, Global Economics and Management Department, Faculty of Economics and Business, University of Groningen, The Netherlands.

Loretta J. Mester, Research Department, Federal Reserve Bank of Philadelphia, USA and Finance Department, The Wharton School, University of Pennsylvania, USA.

Philip Molyneux, Bangor Business School, Bangor University, UK.

Fotios Pasiouras, University of Surrey, UK and Technical University of Crete, Greece.

Ignace Van de Woestyne, Hogeschool-Universiteit Brussel, Belgium.

William L. Weber, Department of Economics and Finance, Southeast Missouri State University, USA.

Jonathan Williams, Bangor Business School, Bangor University, UK.

Constantin Zopounidis, Financial Engineering Laboratory, Department of Production Engineering and Management, Technical University of Crete, Greece.

1

Bank efficiency in Latin America

Philip Molyneux and Jonathan Williams

Bangor Business School, Bangor University, UK

1.1 Introduction

Across Latin America, the period from the mid-1980s is characterised by fundamental
shifts in public policy that have led to a reconfiguration of the industrial structure of
national banking sectors. Policies associated with financial repression, namely interest rate
controls and directed lending, were replaced by liberal policies that sought to increase
competition and bank efficiency. Amendments to entry and exit conditions, the privatisation of state-owned banks and repeal of restrictions on foreign bank entry led to changes in
bank ownership and the reform of governance (Carvalho, Paula and Williams, 2009).
Improvements to bank governance that temper the risk-taking behaviour of bank owners
are expected to lead to increases in bank efficiency (Caprio, Laeven and Levine, 2007;
Laeven and Levine, 2009).

Latin America was badly affected by regional banking crises during the mid-1990s. The
resolution of the crises required extensive government intervention that led to increases in
market concentration (Domanski, 2005). Intervention involved a restructuring process that
included the nationalisation of banks; transfer of ownership to healthy institutions; liquida-
tion of bankrupts; and use of public funds to recapitalise and give liquidity to distressed
banks. As a result, Latin American banking sectors operate under conditions of monopolistic
competition (Gelos and Roldós, 2004). In terms of efficiency, increases in concentration may
stifle competition because concentrated markets lack market discipline, which leads to lower
efficiencies (Berger and Hannan, 1998). In spite of this concern, across the region there was
an implicit assumption that private ownership would lead to a more efficient outcome, especially as public banks had served political and social purposes (Carvalho, Paula and Williams,
2009). Public ownership of banks is a feature of institutional and financial underdevelopment
(La Porta et al., 2002), and the state's share of banking sector assets had been around 45% and
50% in Argentina and Brazil in the early 1990s (Carvalho, Paula and Williams, 2009). At this

Efficiency and Productivity Growth: Modelling in the Financial Services Industry, First Edition. Edited by Fotios Pasiouras.
© 2013 John Wiley & Sons, Ltd. Published 2013 by John Wiley & Sons, Ltd.

time, public ownership of banking sector assets amounted to 100% in Mexico following the 1982 nationalisation in response to the debt crisis (Haber, 2005). According to Ness (2000), public ownership created moral hazards between the government's economic and political goals and bank's business goals, and the relatively large size of public banks conferred a too-big-to-fail status that required frequent use of public funds to support ailing institutions.

To facilitate competition and improve efficiency, and to recapitalise distressed banks, governments repealed restrictions on foreign bank entry. The sale of local banks to foreigners is based on an assumption that private ownership is more effective in resolving agency problems (Megginson, 2005), and foreign banks possess superior management skills and technological capabilities that let them export efficiencies from home to host. Foreign bank entry is expected to boost banking sector efficiency because incumbent domestic banks must improve efficiencies or face losing market share. Operational diseconomies associated with distance from the home headquarters and cultural difference between the home and host countries can raise costs and lessen efficiencies at foreign banks (Berger et al., 2000; Mian, 2006). There is evidence to suggest foreign bank penetration in the post-restructuring period in Latin America did improve competition particularly when more efficient and less risky foreign banks entered the market (Jeon, Olivero and Wu, 2010). Efficiencies may be adversely impacted because foreign banks could 'cherry pick' the best customers and force local banks to service higher risk customers; foreign banks face information constraints and are less effective at monitoring soft information, which suggests credit to the private sector may be lower and certain sectors could face financial exclusion under conditions of increasing foreign bank penetration.

In order to investigate bank inefficiency in Latin America, we use a relatively new approach that deals with the problems associated with firm heterogeneity over time. Bank inefficiency is measured in terms of a bank's deviation from a best-practice frontier that represents the underlying production technology of a banking industry. Best-practice or efficient frontiers can be estimated by parametric and/or non-parametric methods. The most popular approaches are stochastic frontier analysis (Aigner, Lovell and Schmidt, 1977; Meeusen and van den Broeck, 1977) and data envelopment analysis (Farrell, 1957; Banker, Charnes and Cooper, 1984). The results reported later in this chapter apply the former approach to estimate bank efficiency and utilise methodological advances in efficiency modelling to account for an anomaly that can 'seriously distort' estimated inefficiency (Greene, 2005a, 2005b; Bos et al., 2009). As just noted, the anomaly is how to treat cross-firm heterogeneity. Standard panel data approaches confound any time-invariant cross-firm heterogeneity with the inefficiency term. The problem may be resolved using so-called true effects models and random parameters models that are adapted to stochastic frontier analysis. This class of model is attractive because it relaxes the restrictive assumption of a common production technology across firms (Tsionas, 2002).

The remainder of this chapter outlines the bank efficiency literature on Latin America and then briefly presents our results on four systems – Argentina, Brazil, Chile and Mexico – from 1985 to 2010 using modelling approaches that deal with the problem of firm heterogeneity in panel estimations.

1.2   Privatization and foreign banks in Latin America

Bank privatisation and foreign bank penetration altered the market structure of national banking sectors and transformed the governance structure of banks as new, private owners (domestic and foreign) assumed control of banks. Formerly, Argentina and Brazil had extensive state-owned banking sectors, but privatisation offloaded banking sector assets onto the private sector that was expected to manage the assets more efficiently (Carvalho, Paula and Williams, 2009). State-owned banks had served political and social purposes but their characteristics included weak loan quality, underperformance, and poor cost control. Yet, privatisation outcomes are variable. In Argentina and Brazil, privatised bank performance improved post-privatisation (Berger et al., 2005; Nakane and Weintraub, 2005). In contrast, the failed 1991 Mexican bank privatisation programme cost an estimated $65 billion (Haber, 2005). Across the region, foreign banks have acquired large, local banks, many under temporary government control for restructuring. Some evidence finds a positive association between foreign bank penetration and bank efficiency. There are caveats: the need to distinguish between the performance of existing foreign banks and local banks acquired by foreign banks; and to disentangle the effects of foreign bank entry from other liberalisation effects that could impact bank efficiency.

Studies report differences in performance between local, private-owned and foreign-owned banks. Foreign banks achieved higher average loan growth (in Argentina and Chile) with loan growth stronger at existing foreign banks compared to acquired foreign banks. This suggests management at foreign bank acquisitions focused on restructuring their acquisitions and integrating operations with the parent (foreign) bank. The cautious nature of foreign bank strategies explains why foreign banks, and foreign bank acquisitions in particular, achieved better loan quality than local banks (Clarke, Crivelli and Cull, 2005), although stronger provisioning and higher loan recovery rates translated into weaker profitability at foreign banks. Foreign banks are relatively more liquid, rely less on deposit financing and produce stronger loan growth during episodes of financial distress than domestic banks. It is suggested that the greater intermediation efficiency of foreign banks arose because they were more able to evaluate credit risks and allocated resources at a faster pace than their local competitors (Crystal, Dages and Goldberg, 2002).

Evidence from Argentina shows state-owned banks underperformed against private-owned and foreign-owned banks due partly to poor loan quality associated with direct lending and subsidised credit. Bank privatisation produced efficiency gains because of falling non-performing loans and higher profit efficiencies. However, local M&A activity and foreign bank entry exerted little effect on bank performance (Berger et al., 2005). These findings do not generalize to Brazil where foreign banks faced difficulties in adapting to the peculiarities of the Brazilian banking sector, which is dominated by local, private-owned banks (Paula, 2002). The empirical record offers no support to suggest foreign banks are more or less efficient than domestic banks (Guimarães, 2002; Vasconcelos and Fucidji, 2002). This is unsurprising in the light of evidence that the operational characteristics and balance sheets of domestic and foreign banks are similar (Carvalho, 2002). Hence, the expected benefits of foreign bank penetration have been slow to emerge because foreign banks follow operational characteristics similar to large domestic, private-owned banks (Paula and Alves, 2007).

Although foreign bank penetration and foreign banks' share of bank lending are positively related, evidence suggests that foreign banks engage in cherry-picking behaviour. In Argentina and Mexico, foreign banks concentrated lending in the commercial loans market and limited exposure to the household and mortgage sectors (Dages, Goldberg and Kinney, 2000; Paula and Alves, 2007). Foreign bank acquisitions in Argentina used growth in lending to diversify away from manufacturing and target consumer markets. In addition, foreign banks aggressively penetrated regional markets that eliminated concerns over geographic concentration and increased regional lending to offset changes in local banks' lending. Lastly, foreign banks are an important source of finance. Their loan growth is higher (better quality and less volatile) than local (especially state-owned) banks (Dages, Goldberg and Kinney, 2000). Foreign banks – and private local banks – responded to market signals with pro-cyclical lending that is sensitive to movements in GDP and interest rates. Foreign banks' loan growth and lower volatility – even during crisis periods – suggests they can help to stabilise bank credit (Dages, Goldberg and Kinney, 2000).

Whereas policymakers expect consolidation to lead to greater competition and efficiency improvements, there is the possibility that competitive gains would not materialise, and instead bank market power would increase. The latter implies that the evolution of highly concentrated market structures could limit the deepening of financial intermediation and the development of more efficient banking sectors (Rojas Suarez, 2007). Since a non-competitive market structure often produces oligopolistic behaviour by banks, the suggestion is that more consolidation may incentivise banks to exploit market power rather than become more efficient. In general, the literature rejects the notion of collusion between banks, but evidence from Brazil suggests that banks possess some degree of market power (Nakane, Alencar and Kanczuk, 2006). Other Brazilian evidence illustrates the complexities associated with identifying competition effects: whilst the banking sector operates under monopolistic competition, this finding cannot be generalised across ownership and size.

1.3   Methodology

The stochastic frontier production function (see Aigner, Lovell and Schmidt, 1977; Meeusen and van den Broeck, 1977) specifies a two-component error term that separates inefficiency and random error. In the composed error, a symmetric component captures random variation of the frontier across firms, statistical noise, measurement error and exogenous shocks beyond managerial control. The other component is a one-sided variable that measures inefficiency relative to the frontier. In its general form, the stochastic frontier cost function is written as

$$C_{it} = \left(\mathbf{X}_{it}\boldsymbol{\beta} \right) \cdot e^{v_{it}+u_{it}}; \quad i = 1,2,\ldots,N, \quad t = 1,2,\ldots,T, \tag{1.1}$$

where C_{it} is a scalar of the variable cost of bank i in period t; $\mathbf{X}_{it}$ is a vector of known inputs and outputs; $\boldsymbol{\beta}$ is a vector of unknown parameters to be estimated; the v_{it} are independently and identically distributed $N(0,\sigma_v^2)$ random errors that are independently distributed of the u_{it}'s, which are non-negative random variables that account for the cost of inefficiency in production; the u_{it} are assumed to be positive and distributed normally with zero mean and variance σ_u^2.

The total variance is defined as $\sigma^2 = \sigma_v^2 + \sigma_u^2$. The contribution of the error term to the total variation is as follows: $\sigma_v^2 = \sigma^2 / (1+\lambda^2)$. The contribution of the inefficiency term is $\sigma_u^2 = \sigma^2 \lambda^2 / (1+\lambda^2)$. Where σ_v^2 is the variance of the error term v, σ_u^2 is the variance of the inefficiency term u and λ is defined as σ_u/σ_v, providing an indication of the relative contribution of u and v to $\varepsilon = u+v$.

Estimation of Equation (1.1) yields the residual ε_{it}, meaning that the inefficiency term u_{it} must be calculated indirectly. The solution is proposed by Jondrow et al. (1982): the estimator uses the conditional expectation of u_{it}, conditioned on the realised value of the error term

$\varepsilon_{it} = (v_{it} + u_{it})$, as the estimator of u_{it}. In other words, $E|u_{it}/\varepsilon_{it}|$ is the mean inefficiency for the ith bank at time t. The Jondrow et al. (1982) estimator for panel data is shown in Equation (1.2):

$$E\left[u_{it}|\varepsilon_{it}\right] = \frac{\sigma\lambda\left(f\left(a_{it}\right)/\left[1-\Phi\left(a_{it}\right)\right]-a_{it}\right)}{1+\lambda^2}, \quad (1.2)$$

where $\sigma = \left(\sigma_v^2 + \sigma_u^2\right)^{1/2}$, $\lambda = \sigma_u/\sigma_v$, $a_i = \pm\varepsilon_i\lambda/\sigma$, and $\phi(\cdot)$ and $\Phi(\cdot)$ are the density and distribution of the standard normal, respectively; $v_{it} \sim N\left(0,\sigma_v^2\right]$, $u_{it} = |U_{it}|$, $U_{it} \sim N\left[0,\sigma_{ui}^2\right]$ and v_{it} is independent of u_{it}.

The availability of panel datasets boosted developments in the frontier literature. Early approaches modelled inefficiency as time invariant, a very restrictive assumption particularly in long datasets. Later panel data methods removed this limitation. Other challenges remained, including the key issue of how to treat observed and unobserved heterogeneity. Observed heterogeneity can be incorporated into the stochastic frontier cost function by specifying variables such as a time trend and/or other control factors. Their inclusion will affect measured inefficiency, however. Should the variables be specified as arguments in the cost function or as determinants of inefficiency in a second-stage analysis? Whilst arguments exist in both directions, ultimately, the decision is arbitrary.[1]

Unobserved heterogeneity presents more of a challenge. Generally, it enters the stochastic frontier through the form of either fixed or random effects. This approach can confound cross-firm heterogeneity with the inefficiency term that will bias estimated inefficiency. Greene (2005a) and Greene (2005b) solve this problem by extending both fixed effects and random effects models to account for unobserved heterogeneity. The literature refers to them as 'true' effects models.

The research strategy is to estimate alternative specifications of the stochastic frontier cost function in Equation (1.1). In the base case, we estimate a standard panel data cost function that assumes the inefficiency term follows a half-normal distribution with the following features: $U \sim N\left(0,\sigma_u^2\right)$ and $v \sim N\left(0,\sigma_v^2\right)$, where σ_v^2 is constant. This is Model 1. Model 2 is the true fixed effects model (Greene, 2005a, 2005b). An advantage of the effects models over the standard panel data approaches is that the former models relax the restrictive assumption of a common production technology across firms (Tsionas, 2002).

Models 3–6 belong to the random parameters class of models and the estimations we report are based on the general framework developed by Greene (2005a). In the most general cost function specification that is reported later, the coefficients on the linear terms in the output, input and time variables and the constant term are assumed to be random with heterogeneous means. The heterogeneous means of these random coefficients are linear in average asset size. The coefficients on the remaining cost function covariates (the control variables) are assumed to be constant. The log standard deviation of the half-normal distribution that is

---

[1] Alternative estimations approaches include those of Battese and Coelli (1995) (see for example, Pasiouras, Tanna, Zopounidis, 2009; and Lozano-Vivas and Pasiouras, 2010). The Battese and Coelli 1995 approach allows for estimation of inefficiency in a single step while controlling for country differences, and you can also include the same variables (if needed) in both the frontier function and the inefficient term without a problem. Another approach used to deal with cross-country frontier estimation is the use of the meta-frontier where individual country best-practice frontiers are enveloped by meta-frontiers and differences between country frontiers and the meta-frontier are gauged by technology gaps. See Bos and Schmiedel (2007) and Kontolaimou and Tsekouras (2010) for explanation of the meta-frontier approach.

used to define the inefficiency term in the cost function is assumed to be linear in asset size. The coefficient on asset size is assumed to be random with a constant mean. The most general specification is as follows:

$$c_{it} = a_i + \boldsymbol{\beta}_i' \mathbf{x}_{it} + \boldsymbol{\phi}' \mathbf{y}_{it} + v_{it} + u_{it}$$

$$v_{it} \sim N(0, \sigma_v^2), \quad \text{where} \quad \sigma_v^2 \text{ is constant}$$

$$u_{it} = |U_{it}|, \quad \text{where} \quad U_{it} \sim N(0, \sigma_{ui}^2) \quad \text{and} \quad \sigma_{ui} = \sigma_u \exp(\theta_i)$$

$$(\alpha_i \boldsymbol{\beta}_i') = (\overline{\alpha} \overline{\boldsymbol{\beta}}')' + \Delta_{\alpha,\beta} s_i + \Gamma_{\alpha,\beta} (\mathbf{w}_{\alpha i} \mathbf{w}'_{\beta i})'$$

$$\theta_i = \overline{\theta} + \delta_\theta s_i + \gamma_\theta w_{\theta i}, \tag{1.3}$$

where $\mathbf{x}_{it}$ is a (27×1) vector of output, input and time variables; $\mathbf{y}_{it}$ is a (9×1) vector of other cost function covariates; and s_i is the average asset size of bank i. The coefficient vectors are as follows: $(\alpha_i \boldsymbol{\beta}_i')'$ is a (28×1) vector of random coefficients; $(\overline{\alpha} \overline{\boldsymbol{\beta}}')'$ and $\Delta_{\alpha,\beta}$ are (28×1) vectors of (fixed) coefficients; $\Gamma_{\alpha,\beta}$ is a free (4×4) lower-triangular matrix of (fixed) coefficients; $\boldsymbol{\phi}$ is a (9×1) vector of (fixed) coefficients; θ_i is a random coefficient; $\overline{\theta}$ and $\gamma\theta$ are (fixed) coefficients. $(\mathbf{w}_{\alpha i} \mathbf{w}'_{\beta i})'$ is a (28×1) vector of NIID random disturbances, where $w'_{\beta i} = \{w_{\beta j i}\}$ for $j = 1, \ldots, 27$; and $w\theta_i$ is a NIID random disturbance. The individual elements of the coefficient vectors are denoted as follows: $\boldsymbol{\beta}' = \{\beta_{ji}\}$, $\overline{\boldsymbol{\beta}} = \{\overline{\beta}_j\}$ for $j = 1, \ldots, 3$; $\Delta_{\alpha,\beta} = \{\delta_j\}$ for $j = 0, \ldots, 27$; $\boldsymbol{\phi}' = \{\phi_j\}$ for $j = 1, \ldots, 9$; and $\Gamma_{\alpha,\beta} = \{\gamma_{jk}\}$ for $j = 0, \ldots, 27$ and $k = 0, \ldots, j$. The specification of $\Gamma_{\alpha,\beta}$ implies the variances of the random coefficients conditional on s_i are $\text{var}(\alpha_i | s_i) = \gamma_{00}^2 v$, $\text{var}(\beta_{ji} | s_i) = \sum_{k=0}^{i} \gamma_{jk}^2$, $\text{var}(\theta_i | s_i) = \gamma_\theta^2$. The corresponding conditional standard deviations are denoted as $\sigma(\alpha_i | s_i)$ and so on. The specification of $\Gamma_{\alpha,\beta}$ allows for non-zero conditional covariances between the elements of $(\alpha_i \boldsymbol{\beta}_i')'$.

We estimate various restricted versions of Equation (1.3). For a stochastic cost frontier with no random coefficients (Model 2), the parameter restrictions are: $\{\delta_j\} = 0$ for $j = 0, \ldots, 27$; $\{\gamma_{jk}\} = 0$ for $j = 0, \ldots, 27, k = 0, \ldots, j$; and $\overline{\theta} = \delta_\theta = \gamma_\theta = 0$. For random (individual) effects with homogeneous means (Model 3), the restrictions are: $\{\delta_j\} = 0$ for $j = 0, \ldots, 27$; $\{\gamma_{jk}\} = 0$ for $j = 1, \ldots, 27, k = 0, \ldots, j$; and $\overline{\theta} = \delta_\theta = \gamma_\theta = 0$. For random effects and random coefficients on the output, input and time variables with homogeneous means (Model 4), the restrictions are: $\{\delta_j\} = 0$ for $j = 0, \ldots, 27$; and $\overline{\theta} = \delta_\theta = \gamma_\theta = 0$. For random effects and random coefficients on the 28 output, input and time variables with heterogeneous means (Model 5), the restrictions are $\overline{\theta} = \delta_\theta = \gamma_\theta = 0$. For random effects and random coefficients on the 28 output, input and time variables with homogeneous means, and a random coefficient with a heterogeneous mean in the equation for the log standard deviation of the half-normal distribution used to define the inefficiency term (Model 6), the restrictions are $\{\delta_j\} = 0$ for $j = 0, \ldots, 28$. The heterogeneous means of the random coefficients in Models 5 and 6 are linear in average asset size.

The random coefficient stochastic frontier cost function is estimated by maximum simulated likelihood. In the estimation procedure, we use 500 Halton draws to speed up estimation and achieve a satisfactory approximation to the true likelihood function. u_{it} has a half-normal distribution truncated at zero to signify that each bank's cost lies either on or above the cost frontier, and deviations from the frontier are interpreted as evidence of the quality of bank management. The choice of distribution for the inefficiency term is arbitrary and other distributions are employed elsewhere (Greene, 2008). Efficiency analysis is characterised by arbitrary assumptions, and it is not always possible to carry out formal statistical tests between alternatives; for instance, the random coefficient models we estimate are not nested.

1.4   Model specification and data

We model the bank production process using the intermediation approach that assumes banks purchase funds from lenders and transform liabilities into the earning assets demanded by borrowers (Sealey and Lindley, 1977). The underlying cost structure of the banking sector is represented by the translog functional form. A unique feature of this study is the construction of a panel dataset covering over a quarter of a century from 1985 to 2010 for banks from Argentina, Brazil, Chile and Mexico. Financial statements data is sourced from the IBCA and BankScope databases. Data is deflated by national GDP deflators and converted in US$ millions at 2000 prices. The dimension of the dataset is 419 banks and 4571 observations over 26 years. Bank ownership is identified using BankScope, central bank reports, academic papers, newswire services, and bank websites. The macroeconomic data is from the World Bank Financial Indicators and World Economic Outlook databases. Table 1.1 shows the descriptive statistics of the sample banks.

Table 1.1   Descriptive statistics for the stochastic frontier cost function.

Variables[a]	Mean	Std. dev.	Minimum	Maximum
Variable cost ($m)	773.3	2 433.8	0.07	59 790.1
Loans ($m)	1963.9	5 517.8	0.04	83 773.5
Customer deposits ($m)	1943.1	5 655.3	0.02	83 653.6
Other earning assets ($m)	1748.8	5 817.4	0.00	85 888.7
Total assets ($m)	4425.0	13 446.2	2.76	204 730.0
Price of financial capital	0.1777	0.2030	0.0014	1.0789
Price of physical capital	0.8205	0.7816	0.0309	5.0234
Price of labour	0.0304	0.0233	0.0005	0.1222
Equity-to-assets[b]	0.0943	0.0214	0.0333	0.2330
Z score (rolling four years)[b]	20.028	12.742	3.280	81.144
Herfindahl index	1144.8	770.0	584.3	7591.4
Loan loss reserves-to-loans[b]	0.1263	0.0938	0.0113	0.3628
Diversification index[b,c]	0.3554	0.0748	0.1151	0.4716
GDP per capita ($m)	5228.1	2 211.9	2606.4	10 418.1
GDP growth	0.0333	0.0400	−0.1089	0.1228
CR – bank credit-to-GDP	0.6303	0.3285	0.2248	2.1292
SO – state-owned assets/ total assets	0.1363	0.3431	0.0000	1.0000

[a]The data is expressed as ratios unless otherwise indicated.
[b]The data is weighted annual averages where the weight is the share of bank i in total assets in country j at time t.
[c]The diversification index is calculated for bank income as in Sanya and Wolfe (2011).

We employ stochastic frontier cost function and translog functional form methodologies to estimate cost inefficiency. The cost function specifies three outputs in value terms (loans, deposits and other earning assets) and three inputs expressed as prices (the prices of financial capital, physical capital and labour). The specification of customer deposits as an output is a contentious issue in the literature. We take the view that customers purchase deposit accounts for the services that they offer, such as cheque clearing, record keeping and safe keeping. Customers do not pay for these services explicitly and banks must incur implicit costs, such as labour and fixed capital costs, in the absence of a direct revenue stream. Fixler and Zieschang (1992, p. 223) suggest banks cover these costs by setting lending rates in excess of deposit rates and propose that 'deposits ... are simultaneously an input into the loan process and an output, in the sense that they are purchased as a final product providing financial services'. Berger and Humphrey (1992) treat deposits as an output because of the large share of bank added value that they generate.

The cost function by construction assumes a common production technology across banks. This assumption is unrealistic given the rate of technological progress over such a long time period. Our cost function is common to banks from four countries and we should account for the effect of cross-country differences as well as inter-temporal differences on bank cost. We control for inter-temporal variation in cost by specifying a time trend, its quadratic term (T^2) and interaction terms between time and outputs, and time and inputs. The sum of the estimated coefficients on the time variables measures the effect of technical change in production on bank cost.[2] We control for the impact of cross-country differences on bank cost by specifying a vector of banking sector and economic variables at country level.

To mitigate potential endogeneity issues, we construct weighted annual averages of four banking sector variables to proxy for underlying conditions, where the weight is the share of bank i in total assets in country j at time t. The variables are as follows:

- The ratio of equity-to-assets (ETA) or capitalisation that is positively associated with prudence or risk aversion. We expect capitalisation is positively related to stability because better capitalized banks are less susceptible to losses arising from unanticipated shocks.

- The Z score (Z) is constructed for each bank as $Z = RoA + ETA/\sigma_{RoA}$ which combines a performance measure (RoA, return on assets), a volatility measure to capture risk (σ_{RoA}) over a four-year rolling window and book capital (ETA, equity-to-assets) as a proxy for soundness or prudence of bank management. Z is expressed in units of standard deviation of RoA and shows the extent to which earnings can be depleted until the bank has insufficient equity to absorb further losses. Lower values of Z imply a greater probability of bankruptcy with larger values implying stability. Our measure of Z is the natural logarithm of Z plus 100.

- It is common to control for differences in the risk appetite of management across banks by specifying variables like the stock of loan loss reserves (LLR)-to-gross loans to proxy asset quality. However, this variable is not strictly exogenous if managers are inefficient at portfolio management or skimp on controlling costs. Hence, we use the

---

[2] An alternative approach specifies fixed time effects using dummy variables to control for the impact of changes in bank regulations and other government policies upon bank cost.

weighted annual average to proxy the underlying level of risk facing the banking sector (Berger and Mester, 1997).

- We measure income diversification (DIV) using a Herfindahl type index that is calculated as $\sum_{i=1}^{n}\left(X_i / Q\right)^2$ where the X variables are net interest revenue and net non-interest income and Q is the sum of X (Acharya, Hasan and Saunders, 2006). Income diversification is proxy for a bank's business model (Fiordelisi, Marques-Ibanez and Molyneux, 2011). The literature focuses on establishing the benefits of diversification in terms of reducing the potential for systemic risk (Demsetz and Strahan, 1997), though the empirical evidence on this point is mixed (Stiroh and Rumble, 2006). The expected relationship between diversification and bank cost is less clear cut although some studies find an inverse link.

- The Herfindahl–Hirschman index of assets concentration in each country by year is specified to control for the effects of increases in market concentration on bank cost. Under the franchise value hypothesis, there is less incentive for banks to assume unnecessary risks in more concentrated markets.

- The natural logarithm of GDP per capita is a proxy for country-level wealth effects.

- We capture business cycle effects by the annual growth in GDP (GDPCHA).

- The ratio of banking sector credit-to-GDP indicates financial deepening, which Levine (2005) suggests is important in exerting corporate governance on bank borrowers. Incremental credit provision requires further screening and monitoring costs for banks that could reduce cost efficiencies.

- The ratio of state-owned bank assets-to-banking sector assets in country j at time t is proxy for the level of financial repression. State-ownership is reported to result in poorly developed banks (Barth, Caprio and Levine, 2001) and less cost efficient banks (Megginson, 2005). State-owned banks may face a soft budget constraint, which implies that incentives for managers to behave in a cost-minimising manner are absent (Altunbas, Evans and Molyneux, 2001). Hence, the underperformance of state-owned banks is correlated with the level of government involvement and the perverse incentives of political bureaucrats (Cornett et al., 2010).

The stochastic frontier cost function is written in Equation (1.4) as

$$
\begin{aligned}
\ln\left(\frac{VC}{P_3}\right) = {} & (\alpha + w_i) + \sum_{i=1}^{3}\beta_i \ln\left(Q_i\right) + \sum_{k=1}^{3}\varphi_l \ln\left(\frac{P_k}{P_3}\right) \\
& + \frac{1}{2}\left[\sum_{i=1}^{3}\sum_{j=1}^{3}\theta_{ij}\ln\left(Q_i\right)\ln\left(Q_j\right) + \sum_{k=1}^{3}\sum_{l=1}^{3}\varphi_{kl}\ln\left(P_k\right)\ln\left(P_l\right)\right] + \sum_{ik=1}^{3}\Omega_{ik}\ln\left(Q_i\right)\ln\left(\frac{P_k}{P_3}\right) \\
& + \kappa_t T + \frac{1}{2}\kappa_{tt}T^2 + \sum_{i=1}^{3}\rho_{ti}\ln\left(Q_i\right)*T + \sum_{k=1}^{3}\varsigma_{tk}\ln\left(P_k\right)*T + \eta_k\sum_{k=1}^{9}\text{controls}_{kt} + \ln\varepsilon_c + \ln\mu_c,
\end{aligned}
$$

(1.4)

where

$\ln(VC/P_3)$ is the natural logarithm of variable cost (the sum of interest paid, personnel expense and non-interest expense) normalised by P_3;

ln Q_i is the natural logarithm of i output values (loans, deposits and other earning assets);

ln P_k is the natural logarithm of k input prices (P_1 is the price of financial capital, that is the ratio of interest paid-to-purchased funds; P_2 is the price of physical capital, that is the ratio of non-interest expenses-to-fixed assets; and P_3 is the price of labour, that is the ratio of personnel expense-to-total assets); and

T is a time trend where 1985 is equal to 1...2010 is equal to 26.

Controls comprise a vector of the following variables:

ETA is the weighted annual average of the ratio of equity-to-assets to proxy capitalisation.

Z is the weighted annual average of the Z score expressed as the log of Z4 plus 100.

HHI is the natural logarithm of the Herfindahl–Hirschman index of total assets.

LLR is the weighted annual average of the ratio of loan loss reserves-to-gross loans.

DIV is the weighted annual average of the diversification index.

GDP is the natural logarithm of real gross domestic product per capita.

GDPCHA is the rate of annual GDP growth.

CR is the ratio of bank credit to the private sector-to-GDP.

SO is the ratio of state-owned bank assets-to-banking sector assets.

ε_i are identical and independently distributed random variables, which are independent of the μ_i, which are non-negative random variables that are assumed to account for inefficiency.

$\alpha, \beta, \psi, \theta, \phi, \Omega, \kappa, \rho, \varsigma$ and η are the parameters to be estimated using maximum likelihood methods.

Standard restrictions of linear homogeneity in input prices and symmetry of the second-order parameters are imposed on the cost function. Whilst the cost function must be non-increasing and convex with regard to the level of fixed input and non-decreasing and concave with regard to prices of the variable inputs, these conditions are not imposed, but may be inspected to determine whether the cost function is well-behaved at each point within a given dataset.

1.5   Estimated parameters and cost efficiency

We begin by considering the estimated coefficients of the stochastic frontier cost functions. In line with expectations, the coefficients on the output and input terms are significantly positive across the different specifications. In the standard panel data model, estimations of the inefficiency term are much larger compared with estimates obtained from 'true' random effects and random parameter models, which suggests that the standard model confounds heterogeneity and inefficiency. For example, in the true random effects and random parameter models, σ_u is more than half the size of the corresponding estimate in the standard model. This key finding, here, therefore suggests that failing to take account of firm heterogeneity results in efficiency underestimation – it also suggest that virtually all the previous evidence on bank efficiency is biased.

The dispersion in mean cost inefficiency that is reported in the literature suggests that different estimation techniques – samples and time periods – have an important bearing on measured inefficiency (Bauer et al., 1998). The distributional properties of the estimated cost efficiencies from each model are shown in Table 1.2. The mean cost efficiency for the true

Table 1.2 Descriptive statistics: variable cost efficiency by model (U_{it} = half-normal distribution).

Model number and type	Mean	Std. dev.	Minimum	Maximum	Skewness	Kurtosis
(1) Standard panel (SP) data model	0.7441	0.1172	0.4108	0.9760	−0.936	3.756
(2) True fixed effects	0.5577	0.0928	0.0579	0.9412	−1.046	6.719
(3) True random effects	0.8696	0.0395	0.3734	0.9776	−4.094	34.783
(4) Random parameters	0.8526	0.0627	0.2137	0.9802	−3.668	25.726
(5) RP heterogeneity in means of RPs	0.8547	0.0611	0.2344	0.9805	−3.510	25.200
(6) RP heterogeneity in variance of U_{it}	0.8300	0.0802	0.2281	0.9860	−1.950	9.690

RP refers to random parameter models.

Table 1.3 Spearman rank-order correlations of variable cost efficiency.

Model number and type	(1) Standard panel	(2) Fixed effects	(3) Random effects	(4) Random parameters	(5) RPM heterogeneity 1
(1) Standard panel					
(2) Fixed effects	0.6400				
(3) Random effects	0.6891	0.8150			
(4) Random parameters	0.5626	0.6772	0.7984		
(5) RPM heterogeneity 1	0.5531	0.6593	0.7862	0.8299	
(6) RPM heterogeneity 2	0.8262	0.4917	0.5793	0.4801	0.4777

(1) All coefficients are significant at the 1% level.

random effects and random parameter models ranges from 83% to 87%. The mean cost efficiency drawn from the standard model is just over 74% and its standard deviation is up to two times larger than the comparative figures for the random parameter models. The 'true' fixed effects model yields the lowest mean cost efficiency at less than 56%, and though standard deviation is less than the standard model, it is larger than in the random parameter models.

Bauer et al. (1998) suggest that measured efficiencies derived from alternative approaches should comply with a set of consistency conditions, such as efficiency should have comparable means, standard deviations and other distributional properties, and the different approaches should rank the banks in approximately the same order. Spearman rank-order correlation coefficients test the null of independence between two sets of rank efficiencies. After ranking cost efficiencies, we calculate the Spearman correlation coefficients for each pair of ranks and present the results in Table 1.3. The random parameter models produce the highest rank-order correlations. The highest correlations are between the 'true' random effects and random parameter models. Whereas each correlation coefficient is statistically significant at the 1% level, Table 1.3 demonstrates the variation in the size of the coefficients.

In this section, we report cost efficiency estimates drawn from the random parameters model that allows for heterogeneity to enter the means of the random parameters. We convert

Table 1.4   Rank cost efficiency by country, 1985–2010.

Year	Argentina	Brazil	Chile	Mexico	Year	Argentina	Brazil	Chile	Mexico
1985	0.2645	0.3701	0.3882	0.7882	1998	0.5117	0.5101	0.4617	0.4306
1986	0.9446	0.5779	0.5367	0.5595	1999	0.5194	0.4930	0.6295	0.3222
1987	0.8099	0.2956	0.4570	0.2891	2000	0.4235	0.5235	0.6330	0.3829
1988	0.1469	0.5028	0.7632	0.4013	2001	0.6451	0.5198	0.4653	0.3845
1989	0.0981	0.3199	0.5471	0.5582	2002	0.6596	0.4963	0.5898	0.5169
1990	0.0547	0.6525	0.5570	0.7015	2003	0.5771	0.5403	0.5556	0.6067
1991	0.8096	0.6006	0.6585	0.6434	2004	0.5663	0.4860	0.5717	0.5678
1992	0.6358	0.4812	0.5594	0.5860	2005	0.5988	0.5127	0.5261	0.5447
1993	0.5292	0.5012	0.5180	0.5556	2006	0.4803	0.4573	0.4520	0.5055
1994	0.4280	0.5548	0.4417	0.5567	2007	0.4420	0.4678	0.4985	0.4294
1995	0.6038	0.4346	0.4042	0.5549	2008	0.3728	0.4008	0.5367	0.2468
1996	0.5525	0.4118	0.3585	0.4613	2009	0.4111	0.5703	0.5483	0.4683
1997	0.4738	0.3983	0.3744	0.5326	2010	0.3941	0.5711	0.4866	0.6688

the cost efficiencies into average rank-order efficiencies following Berger, Hasan and Klapper (2004). The rank efficiencies are interpreted as follows. In Table 1.3, the average rank cost efficiency of Argentine banks is 0.2645 in 1985. It means that the average Argentine bank is more cost efficient than 26% of banks operating in Latin American over 1985–2010. As a test of robustness, we calculated rank cost efficiency at country level and correlated the ranks with the regional ranking. The correlation coefficient exceeds 0.9.

Table 1.4 shows average rank-order cost efficiency by country and year. The period from the mid-1980s to the early 1990s reveals a greater variation in mean cost efficiency, which is to be expected given the ongoing troubles associated with the international debt crisis. Nevertheless, if we compute average rank cost efficiency for 1985–1993, the value for each country is greater than the comparative data for 1994–2000 that covers the period of regional banking crises. For instance, for Mexico the averages are 0.5814 and 0.4573, respectively. We compute average rank cost efficiency for 2001–2006 and 2007–2010 to disentangle the impact of the global banking crisis upon regional cost efficiency. Average rank cost efficiency improves in each country between 1994–2000 and 2001–2006. The biggest improvement is in Argentina from 0.5042 to 0.5900, which made the average Argentine bank more cost efficient than 59% of Latin American banks during this period. However, the impact of the global crisis sees average rank cost efficiency decrease by around 31% in Argentina and 20% in Mexico (from 0.5900 to 0.4065, and 0.5177 to 0.4130, respectively). On the contrary, average rank cost efficiencies held up relatively well in Brazil and Chile with only minor reductions in comparison to the previous period: for Brazil, the mean rank cost efficiency equals 0.4909 in 2007–2010 and it is 0.5180 in Chile.

In Table 1.5 and Table 1.6, we present the average rank cost efficiencies by bank ownership in each country and by year. Table 1.5 shows private-owned domestic banks as well as state banks, and Table 1.6 shows the efficiency performance of foreign-owned banks divided into *de novo* entrants and entry made via mergers and acquisitions (M&A), respectively. Using the same sub-period classification to discuss the results, the average rank cost efficiency of state-owned banks fell between 1985–1993 and 1994–2000 (except in Chile) from 0.5332 to 0.4988 in Argentina, from 0.5245 to 0.4597 in Brazil and 0.6068 to 0.5224 in Mexico. Whilst this feature is unsurprising given the incidence of banking sector crises over

Table 1.5 Rank cost efficiency by country, 1985–2010 – state- and private-owned banks.

Year	Argentina	Brazil	Chile	Mexico	Year	Argentina	Brazil	Chile	Mexico
State-owned banks									
1985	0.8333	0.2381	0.2213	0.7882	1998	0.5828	0.4492	0.7103	—
1986	0.7318	0.5655	0.3470	0.5595	1999	0.4990	0.6515	0.7502	0.4631
1987	0.8077	0.2908	0.3758	0.2891	2000	0.3919	0.6443	0.7696	0.3916
1988	0.5049	0.6411	0.4844	0.4013	2001	0.5493	0.5981	0.6841	0.9005
1989	0.4771	0.3404	0.8289	0.6054	2002	0.5618	0.4460	0.8606	0.2590
1990	0.5896	0.7516	0.6655	0.7154	2003	0.6143	0.5673	0.6515	0.6165
1991	0.6248	0.5530	0.6471	0.7329	2004	0.4984	0.6391	0.5649	—
1992	0.5043	0.6114	0.7206	0.7633	2005	0.5866	0.6218	0.4603	—
1993	0.4871	0.4611	0.7852	0.6488	2006	0.4771	0.5259	0.2660	0.3540
1994	0.4444	0.4844	0.7311	0.7970	2007	0.4659	0.4815	0.1483	—
1995	0.5277	0.4425	0.6257	0.4868	2008	0.4597	0.5147	0.1755	0.6213
1996	0.5170	0.3700	0.4975	0.1989	2009	0.4901	0.6482	0.3448	0.4756
1997	0.3796	0.3349	0.5749	—	2010	0.4311	0.6207	0.2964	—
Private-owned banks									
1985	0.1507	0.4237	0.5218	—	1998	0.4879	0.5178	0.3860	0.3634
1986	0.9750	0.5802	0.5604	—	1999	0.4926	0.4697	0.7010	0.2391
1987	0.8101	0.3015	0.4672	—	2000	0.4220	0.5236	0.6327	0.3521
1988	0.0447	0.4019	0.7589	—	2001	0.6834	0.4570	0.5314	0.4199
1989	0.0666	0.2622	0.6208	0.3008	2002	0.7781	0.4949	0.5893	0.6111
1990	0.0190	0.6174	0.5976	0.2689	2003	0.5711	0.5639	0.5511	0.6610
1991	0.8482	0.6333	0.6428	0.6217	2004	0.5771	0.4963	0.4932	0.5847
1992	0.6768	0.4269	0.4991	0.6062	2005	0.5976	0.4782	0.5376	0.5862
1993	0.5300	0.5526	0.5043	0.5622	2006	0.4425	0.4574	0.4381	0.5243
1994	0.4510	0.5756	0.3633	0.5241	2007	0.3977	0.4536	0.4068	0.4063
1995	0.6530	0.4578	0.3857	0.4291	2008	0.3308	0.3673	0.6744	0.2382
1996	0.5619	0.4208	0.3178	0.4088	2009	0.3501	0.5281	0.6360	0.4449
1997	0.4864	0.4450	0.4220	0.4536	2010	0.3770	0.5632	0.5495	0.5471

1994–2000, we find that this pattern of performance is not common to private-owned banks in Argentina and Brazil for whom average rank cost efficiency is relatively stable at around 0.51 and 0.48 in each sub-period. In contrast, the reduction in average rank cost efficiency for private-owned banks in Mexico deteriorated by 31% compared to 14% for state-owned banks. In Mexico, the cost efficiency performance of banks remains relatively constant in the two subsequent sub-periods. In Brazil, rank cost efficiency improves for state-owned banks over 2001–2006 and it remains relatively constant during the 2007–2010 sub-period at 0.5607. Whereas state-owned bank efficiency performance in Argentina improves over 2001–2006 (to 0.5491), it deteriorates by 15% to 0.4662 in 2007–2010. Across the region, the cost efficiency performance of private-owned improves between 1994–2000 and 2001–2006: in ascending order, by 41% in Mexico, 18% in Argentina, 15% in Chile and less than 2% in Brazil. Only in Chile did private-owned banks improve cost efficiency performance over 2007–2010 (by 8%); cost efficiency deteriorated mildly in Brazil (by 6%) and substantially in Argentina and Mexico (by 40% and 33%).

Table 1.6 Rank cost efficiency by country; 1985–2010 – foreign-owned banks (*de novo* entry and via acquisition).

Year	Argentina	Brazil	Chile	Mexico	Year	Argentina	Brazil	Chile	Mexico
Foreign-owned banks (*de novo* entry)									
1985	—	0.2553	—	—	1998	0.5265	0.4580	0.4972	0.5430
1986	—	0.5977	—	—	1999	0.5223	0.4850	0.5351	0.4522
1987	—	0.2387	—	—	2000	0.3996	0.5045	0.7280	0.5042
1988	—	0.6474	0.8403	—	2001	0.6254	0.6738	0.3243	0.3454
1989	—	0.4870	0.3649	0.0120	2002	0.5989	0.5655	0.5581	0.4199
1990	—	0.6665	0.4546	0.8963	2003	0.5265	0.4845	0.5522	0.3622
1991	0.3292	0.5385	0.6320	0.0333	2004	0.6097	0.3860	0.6341	0.4408
1992	0.3250	0.5149	0.6816	0.0654	2005	0.5986	0.5374	0.4398	0.4180
1993	0.5677	0.3892	0.5361	0.2507	2006	0.5628	0.3880	0.4298	0.4189
1994	0.3642	0.5575	0.4818	0.7937	2007	0.5274	0.4326	0.6829	0.3290
1995	0.5862	0.3588	0.4187	0.8270	2008	0.4094	0.4363	0.3467	0.2504
1996	0.5493	0.4146	0.4074	0.5488	2009	0.4616	0.6333	0.4824	0.4097
1997	0.4925	0.3297	0.3384	0.6474	2010	0.3551	0.6154	0.4938	0.6681
Foreign-owned banks (acquisitions)									
1985	—	—	—	—	1998	0.5042	0.6542	0.4773	0.0807
1986	—	—	—	—	1999	0.6133	0.4972	0.6701	0.0665
1987	—	—	—	—	2000	0.5157	0.4885	0.4356	0.1255
1988	—	—	0.9438	—	2001	0.6634	0.3824	0.5150	0.2935
1989	—	—	0.2186	—	2002	0.5275	0.4233	0.5790	0.4896
1990	—	—	0.4332	—	2003	0.6965	0.5070	0.5509	0.7917
1991	—	—	0.8325	—	2004	0.5452	0.4715	0.6646	0.6661
1992	—	—	0.5658	—	2005	0.6435	0.5187	0.6861	0.5817
1993	—	—	0.4004	—	2006	0.4552	0.5396	0.5776	0.5941
1994	—	—	0.4472	—	2007	0.3975	0.6099	0.4792	0.6079
1995	—	—	0.3241	0.2719	2008	0.2973	0.3870	0.5805	0.2024
1996	—	—	0.2196	0.5110	2009	0.4198	0.5402	0.4927	0.5629
1997	0.4784	0.3312	0.3009	0.4229	2010	0.4416	0.4835	0.4190	0.8216

Table 1.6 shows the average rank cost efficiency of foreign-owned banks segmented by entry status, that is, *de novo* entrants and entrants through M&A mainly following the bank restructuring programmes of the mid-1990s. Some patterns of performance we describe earlier for state and private-owned banks are visible for *de novo* entrants; namely, the deterioration in performance between 1985–1993 and 1994–2000 (except Mexico). In Argentina, Brazil and Chile, average cost efficiency improves in 2001–2006 and remains relatively stable over 2007–2010 in Brazil and Chile, whereas it deteriorates by 22% in Argentina. In each country, except Mexico, the average *de novo* foreign entrant is more cost efficient than the average M&A entrant.

Foreign acquisitions of local banks take off over 1994–2000. Comparing the average rank cost efficiency of this cohort for 1994–2000 and 2001–2006 we observe improvements in performance in Argentina (by 10%), Chile (by 37%) and Mexico (by 185%), whilst a decline occurs in Brazil (by 11%). It is interesting to consider if foreign-acquired banks' cost efficiency performance held up during the 2007–2010 sub-period. Only in Brazil did foreign-acquired banks improve average rank cost efficiency (by 9%): performance falls of the magnitude of 36%, 17% and 9% are recorded in Argentina, Chile and Mexico, respectively.

1.6   Conclusion

This chapter contributes to the bank efficiency literature through its application of recently developed effects models for stochastic frontier analysis. We estimate several variants of this class of model including fixed and random effects models, and alternative specifications of random parameters models that accommodate heterogeneity in different ways. We find that estimated mean efficiency drawn from the effects models is greater and arguably more precise because heterogeneity is not confounded with inefficiency. Or to put another way, previous studies on bank cost efficiency provide underestimates of the 'true' efficiency of banking markets if they have used panel data and not controlled for firm heterogeneity.

We then chose one of these random effects and parameters model to analyse the evolution of average rank cost efficiency for Latin American banks between 1985 and 2010 and show the results by country, year and bank ownership. Here we find that bank cost efficiency generally deteriorated between 1985–1993 and 1994–2000 and this was particularly pronounced for state-owned institutions. The period up to 2006 experienced widespread foreign bank expansion in the region, reflecting a strengthened economic operating environment with cost efficiency generally improving over this period (even for state-owned Brazilian and Argentine banks). Since the 2007 crisis, cost efficiency appears to have either stabilised (foreign-owned banks) or mildly fallen (private banks). An interesting feature of our findings is that *de novo* foreign bank entry appears to be more cost efficient compared to entry via M&A.

References

Acharya, V., Hasan, I. and Saunders, A. (2006) Should banks be diversified? Evidence from individual bank loan portfolios. *Journal of Business*, **79** (3), 1355–1412.

Aigner, D.J., Lovell, C.A.K. and Schmidt, P. (1977) Formulation and estimation of stochastic frontier production function models. *Journal of Econometrics*, **6** (1), 21–37.

Altunbas, Y., Evans, L. and Molyneux, P. (2001) Bank ownership and efficiency. *Journal of Money, Credit and Banking*, **33** (4), 926–954.

Banker, R.D., Charnes, A. and Cooper, W.W. (1984) Some models for estimating technical and scale inefficiencies in data envelopment analysis. *Management Science*, **30** (9), 1078–1092.

Barth, J.R., Caprio, G. Jr and Levine, R. (2001) Banking systems around the globe: do deregulation and ownership affect performance and stability? in *Prudential Supervision: What Works and What Doesn't?* National Bureau of Economic Research Conference Report Series (ed. F. Mishkin), University of Chicago Press, Chicago, pp. 31–96.

Battese, G.E. and Coelli, T.J. (1995) A model for technical inefficiency effects in a stochastic frontier production for panel data. *Empirical Economics*, **20** (2), 325–332.

Bauer, P.W., Berger, A.N., Ferrier, G.D. and Humphrey, D.B. (1998) Consistency conditions for regulatory analysis of financial institutions: a comparison of frontier efficiency methods. *Journal of Economics and Business*, **50** (2), 85–114.

Berger, A.N. and Hannan, T.H. (1998) The efficiency cost of market power in the banking industry: a test of the 'quiet life' and related hypotheses. *The Review of Economics and Statistics*, **80** (3), 454–465.

Berger, A.N. and Humphrey, D.B. (1992) Measurement and efficiency issues in commercial banking, in *Output Measurement in the Service Sectors* (ed. Z. Griliches), National Bureau of Economic Research, University of Chicago Press, Chicago, pp. 245–279.

Berger, A.N. and Mester, L.J. (1997) Inside the black box: what explains differences in the efficiencies of financial institutions? *Journal of Banking and Finance*, **21** (7), 895–947.

Berger, A.N., DeYoung, R., Genay, H. and Udell, G.F. (2000) Globalisation of financial institutions: evidence from cross-border banking performance. *Brookings-Wharton Papers on Financial Services*, **3**, 23–158.

Berger, A.N., Hasan, I. and Klapper, L. (2004) Further evidence on the link between finance and growth: an international analysis of community banking and economic performance. *Journal of Financial Services Research*, **25** (2–3), 169–202.

Berger, A.N., Clarke, G.R.G., Cull, R. et al. (2005) Corporate governance and bank performance: a joint analysis of the static, selection, and dynamic effects of domestic, foreign, and state ownership. *Journal of Banking and Finance*, **29** (8–9), 2179–2221.

Bos, J.W.B. and Schmiedel, H. (2007) Is there a single frontier in a single European banking market? *Journal of Banking and Finance*, **31** (7), 2081–2102.

Bos, J.W.B., Koetter, M., Kolari, J.W. and Kool, C.J.M. (2009) Effects of heterogeneity on bank efficiency scores. *European Journal of Operational Research*, **195** (1), 251–261.

Caprio, G., Laeven, L. and Levine, R. (2007) Governance and bank valuation. *Journal of Financial Intermediation*, **16** (4), 584–617.

Carvalho, F.J.C. (2002) The recent expansion of foreign banks in Brazil: first results. *Latin American Business Review*, **3** (4), 93–119.

Carvalho, F.J.C., Paula, L.F. and Williams, J. (2009) Banking in Latin America, in *The Oxford Handbook of Banking* (eds A.N. Berger, P. Molyneux, and J. Wilson), Oxford University Press, Oxford, pp. 868–902.

Clarke, G.R.G., Crivelli, J.M. and Cull, R. (2005) The direct and indirect impact of bank privatization and foreign entry on access to credit in Argentina's provinces. *Journal of Banking and Finance*, **29** (1), 5–29.

Cornett, M.M., Guo, L., Khaksari, S. and Tehranian, H. (2010) Performance differences in privately-owned versus state-owned banks: an international comparison. *Journal of Financial Intermediation*, **19** (1), 74–94.

Crystal, J.S., Dages, B.G. and Goldberg, L. (2002) Has foreign bank entry led to sounder banks in Latin America? *Current Issues in Economics and Finance*, **8** (1), 1–6.

Dages, B.G., Goldberg, L. and Kinney, D. (2000) Foreign and domestic bank participation in emerging markets: lessons from Mexico and Argentina. Federal Bank of New York. Economic Policy Review (Sep), pp. 17–36.

Demsetz, R. and Strahan, E. (1997) Diversification, size, and risk at bank holding companies. *Journal of Money, Credit and Banking*, **29** (3), 300–313.

Domanski, D. (2005) Foreign banks in emerging market economies: changing players, changing issues. Bank for International Settlements Quarterly Review (Dec), pp. 69–81.

Farrell, M.J. (1957) The measurement of productive efficiency. *Journal of the Royal Statistical Society, Series A*, **120** (3), 253–290.

Fiordelisi, F., Marques-Ibanez, D. and Molyneux, P. (2011) Efficiency and risk in European banking. *Journal of Banking and Finance*, **35** (5), 1315–1326.

Fixler, D.J. and Zieschang, K.D. (1992) User costs, shadow prices and the real output of banks, in *Output Measurement in the Service Sectors* (ed. Z. Griliches), National Bureau of Economic Research, University of Chicago Press, Chicago, pp. 218–243.

Gelos, R.G. and Roldós, J. (2004) Consolidation and market structure in emerging market banking systems. *Emerging Markets Review*, **5** (1), 39–59.

Greene, W. (2005a) Reconsidering heterogeneity in panel data estimators of the stochastic frontier model. *Journal of Econometrics*, **126** (2), 269–303.

Greene, W. (2005b) Fixed and random effects in stochastic frontier models. *Journal of Productivity Analysis*, **23** (1), 7–32.

Greene, W.M. (2008) The econometric approach to efficiency analysis, in *The Measurement of Productive Efficiency: Techniques and Applications* (eds H.O. Fried, C.A.K. Lovell and P. Schmidt), Oxford University Press, Oxford, pp. 92–251.

Guimarães, P. (2002) How does foreign entry affect domestic banking market? The Brazilian case. *Latin American Business Review*, **3** (4), 121–140.

Haber, S. (2005) Mexico's experiments with bank privatization and liberalization, 1991–2003. *Journal of Banking and Finance*, **29** (8–9), 2325–2353.

Jeon, B.N., Olivero, M.P. and Wu, J. (2011) Do foreign banks increase competition? Evidence from emerging Asian and Latin American banking markets. *Journal of Banking and Finance*, **35** (4), 856–875.

Jondrow, J., Lovell, C.A.K., Materov, I.S. and Schmidt, P. (1982) On estimation of technical inefficiency in the stochastic frontier production function model. *Journal of Econometrics*, **19** (2–3), 233–238.

Kontolaimou, A. and Tsekouras, K. (2010) Are cooperatives the weakest link in European banking? A non-parametric metafrontier approach. *Journal of Banking and Finance*, **34** (8), 1946–1957.

La Porta, R., Lopez-de-Silanes, F., Shleifer, A. and Vishny, R. (2002) Investor protection and corporate valuation. *Journal of Finance*, **57**, 1147–1170.

Laeven, L. and Levine, R. (2009) Bank governance, regulation and risk taking. *Journal of Financial Economics*, **93** (2), 259–275.

Lozano-Vivas, A. and Pasiouras, F. (2010) The impact of non-traditional activities on the estimation of bank efficiency: international evidence. *Journal of Banking and Finance*, **34** (7), 1436–1449.

Meeusen, W. and van den Broeck, J. (1977) Efficiency estimation from a Cobb-Douglas production function with composed error. *International Economic Review*, **18** (2), 435–444.

Megginson, W. (2005) The economics of bank privatisation. *Journal of Banking and Finance*, **29** (8–9), 1931–1980.

Mian, A. (2006) Distance constraints: the limits of foreign lending in poor economies. *Journal of Finance*, **61** (3), 1465–1505.

Nakane, M.I. and Weintraub, D.B. (2005) Bank privatization and productivity: evidence for Brazil. *Journal of Banking and Finance*, **29** (8–9), 2259–2289.

Nakane, M.I., Alencar, L.S. and Kanczuk, F. (2006) Demand for bank services and market power in Brazilian banking. Banco Central do Brasil working paper series 107, June.

Ness, W.L. (2000) Reducing government bank presence in the Brazilian financial system: why and how. *The Quarterly Review of Economics and Finance*, **40** (1), 71–84.

Pasiouras, F., Tanna, S. and Zopounidis, C. (2009) The impact of banking regulations on banks' cost and profit efficiency: cross-country evidence. *International Review of Financial Analysis*, **18** (5), 294–302.

Paula, L.F. (2002) Expansion strategies of European banks to Brazil and their impacts on the Brazilian banking sector. *Latin American Business Review*, **3** (4), 59–91.

Paula, L.F. and Alves, A.J., Jr (2007) The determinants and effects of foreign bank entry in Argentina and Brazil: a comparative analysis. *Investigación Económica*, **66** (259), 63–102.

Rojas-Suarez, L. (2007) The provision of banking services in Latin America: obstacles and recommendations. Center for Global Development, Washington, DC, Center for Global Development working paper no. 124, June.

Sanya, S. and Wolfe, S. (2011) Can banks in emerging economies benefit from revenue diversification? *Journal of Financial Services Research*, **40** (1), 79–101.

Sealey, C. and Lindley, J.T. (1977) Inputs, outputs and a theory of production and cost at depository financial institution. *Journal of Finance*, **32** (4), 1251–1266.

Stiroh, J. and Rumble, A. (2006) The dark side of diversification: the case of US financial holding companies. *Journal of Banking and Finance*, **30** (8), 2131–2161.

Tsionas, E. (2002) Stochastic frontier models with random coefficients. *Journal of Applied Econometrics*, **17** (2), 127–147.

Vasconcelos, M.R. and Fucidji, J.R. (2002) Foreign entry and efficiency: evidence from the Brazilian banking industry. State University of Maringá, Brazil.

2

A primer on market discipline and governance of financial institutions for those in a state of shocked disbelief*

Joseph P. Hughes[1] and Loretta J. Mester[2]

[1] *Department of Economics, Rutgers University, USA*
[2] *Research Department, Federal Reserve Bank of Philadelphia, USA and Finance Department, The Wharton School, University of Pennsylvania, USA*

Except where market discipline is undermined by moral hazard, for example, because of federal guarantees of private debt, private regulation generally has proved far better at constraining excessive risk-taking than has government regulation.
—Alan Greenspan, former Federal Reserve Board Chairman, in a speech to the Forty-First Annual Conference on Bank Structure at the Federal Reserve Bank of Chicago, May 2005

When the music stops, in terms of liquidity, things will be complicated. But as long as the music is playing, you've got to get up and dance. We're still dancing.
—Charles O. Prince, former CEO and Chairman of Citigroup, in an interview by Nakamoto and Wighton in the *Financial Times*, July 2007

...those of us who have looked to the self-interest of lending institutions to protect shareholders' equity (myself especially) are in a state of shocked disbelief.
—Alan Greenspan, former Federal Reserve Board Chairman, in testimony to the House Committee on Oversight and Government Reform, October 2008

---

*The views expressed here are those of the authors and do not necessarily reflect those of the Federal Reserve Bank of Philadelphia or of the Federal Reserve System.

2.1   Introduction

Self-regulation encouraged by market discipline constitutes a key component of the third pillar of Basel II. As implied by the third pillar, markets are thought to punish the banks that imprudently take risk and reward those that do not. As former Federal Reserve Board Chairman Greenspan suggested, market discipline has traditionally been thought to enhance managerial performance and shareholder wealth, and to constrain excessive risk-taking. Empirical research has generally confirmed that, where market discipline is not impeded by managerial entrenchment, it has promoted efficiency and enhanced value. However, the comment of former Citigroup CEO Charles Prince, 'we're still dancing', and his worry about the liquidity problems that will arise when the music stops suggest that high-risk investment strategies may maximize the *expected* value of some banks. If so, does market discipline in these cases encourage risk-taking that erodes the stability of banks in economic downturns? And, what are the sources of these risk-taking incentives?

Compared with nonfinancial firms, commercial banks face unique risk-taking incentives. Marcus (1984) shows that regulatory limitations on entry and the mispriced federal safety net create dichotomous incentives for risk-taking. For banks with valuable investment opportunities, protecting their charters from episodes of financial distress by pursuing relatively less risky investment strategies maximizes their expected value. On the other hand, for banks with less valuable investment opportunities, say, because they operate in very competitive markets, exploiting the cost-of-funds subsidy due to implicit and explicit deposit insurance by pursuing relatively more risky investment strategies maximizes their expected value. For this latter type of financial institution, market discipline encourages risk-taking and may work against financial stability.

Managers with substantial undiversified investments of human capital and ownership stakes in their firms and managers who enjoy substantial private benefits of control may protect their advantages by avoiding higher risk investment strategies. However, diversified outside owners may prefer that managers pursue these risky investments.[1] When they own enough of the firm to overcome managerial resistance, they can induce managers to adopt higher risk strategies that tend to maximize expected value. And, there are a variety of other sources of discipline, internal as well as external to the firm, that can ameliorate agency problems and improve individual firm performance, but not necessarily the stability of the financial system as a whole.

Competition among firms is thought to be one source of discipline. Many studies find an important role for competition in promoting efficiency. Competition among firms in markets for products and services enhances managerial efficiency (Berger and Hannan, 1998). The efficiency of competitive labor markets in banking appears sufficient to distinguish poor managerial performance from poor firm performance and to hold senior managers accountable (Cannella, Fraser, and Lee, 1995). Moreover, relaxation of restrictions on interstate banking in recent years has increased competition in the market for corporate control and led to improved performance among underperforming banks whose management is not entrenched by means of higher insider ownership, lower outside block ownership, or less

---

[1] Gorton and Rosen (1995), however, show that the conflict between managers' and equity holders' risk-taking incentives also depends on the investment opportunities facing the bank. In an environment of declining investment opportunities, when bank managers receive private benefits of control and outside shareholders cannot perfectly control them, managers will tend to take on excessive risk. In contrast, when the industry has increasing investment opportunities, managers act too conservatively.

independent boards (Brook, Hendershott, and Lee, 1998). In banking, however, competition may be a two-edged sword: it can change risk-taking incentives, which then flows through to performance. For example, Keeley (1990) found that competition reduces the value of banks' charters and creates risk-taking incentives. Grossman (1992) offered evidence that cost-of-funds subsidies that result from mispriced deposit insurance as well as lax supervision encourage bank risk-taking.

Ownership of stock by officers and directors can align the interests of insiders with that of outside owners (Jensen and Meckling, 1976; Fama and Jensen, 1983), but it can also entrench insiders and lead to poorer performance (Morck, Shleifer, and Vishny, 1988). Similarly, ownership of stock by blockholders whose economic stake in the firm is large enough to overcome free-rider problems can improve monitoring of insiders and, consequently, better align the interests of insiders with those of outside owners. However, when high-risk investment strategies maximize expected value, the influence of blockholders can increase bank risk-taking and threaten banking system stability in troubled economic times (Laeven and Levine, 2009).

In addition to market sources of discipline, arrangements internal to the firm can also ameliorate agency problems and improve performance. The board of directors monitors management, sets compensation for senior managers, and hires and fires the CEO. The board can structure managerial compensation contracts so that they lessen agency conflicts between managers and outside stakeholders. However, boards themselves may have agency conflicts and fail to put the optimal compensation structure in place. Core, Holthausen, and Larcker (1999) find that CEOs of firms with weaker governance structures earn higher compensation and that their firms perform worse than those with stronger governance. In contrast, Cheng, Hong, and Scheinkman (2010) find that financial institutions with institutional investors often provide unusually large compensation incentives to adopt high-risk investment strategies that typically perform well above average in good economic times and well below average in poor economic times.

Market discipline and internal governance interact with banking regulations and supervision to influence the performance and stability of banks. The components of market discipline and internal governance in the context of regulation are considered in the sections that follow. Section 2.2 describes a variety of techniques for assessing bank performance that are found in the literature on discipline and governance. Section 2.3 considers Chairman Greenspan's caveat on private regulation: how public regulation and the federal safety net interact with market discipline to influence risk-taking incentives and bank stability. Section 2.4 examines sources of market discipline: ownership structure, capital market discipline, product market competition, labor market competition, boards of directors, and compensation. Section 2.5 concludes.

2.2   Assessing the performance of financial institutions

Investigations into the relationship of banks' financial performance to sources of market discipline and governance arrangements use accounting data and data on market value.[2] Accounting data permits the construction of various measures of historical cost and profit. Unlike accounting data, market-value data includes the market's valuation of expected *future* cash flows as well as current cash flow. The market's calculation of present value also

---

[2] Hughes and Mester (2010) provide a more detailed discussion of measuring performance in banking.

contains its evaluation of a firm's discount rate – that is, its exposure to market-priced risk. Thus, performance measured by market value offers two advantages over accounting data: the evaluation of market-priced risk and future expected earnings. While some studies seek to evaluate banking performance in terms of quantities of inputs used to produce the outputs, the focus on quantities rather than value makes incorporating risk into the analysis extremely difficult. In fact, many studies that use accounting data as well as production data often ignore risk and reach misleading conclusions. This is explained further in Berger and Mester (1997), Hughes, Mester, and Moon (2001), and Hughes and Mester (2010). Consequently, we do not review the quantity-based approach.

Bank performance can be measured using either a structural or a nonstructural approach. Let y_i represent the measure of the ith bank's performance. Let z_i be a vector of variables that represent components of the ith bank's technology such as output levels and input prices. Let τ_i be a vector of variables affecting the technology, such as the number of branches and measures of asset quality. A number of studies reviewed later in the chapter include a vector, θ_i, which characterizes the property rights system, contracting, and regulatory environment in which the ith firm operates. This vector can include the characteristics of deposit insurance and legal protection of investors. In addition, the organizational form and characteristics of market discipline and governance of the ith firm are included in a vector, φ_i, which might include the degree of market concentration, the status of the firm as a mutual or stock-owned firm, the size of its board of directors, and the proportion of the bank's outstanding shares owned by officers and directors.

Letting ε_i represent random error, the performance equation to be estimated takes the form,

$$y_i = f\left(z_i, \tau_i, \varphi_i, \theta_i \mid \beta\right) + \varepsilon_i. \tag{2.1}$$

The nonstructural approach specifies the performance equation in terms of either an accounting measure of performance, such as return on assets, or a measure based on market value, such as Tobin's q ratio or cumulative abnormal return from an event-study model. It is less likely to focus on a detailed vector of input prices, output levels, or output prices in the specification of the vector z_i. Instead, it might consider how performance is related to the degree of market discipline and the quality of governance.

In contrast, the structural model incorporates an optimization assumption, such as profit maximization, cost minimization, or utility maximization. In the structural model of cost minimization, the vector z_i characterizes the outputs banks produce and the prices of inputs used in bank production. In addition, the vector τ_i might include the level of equity capital and various measures of asset quality, such as the ratio of nonperforming loans to total assets. In measuring performance, the structural model is usually estimated as a frontier – a lower envelope in the case of cost and an upper envelope in the case of profit. Various parametric and nonparametric techniques have been developed to identify the best-practice frontier.[3] The difference between the best-practice frontier and the observed practice represents, in the

---

[3] Berger and Mester (1997) discuss several of these techniques and point out the advantages of the parametric techniques, such as stochastic frontier estimation and the distribution-free approach, over nonparametric techniques like data envelopment analysis. The nonparametric techniques typically focus on technological optimization rather than economic optimization. Since they generally ignore prices, the nonparametric methods can account only for technical inefficiency in using too many inputs or producing too few outputs and cannot account for allocative inefficiency in which firms inefficiently choose inputs and outputs given their relative prices.

case of the cost function, excessive cost relative to best practice and, in the case of the profit function, lost profit relative to best practice. Having estimated cost or profit efficiency from the structural model, studies typically regress the efficiency estimate on a set of explanatory variables that could include measures of market discipline and the quality of governance.

When profit and cost are estimated by a frontier technique, the goal is to measure best-practice technology and the failure to achieve it. Since the frontiers that are estimated are obtained by minimizing cost and maximizing profit, they fit the data for these best practices and, thus, do not provide a theoretical model to explain the inefficient behavior in the data captured by the frontiers. In a series of papers – Hughes et al. (1996), Hughes et al. (2000), Hughes and Mester (2010) and Hughes and Mester (2011), and Hughes, Mester, and Moon (2001) – the authors develop and estimate a model of managerial utility maximization that is sufficiently general to subsume profit maximization and cost minimization and, more generally, managerial objectives that trade profit for other objectives, such as risk reduction and the consumption of agency goods. Thus, the objective function that yields the equations they estimate allows for agency problems: it explains each bank's utility-maximizing expected return and return risk. It is a behavioral model that explains inefficiency. To estimate the inefficiency present in the data, the authors fit a stochastic frontier of expected return to return risk. The frontier yields the best-practice risk versus expected-return frontier and each bank's lost return at its estimated risk exposure.

The utility-maximizing expected return and return risk are estimated from a structural model of managerial behavior that allows for risk versus expected-return inefficiency; however, the frontier is fitted as a nonstructural model and estimates the degree of inefficiency of each bank's predicted return given its return risk. Thus, this approach represents a hybrid of the standard model in which a minimum cost or maximum profit function is fitted as a best-practice frontier. Finally, the estimated inefficiency is explained by estimating Equation (2.1) with the lost return as the dependent variable.

This nonstructural approach specifies the performance equation in terms of an accounting measure of performance; however, the accounting measure was ultimately derived from a structural model of banking. Other structural models, as noted earlier in the chapter, might include other accounting measures of inefficiency derived directly from a maximum profit function or a minimum cost function fitted as a frontier.

Hughes and Moon (2003) develop a structural model of managerial behavior to explain the market value managers produce and, given their firm's potential value, the market value they fail to produce, a measure of agency costs. Managers' choice of the value they produce and the value they consume as agency goods maximizes their utility. The authors use this framework to derive a utility-maximizing managerial demand function for agency goods (inefficiency) and apply the structural properties of utility-maximizing demand to decompose the effect of ownership changes into substitution and wealth effects.

Many nonstructural models simply begin their specification of Equation (2.1) with either an accounting measure of performance, such as return on assets, return on equity, or the ratio of noninterest expense to total expense, which gauges operating cost efficiency. Alternatively, they may use a measure of performance derived from the market value of assets, such as Tobin's q ratio, the market value of assets divided by the replacement cost of assets. Tobin's q ratio, which is commonly proxied by the market value of equity plus the book value of liabilities divided by the book value of assets, measures how much market value is created from a particular investment in assets. For example, Morck, Shleifer, and Vishny (1988) regress Tobin's q on the proportion of outstanding shares owned by officers and directors to look for evidence that ownership aligns the interests of insiders with those of outside owners.

Hughes et al. (1997) proposed using the stochastic frontier technique to measure the highest potential value of banks' assets across all markets in which they operate. This technique was also used in Hughes et al. (2003). The difference between a bank's potential and achieved values, as a proportion of its potential value, represents the bank's market-value efficiency. The stochastic frontier technique eliminates the influence of statistical noise and estimates the systematic failure to achieve potential value. Appendix 2.A describes the technique in more detail.

Other nonstructural models that gauge performance from market value rely on the Sharpe ratio (the ratio of the firm's expected excess return over the risk-free return to the standard deviation of the excess return $= (R - R_f) / (\sigma_{R-R_f})$ and on event studies, which investigate the response of the market's valuation of banks when an unanticipated event occurs. An asset pricing model separates the systematic movement of a stock's price from the unexplained 'abnormal' return. Summed over the event window, the cumulative abnormal return, or CAR, is then regressed on factors thought to explain it. For example, Brook, Hendershott, and Lee (1998) considered the reaction of bank stock prices to the passage of the Interstate Banking and Branching Efficiency Act and identified a statistically significant positive CAR. They hypothesized that the act would increase the probability of takeovers for inefficient banks. When they regressed the CAR of each bank in their sample on banks' performance and ownership structure, they found that underperforming banks whose management was least entrenched received the strongest price reaction.

2.3   Market discipline, public regulation, and the federal safety net

Market discipline interacts with banking regulations and supervision to influence the performance of banks and the stability of the banking system. The federal safety net seeks to promote banking system stability. The formal safety net guarantees payments on Fedwire, the large-value payments system, and deposits up to $250 000. The informal safety net applied to institutions considered too big to fail provides an implicit guarantee of formally uninsured liabilities of commercial banks. Both the formal and informal safety nets imply that the depositor and creditor discipline of bank management will be significantly eroded. In addition, to obtain a charter to gain entry into commercial banking markets, a start-up bank must demonstrate that its management is experienced and that it commences operation with adequate capitalization. Limitations on entering commercial banking markets through the chartering process promote bank safety, but they also create market power for banks in some local markets. Market power is especially valuable when markets are experiencing economic growth. These regulatory features of banking, explicit and implicit deposit insurance and restrictions on entry, create contrasting incentives for risk-taking and value maximization. In the case of banks for which high-risk investment strategies maximize shareholder value, market discipline that promotes value maximization can threaten the stability of the banking system.

Marcus (1984) shows that value-maximizing banks face dichotomous incentives for risk-taking that result from regulatory limitations on entry and from the mispriced federal safety net. He finds that banks with valuable investment opportunities, say, because they operate with market power in growing markets, protect their charters from episodes of financial distress by pursuing relatively less risky investment strategies to maximize the expected value of their assets. On the other hand, banks with less valuable investment opportunities, say, because they operate in very competitive markets, exploit the cost-of-funds subsidy due to implicit and explicit deposit insurance by pursuing relatively more risky investment strategies to maximize the expected value of their assets. For the latter financial institutions, market discipline encourages risk-taking, which may work against financial stability.

Keeley (1990) provides evidence that the liberalization of a number of regulatory restrictions on banking has increased the competition banks face and has caused the value of their charters to fall. In turn, the falling charter values have encouraged bank risk-taking as investment strategies that protect charter value have lost value relative to those that exploit the cost-of-funds subsidy of the federal safety net. Grossman (1992) finds that thrift institutions in the United States adopted more risky investment strategies after securing deposit insurance.

Using two measures of financial performance based on the market value of assets, Tobin's q and an efficiency measure equal to the ratio of achieved market value to the potential market value estimated by a stochastic frontier, Hughes et al. (1997) find evidence of the dichotomous investment strategies Marcus described. They find that high-leverage banks could improve financial performance by lowering equity and low-leverage banks could improve performance by raising their equity ratio. Banks in the third of their sample with the highest capital ratios appear to have exhausted the gains from increasing the capital ratio, while those in the middle third can still improve their q ratio and market-value efficiency by increasing the capital ratio. Banks in the lowest third, though, improve their q ratio and market-value efficiency by reducing their capital ratio. Banks with lower capital ratios, they find, tend to have lower valued investment opportunities and tend to be larger. Thus, among the larger financial institutions in their sample, value enhancement tends to be associated with riskier investment strategies.

A number of studies find that the most profitable banks before the recent financial crisis, which took more risks, were the least profitable during the crisis when the risks led to unexpected losses (e.g., Beltratti and Stulz, 2009; Cheng, Hong, and Scheinkman, 2010). These banks tended to be the largest financial institutions, including Bear Stearns, Citigroup, and AIG. Is a higher-risk investment strategy in a bank's self-interest? These strategies appear to maximize *expected* value, which is realized in good economic conditions.

Other important aspects of bank regulation create additional differences between the market discipline of financial and nonfinancial firms. Banks cannot be owned by nonfinancial firms, and mergers of banks are subject to restrictions and must be approved by the Federal Reserve Board. Until the passage of the Interstate Banking and Branching Efficiency Act in 1994, the McFadden Act and the Douglas Amendment of the Bank Holding Company Act had put banks under the branching laws of the state in which they were chartered. Until 1975, states had prevented out-of-state banks from purchasing in-state banks. Beginning in 1975 in Maine and then in 1982 in Massachusetts, states began to relax some of these restrictions in limited ways. The Interstate Banking and Branching Efficiency Act essentially repealed the McFadden Act and the Douglas Amendment to allow banks to merge across state lines. Thus, the passage of this legislation significantly increased the number of potential buyers of a bank in a takeover and increased the takeover threat faced by all but the most entrenched managers. In short, restrictions on ownership of banks, regulatory approval of mergers, and historical restrictions on mergers across state lines have significantly limited the threat of takeover as a disciplining mechanism of management and have meant that other sources of market discipline would be more important in banking. However, with the passage of the Interstate

Banking and Branching Efficiency Act in 1994, as Brook, Hendershott, and Lee (1998) show, takeover discipline has improved.

Another important difference in the discipline faced by banks is the regulation and supervision imposed on them by federal and state law. Bank operations are much more thoroughly regulated than most nonfinancial firms. Rather than simply focus on protecting shareholder value, regulation promotes bank safety and soundness. For banks with valuable investment opportunities, such regulation may be in the interests of the shareholders, since lower-risk investment strategies maximize value at these banks. In the case of institutions with poorer investment opportunities, safety and soundness regulation may conflict with the adoption of higher-risk strategies that maximize the value of these banks. The evidence of dichotomous strategies to maximize value found by Hughes et al. (1997) suggests that the least levered banks have exploited all the efficiency gains from their capital structure, while the most levered banks have unexploited gains to increasing their leverage further. They hypothesize that regulation prevents these banks from doing so. Most of these banks are very large.

Safety and soundness, of course, are enhanced by efficient management, especially of risk. DeYoung, Hughes, and Moon (2001) find evidence in banks' regulatory assessments, their CAMEL ratings, that bank examiners appear to take bank efficiency into account in assigning ratings. Using the risk versus expected-return frontier developed by Hughes et al. (1996) and Hughes et al. (2000), they find that regulators treat the risk-taking of efficient banks differently than the risk-taking of inefficient banks and afford efficient banks more latitude in their investment strategies than inefficient banks. While their US data are drawn from 1994 and may not shed much light on the years preceding the crisis that began in 2007, it does suggest that supervisors hold large inefficient banks (i.e., banks whose safety and soundness is most likely to have implications for the stability of the banking system) to higher standards than large efficient banks.

Barth, Caprio, and Levine (2004) and Barth, Caprio, and Levine (2006) provide comprehensive studies of banks in 107 countries and in over 150 banks, respectively, which assess the relationship between several aspects of bank regulation and supervision and bank performance, efficiency, and stability. The authors surveyed the banks during 2003–2004 and created a large database chronicling multiple aspects of bank regulation, supervision, structure, and performance. The general conclusion from these studies is that market-based discipline, as opposed to government supervision, results in better banking performance along a variety of dimensions.

Other cross-country studies include Pasiouras (2008); Pasiouras, Tanna, and Zopounidis (2009); Delis, Moyneux, and Pasiouras (2011); and Chortareas, Girardone, and Ventouri (forthcoming). These studies generally support the finding that market discipline can improve bank performance, although there are mixed results on whether certain forms of supervision and regulations, for example, capital requirements or restrictions on bank activities, do. Using the survey data from Barth, Caprio, and Levine (2006), Pasiouras (2008) studies the relationship between bank efficiency and bank supervision and regulation using a sample of 715 banks in 95 countries. He estimates technical and scale efficiency using data envelopment analysis and then performs Tobit regressions of efficiency on measures of regulations related to capital adequacy, the degree of private monitoring, bank activities, deposit insurance, the power of banking authorities to discipline banks, and entry restrictions. Private monitoring is measured by an index that indicates the degree to which information is released to officials and the public, the extent of auditing requirements, and whether credit ratings are required. He finds that a higher level of private monitoring is significantly positively related to bank efficiency across all specifications; the other regulatory characteristics are less robust across specifications. Pasiouras, Tanna, and Zopounidis (2009) also investigate the relationship between bank efficiency and regulation using data on a sample of 615 publicly traded commercial banks operating in 74 countries during 2000–2004. Profit and cost efficiency are measured using stochastic frontier analysis. The results indicate that higher market discipline and greater supervisory power are significantly positively related to both profit and cost efficiency, while stricter capital requirements are positively related to cost efficiency and negatively related to profit efficiency, and restrictions on bank activities are negatively related to cost efficiency and positively related to profit efficiency. Delis, Molyneux, and Pasiouras (2011) investigate the relationship between bank productivity growth and supervision and regulation using data on 582 commercial banks in 22 transition countries, including those in the former Soviet Union and others in eastern Europe during 1999–2009. They find that productivity growth is significantly positively related to regulations and incentives that promote private monitoring and to restrictions on bank activities, but not significantly related to capital adequacy requirements or official supervisory power. Chortareas, Girardone, and Ventouri (forthcoming) study the relationship between bank efficiency, measured using data envelopment analysis, and supervision and regulation using data on banks in 22 countries in the European Union over the period 2000–2008. The number of banks included in the sample varies over the years, from a low of 472 to a high of 704. In contrast to the other studies, this chapter finds that a higher degree of private monitoring is related to lower efficiency. It also finds that tighter capital requirements and stronger official supervisory powers are positively related to efficiency, whereas restrictions on activities are negatively related to efficiency.

2.4 Sources of market discipline

2.4.1 Ownership structure

When managers of a firm trade a dollar of firm value for personal benefits such as avoiding effort and consuming perquisites, the cost to them of the dollar of benefits is determined by their ownership stake in the company. If they own 10%, the dollar of personal benefits costs them 10 cents, while outside owners bear 90 cents of the costs. Jensen and Meckling (1976) hypothesized that as the ownership stake of insiders increases, their interests are better aligned with those of outside owners and that agency costs are reduced. They define the firm entirely owned by its manager as the zero-agency-cost case – where there is no principal-agent problem. They define agency cost as the value lost when the owner-manager sells part of the firm to an outsider so that the cost of a dollar of personal benefits is now less than a dollar. Ang, Cole, and Lin (2000) use data on small- to medium-sized businesses in the United States where there are a number of firms that are entirely owned by their managers. They compare various measures of financial performance for these firms with firms where outsiders share in the ownership of the firm. In the limiting case, the primary manager owns none of the firm. They find that a higher ownership stake by managers is associated with improved performance.

Morck, Shleifer, and Vishny (1988) allow for the possibility that as the ownership stake of insiders increases, their ability to resist various forms of market discipline increases: managers become entrenched. Thus, an increasing level of insider ownership not only increases the price of consuming agency goods, which tends to align insiders' interests with those of outside owners, it also increases insiders' control. The relationship of value to insider ownership depends, then, on the relative strength of the alignment-of-interest effect versus the entrenchment effect. Applying a piecewise linear specification of the proportion of the firm owned by officers and directors to data on nonfinancial firms in the United States, the authors find a statistically significant positive relationship between Tobin's q ratio and insider ownership between 0% and 5% ownership, a statistically significant negative relationship between 5% and 25%, and a less significant, positive relationship at ownership exceeding 25%. They interpret alignment as dominant at less than 5% and greater than 25% and entrenchment as dominant in the range of 5–25%. They note that Weston (1979) found that no firm where insiders owned more than 30% of the firm had ever been acquired in a hostile takeover. Weston suggested that the ability of insiders to resist a hostile bid occurs in the ownership range of 20–30%. Morck, Shleifer, and Vishny's finding that entrenchment dominates alignment in the range 5–25% suggests that a firm's ability to resist market discipline may begin at a much lower range of insider ownership.

Gorton and Rosen (1995) use annual call report data on banks for the period 1984–1990 and find that managerial entrenchment and corporate control issues played a more important role than the moral hazard related to mispriced deposit insurance in explaining the increased risk-taking in banking in the 1980s. They find a nonlinear relationship between insider ownership, managerial entrenchment, and bank risk-taking. As they show, the relationship between ownership and control is a complicated one – as the degree of stock ownership by managers increases, it can increase their ability to act on their own behalf rather than aligning their incentives with the majority shareholders. However, that relationship also depends on how healthy the industry is, that is, on the degree of investment opportunities. When investment opportunities are low, managers may be induced to take on more risk because conservative behavior may not be enough to allow them to keep their jobs and perquisites in a declining industry.

Using data on small, closely held US banks in the Tenth Federal Reserve District to study the relationship of performance to ownership, DeYoung, Spong, and Sullivan (2001) note that since these banks are not actively traded, the discipline of the market for corporate control is largely lacking. In addition, outside owners generally are few in number and hold a relative small stake in the company. Consequently, they have little incentive to monitor. The primary owners who are not managing the bank and whose stake is large enough to monitor may lack the skills and inclination to monitor, especially when they have retired from managing the bank or turned to hired managers as the bank's operations grew and became more complex. The authors estimate a stochastic profit frontier to gauge efficiency – achieved profit as a proportion of best-practice potential profit. They estimate the performance Equation (2.1) by regressing profit efficiency on ownership structure and control variables. Rather than specify piecewise continuous insider ownership variables as in Morck, Shleifer, and Vishny (1988), they use a quadratic specification – insider ownership and the square of insider ownership. They find that hired managers are on average slightly more efficient than owner managers. The proportion of the bank owned by the owner manager is not statistically significantly related to the bank's performance, while the proportion owned by the hired manager is significantly positively related to profit efficiency up to 17% insider ownership and is then significantly negatively related. Thus, at less than 17%, the alignment effect of ownership dominates the entrenchment effect, and at greater than 17%, entrenchment dominates. Most of the banks in their sample with hired managers provide them with less than 17% ownership. The quadratic specification of insider ownership has an important advantage over the piecewise linear specification: the quadratic is more flexible in that it does not impose the breakpoints between shifts in slopes. However, without a cubic specification, one cannot investigate whether at a higher ownership stake, greater than 25% in the case of Morck, Shleifer, and Vishny (1988), the alignment effect once again dominates entrenchment.

To consider the possibility that the sign of the relationship between performance and insider ownership changes three times in the bank data and to allow the data to show where these breaks occur, we use the data of Hughes et al. (2003). The sample is all publicly traded, top-tier bank holding companies (BHCs) in 1993 and 1994; there are 169 of these firms. Performance is measured by market-value inefficiency, derived from a market-value frontier and gauging the difference between the achieved market value of assets and the best-practice potential value (the shortfall) as a proportion of potential value. We measure the proportion of outstanding shares owned by officers and directors in the year before, that is, in 1993, and market-value ineffi-ciency at the end of 1994. This attempts to control for the endogeneity of ownership, although admittedly this may not be adequate as ownership does not change very much over time. We control for the size of banks as indicated by the natural log of total assets. Managers of larger banks typically own a smaller proportion because the wealth needed to own large proportions is too great. We also control for the value of a bank's investment opportunities. Managers with much more valuable investment opportunities are, on average, much less efficient – they achieve a smaller proportion of their potential value – than those with poorer investment opportunities, even though both groups have essentially the same average q ratio. The value of a bank's invest-ment opportunities is measured by fitting a stochastic frontier to the market value of assets as a function of the book value of assets and, in the bank's local markets, the market-weighted, 10-year average macroeconomic growth rate and market-weighted Herfindahl index of concen-tration. A bank's investment opportunity index is the best-practice market value of the bank's assets in the local markets in which it operates as a proportion of its book-value investment in assets. Insider ownership is the proportion of the firm owned by officers and directors.

Table 2.1 and Table 2.2 provides summary statistics for the full sample of 169 publicly traded, top-tier BHCs in 1994, and for the more efficient and less efficient halves of the sam-ple, where we gauge performance by the market-value inefficiency measure. (These statistics are also reported in Hughes et al., 2003.) As shown in Panel A, the BHCs in the full sample range in size from $159.86 million to $211.764 billion in assets. The market-value ineffi-ciency ratio indicates that, on average, banks fail to achieve 19.1% of their potential market value while their average q ratio is 1.036. Their mean potential value in the markets in which they operate as a proportion of their book-value investment in assets is 1.073.

As shown in Panels B and C, the more efficient half of the sample holds more total assets and their insiders own less of their banks than the less efficient half. Their lower ownership stake may result from the very large size of the banks they manage. The more efficient half of the sample exhibits more outside blockholder ownership, which may contribute to their effi-ciency through their greater ownership incentive to monitor the performance of insiders. However, the more efficient half also holds proportionately less capital. It is not clear to what extent better diversification and increased risk-taking may contribute to the lower capital ratio.

The potential value of investment opportunities for less efficient banks as a proportion of their book-value investment in assets far exceeds that of the more efficient banks: 1.110 ver-sus 1.037. Recall that a bank's market-value efficiency is measured by the achieved market value of its assets as a proportion of its potential value measured across all markets – not just the markets in which it operates. The less efficient half of the banks in the sample, on average, waste 33.8% of their potential value, while the more efficient half waste only 6% on average. However, their average q ratios are identical. The q ratio fails to capture the stark difference

Table 2.1   Data definitions.

The data are taken from Hughes et al. (2003) and consists of 169 publicly traded, top-tier BHCs in 1993 and 1994.

Market-value inefficiency is derived from a market-value frontier and is the difference between the best-practice potential (frontier) value and the noise-adjusted, observed market value of assets, as a proportion of the potential value. Market-value inefficiency is measured at the end of 1994. (See Appendix 2.A for further information.)

Insider ownership is the proportion of outstanding shares owned by officers and directors in the year before, that is, in 1993.

An *outside blockholder* is defined as a holder of 5% or more of outstanding shares based on 13D filings. Blockholder ownership is the percent of outstanding shares held by outsiders in 1993.

The value of a BHC's *investment opportunities* is measured by fitting a stochastic frontier to the market value of assets as a function of the book value of assets and, in the bank's local markets, the market-weighted, 10-year average macroeconomic growth rate and the BHC's market-weighted Herfindahl index of concentration. A BHC's investment opportunity index is the best-practice market value of the BHC's assets in the local markets in which it operates as a proportion of its book-value investment in assets.

Tobin's q ratio is the market value of assets, net of goodwill, divided by the book value of assets, net of goodwill.

The *capital-to-assets ratio* is the book value of equity including goodwill divided by the book value of total assets goodwill.

The BHC's *Herfindahl index* is given by the weighted sum of the BHC's squared share of deposits in each of its local markets where the weights are the proportion of total deposits found in each market.

The *average interest rate on loans* is total interest earned on loans divided by loans minus nonaccruing loans.

The *price of insured deposits* is the interest expense of deposits in domestic offices excluding time deposits over $100 000 divided by the volume of these deposits.

The *price of uninsured deposits* is the interest expense of domestic time deposits over $100 000 divided by the volume of these deposits.

The *price of other borrowed money* is the expense of foreign deposits, commercial papers, subordinated notes and debentures, mandatory convertible securities, securities sold under agreement to repurchase, federal funds purchased, trading account liabilities, other borrowed money, and mortgage indebtedness divided by the volume of these funds.

Note that Hughes et al. (2003) misreported the definition of this variable as the shortfall as a proportion of the book value of assets net of goodwill rather than as a proportion of potential value. The values of the variable, however, were correct.

in performance between these two groups of banks. The stochastic frontier technique identifies the critical difference in the value of investment opportunities between the two groups of banks and, when used to calculate lost market value, shows that banks with more valuable investment opportunities tend to waste more value than banks with poorer opportunities.

These contrasts between the more efficient and less efficient banks suggest that differences in asset size and the value of investment opportunities play an important role in shaping managerial performance incentives (which is consistent with the results of Gorton and Rosen, 1995).

Table 2.2 Summary statistics.

Variable	Mean	Median	Std. dev.	Minimum	Maximum
Panel A: Full sample of 169 publicly traded, top-tier BHCs					
Book value assets, net of goodwill ($1000)	11796319	1972085	27384208	159860.00	211764250
Market-value inefficiency ratio	0.191	0.149	0.164	0.000961	0.697
Insider ownership = % of outstanding shares held by officers and managers	12.885	7.264	13.449	0.342	66.018
Outside blockholder ownership = % of outstanding shares held by outside blockholders (holders of 5% or more of outstanding shares)	3.307	0.000	6.555	0.000	33.051
Investment opportunity ratio = size of investment opportunity set/book value of assets net of goodwill	1.073	1.057	0.054	1.006	1.319
Tobin's *q* ratio	1.036	1.033	0.033	0.970	1.172
Capital-to-assets ratio = book value of equity including good will/book value of assets including goodwill	0.085	0.082	0.016	0.044	0.135
Herfindahl index	0.238	0.223	0.116	0.059	0.646
Average interest rate on loans = interest earned on loans/volume of loans	0.085	0.084	0.00777	0.0573	0.126
Uninsured deposit interest rate = interest expense for uninsured deposits/volume of uninsured deposits	0.0421	0.0412	0.0102	0.00548	0.107
Insured deposit interest rate = interest expense for insured deposits/volume of insured deposits	0.0254	0.0262	0.00564	0.00938	0.0393
Other borrowed funds rate = interest expense for these funds/volume of these funds	0.0438	0.0413	0.0172	0.0143	0.178
Panel B: More efficient half of the sample (market-value inefficiency ratio < median): 84 publicly traded, top-tier BHCs					
Book value assets, net of goodwill ($1000)	22820516*	8763745	35679798	192094800	211764250
Market-value inefficiency ratio	0.0550*	0.0451	0.0409	0.000961	0.148
Insider ownership	7.577*	4.869	8.632	0.342	55.11

(continued overleaf)

Table 2.2 *(continued)*

Variable	Mean	Median	Std. dev.	Minimum	Maximum
Outside blockholder ownership	4.895*	0.0	7.900	0.0	33.051
Investment opportunity ratio	1.037*	1.037	0.0144	1.006	1.073
Tobin's q ratio	1.036	1.033	0.0269	0.983	1.129
Capital-to-assets ratio	0.080*	0.079	0.014	0.048	0.133
Herfindahl index	0.228	0.221	0.083	0.059	0.521
Average interest rate on loans	0.0823*	0.0818	0.00785	0.0573	0.115
Uninsured deposit interest rate	0.0433	0.0424	0.0123	0.00548	0.107
Insured deposit interest rate	0.0249	0.0251	0.00542	0.00938	0.0393
Other borrowed funds rate	0.0430	0.0419	0.00845	0.0291	0.0916
Panel C: Less efficient half of the sample (market-value inefficiency ratio ≥ median): 85 publicly traded, top-tier BHCs					
Book value assets, net of goodwill ($1000)	901 818*	730 513	510 839	159 860	2 378 657
Market-value inefficiency ratio	0.324*	0.317	0.125	0.149	0.697
Insider ownership	18.131*	11.985	15.235	1.931	66.018
Outside blockholder ownership	1.738*	0.0	4.390	0.0	22.994
Investment opportunity ratio	1.109*	1.095	0.0551	1.028	1.319
Tobin's q ratio	1.036	1.0321	0.0375	0.970	1.172
Capital-to-assets ratio	0.0900*	0.0877	0.0169	0.0442	0.135
Herfindahl index	0.248	0.2310	0.141	0.0593	0.646
Average interest rate on loans	0.0875*	0.0861	0.00681	0.0787	0.126
Uninsured deposit interest rate	0.0408	0.0408	0.00756	0.0235	0.0780
Insured deposit interest rate	0.0260	0.0269	0.00582	0.0110	0.0387
Other borrowed funds rate	0.0447	0.0408	0.0228	0.0143	0.178

* Significantly different from the mean of the other efficiency subsamples at the 10% or better level.

Table 2.3 Regression of market-value inefficiency on insider ownership.

Variable	Parameter estimate	Heteroscedasticity-consistent		
		Standard error	t-Value	Pr>|t|
Dependent variable: market-value inefficiency				
Intercept	−0.85116*	0.22875	−3.72	0.0003
Insider ownership	−0.00504**	0.00217	−2.32	0.0216
(Insider ownership)^2	0.00021254**	0.00009418	2.26	0.0254
(Insider ownership)^3	−0.00000217**	0.00000105	−2.06	0.0408
Investment opportunity ratio	1.66887*	0.14147	11.80	<0.0001
ln(Book value of assets)	−0.04907*	0.00565	−8.69	<0.0001
Number of observations = 169				
R^2 = 0.9225				

The data are taken from Hughes et al. (2003) and consists of 169 publicly traded, top-tier BHCs in 1993 and 1994, and the regressions are used in the course Hughes (2011).
See Table 2.1 for data definitions.
The dependent variable is market-value inefficiency, and the performance equation is estimated by ordinary least squares.
* Significantly different from zero at the 1% level.
** Significantly different from zero at the 5% level.

Hence, we use the natural logarithm of total assets and the investment opportunity ratio as control variables in our performance regressions.

Table 2.3 shows the ordinary least squares estimates of a regression of market-value inefficiency on the cubic specification of insider ownership and the control variables, ln (asset size) and the investment opportunity ratio. Over the range of insider ownership between 0% and 15.6%, market-value inefficiency is significantly negatively related to insider ownership; hence, alignment dominates entrenchment. Over the range 15.6–49.7%, market-value inefficiency is positively related to insider ownership so that entrenchment dominates alignment. Above 49.7%, the relationship becomes negative so that alignment again dominates. While the derivative of market-value inefficiency is significantly negative in this region, there are only five banks with ownership greater than 49.7% (66% is the highest stake).

These results are similar to those of DeYoung, Spong, and Sullivan (2001b) where the positive relationship between performance and insider ownership changes to a negative relationship at 17%; however, the results here differ in that we allow the data to reveal a third regime where the sign of this relationship changes again. In that sense, these results are qualitatively similar to those of Morck, Shleifer, and Vishny (1988); however, the quantitative values of the two turning points for the sign of the relationship of performance to insider ownership for commercial banks differ significantly from those Morck, Shleifer, and Vishny (1988) found by trial and error for nonfinancial firms.

The statistically significant negative coefficient on asset size indicates that larger banks on average achieve more of their best-practice value than smaller banks, and the statistically significant positive coefficient on the investment opportunity ratio indicates that banks with more growth opportunities waste more of this value than those with poorer opportunities.

Large holdings of shares provide another perspective on the relationship of performance to ownership, especially when the large shareholder is unrelated to management (e.g., an outside blockholder). A *blockholder* is defined as a holder of 5% or more of outstanding shares based on 13D filings. Large blocks of stock ownership give their owners a substantial financial stake in the firm, large enough to overcome the free-rider problem of small stakeholders and to monitor managers' performance or perform better when part of management. Holderness (2003) calls the hypothesis that blockholders either monitor insiders better or, when they are insiders themselves, perform better the *shared benefits hypothesis*. On the other hand, blockholders can use their voting power to consume pecuniary and nonpecuniary private benefits of control. Holderness terms this possibility the *private benefits hypothesis*. While the consumption of private benefits by blockholders might be thought to influence firm value negatively, it might also be positive. According to Holderness, the impact of blockholders on firm value has not been firmly established as either positive or negative, and there is little evidence that it has a large effect whatever the sign.

Outside blockholders are thought to be more independent of management and, therefore, better able to monitor and positively influence performance. We consider the relationship of bank performance and ownership by outside blockholders by regressing market-value inefficiency on a quadratic specification of the proportion of the firm owned by outside blockholders to allow for a nonmonotonic relationship that could capture both the shared and private benefits hypotheses at different levels of ownership.[4] We control for the value of a bank's investment opportunities with the investment opportunity ratio and its size with the log of the value of its assets. Table 2.4 reports the results of this regression. The positive coefficient on the linear term, 0.00256, and the negative coefficient on the squared term, −0.00006618, reveal that the

Table 2.4   Regression of market-value inefficiency on outside blockholder ownership.

Variable	Parameter estimate	Heteroscedasticity-consistent		
		Standard error	t-Value	Pr > |t|
Dependent variable: market-value inefficiency				
Intercept	−0.94457*	0.21900	−4.31	<0.0001
Outside blockholder ownership	0.00256**	0.00137	1.87	0.0628
(Outside blockholder ownership)^2	−0.00006618	0.00005078	−1.30	0.1943
Investment opportunity ratio	1.71071*	0.13725	12.46	<0.0001
ln(book value of assets)	−0.04750*	0.00532	−8.93	<0.0001
Number of observations = 169				
Adjusted R^2 = 0.9208				

The data are taken from Hughes et al. (2003) and consists of 169 publicly traded, top-tier BHCs in 1993 and 1994, and the regressions are used in the course Hughes (2011).
See Table 2.1 for data definitions.
The dependent variable is market-value inefficiency, and the performance equation is estimated by ordinary least squares.
* Significantly different from zero at the 1% level.
** Significantly different from zero at the 5% level.
*** Significantly different from zero at the 10% level.

---

[4] In our data set, there are 118 firms in which there is no blockholder ownership and 51 firms with positive blockholder ownership.

positive sign of the derivative of market-value inefficiency with respect to blockholder ownership changes to a negative sign at the level of 19.3% blockholder ownership. Thus, at levels lower than 19.3%, increasing blockholder ownership is associated with higher ineffi- ciency, a result consistent with the private benefits hypothesis. However, at levels above 19.3%, where the opportunity cost of consuming private benefits may be too high, increasing blockholder ownership is associated with lower inefficiency, a result consistent with the shared benefits hypothesis; however, the value of the derivative is not statistically significant in these cases.[5] Again, the negative coefficient on size indicates that larger banks are more efficient, and the positive coefficient on the value of investment opportunities indicates that banks in more valuable markets are less efficient.

The commonly used measure of aggregate blockholder ownership lacks details on the type of blockholders represented in the data. Their identity could be important because the seriousness of incentive misalignments within the block may vary by the type of blockholder ownership. Cronqvist and Fahlenbrach (2009) construct a detailed panel data set over the period 1996–2001 that includes all blockholders of 1919 publicly traded corporations. The data allows the specification not just of time and firm fixed effects but also of unique block- holder fixed effects. They consider how the individual blockholder ownership and type of blockholder are related to various corporate policies and firm performance. They find statisti- cally and economically significant blockholder fixed effects in investment, financial, and compensation policies, which are related to firm performance. On the question of influence versus selection, for activist, pension fund, corporate, individual, and private equity block- holders, they find evidence consistent with influence. But, for large mutual funds, they find evidence consistent with systematic selection. Moreover, their results bear on the incentives of managers of large financial firms that took significant risks leading up to the recent finan- cial crisis. In particular, they find higher return on assets and the q ratio in firms with large shareholders that had aggressive investment styles (including higher levels of investment and M&A activity). They also find that blockholders that are associated with higher total CEO compensation have more aggressive investment styles, higher investment-to-q sensitivity, and fewer diversifying acquisitions.

Cheng, Hong, and Scheinkman (2010) examine risk-taking and executive compensation at financial institutions – banks, insurance companies, and primary dealers – for the period 1992–2008 and find that risk-taking and high executive compensation are strongly positively related to institutional ownership. This suggests that institutional investors want these firms to take more risk (perhaps because of shorter horizons due to agency issues) and give them short-term incentives via compensation to do so.

Ellul and Yerramilli (forthcoming) examine the 74 largest publicly listed US BHCs in a panel spanning the period 2000–2008 and find that a strong risk-management function is associated with lower future risk and better future financial performance. However, institutions with high institutional ownership were found to exhibit more volatile stock returns.

---

[5] The value of the derivative is significant at the 5% level for blockholder ownership values greater than 2.4% and less than or equal to 11.8%; it is significant at the 10% level for blockholder ownership values less than 2.4% and for values greater than 11.8% and less than or equal to 13.1%. There are six observations in the data set for which blockholder ownership is greater than 13.1% and less than 19.3%; for these observations, the derivative is positive but insignificant. There are six observations in the data set for which blockholder ownership is greater than 19.3%; for these observations, the derivative is negative but insignificant.

Laeven and Levine (2009) posit three key hypotheses that they use to examine the relationship of bank risk-taking, ownership structure, and regulatory policies. First, they assert that owners whose stake in a financial institution is a relatively small part of their diversified wealth holdings generally prefer that the institution take more risk than is preferred by debt-holders and nonshareholder managers who may avoid risk to protect their relatively undiversified human capital and their private benefits of control. Second, regulations create risk-taking incentives for diversified owners that differ from those of debt-holders and nonshareholder managers. In particular, while deposit insurance may give nonshareholder managers little incentive to threaten their control by taking extra risk, diversified equity holders may prefer extra risk-taking when exploiting the subsidy of mispriced deposit insurance. Third, ownership structure affects the ability of owners to influence risk-taking. Larger cash flow and voting rights give owners greater risk-taking incentives – both the standard risk-shifting incentives and those created by regulations – and greater ability to influence managerial risk-taking policies.

Laeven and Levine (2009) collect data on the 10 largest publicly listed banks in 2001 in each of 48 countries, which, because of limits on data availability, results in a sample of 279 banks once state-owned banks are eliminated. The resulting sample accounts for over 80% of the assets in each country. The authors consider the relationship of the z-score, a measure of the distance of the bank from insolvency, to various regulatory policies. They create a capital stringency index that measures the degree of regulatory oversight of bank capital and find that the z-score is significantly positively related to the capital stringency index. However, the coefficient on a term that interacts capital stringency with the proportion of cash-flow rights of the largest shareholder is significantly negative. As cash-flow rights become more concentrated, the sign of the relationship of the z-score and capital stringency reverses. In other words, more stringent oversight of bank capital becomes less effective at stabilizing a bank when the bank has a large owner, and with a sufficiently large owner, more stringent oversight of capital regulations increases bank risk. As predicted by Koehn and Santomero (1980), the intensification of capital regulation provides diversified bank equity holders with the incentive to adopt more risky investment for higher expected return to compensate for expected return lost to stricter capital regulation.

Countries often attempt to enhance bank stability by imposing activity restrictions. Laeven and Levine (2009) create an index of activity restrictions and find a statistically significant negative relationship of bank stability to the interaction of the index with the cash-flow rights of the largest shareholder. When banks in a country are widely held, there is no statistically significant relationship of bank stability and activity restrictions; however, when bank ownership is concentrated, activity restrictions are associated with more risky investment strategies and less stable banks. They also find evidence that the moral hazard effect of deposit insurance appears when there is concentrated ownership but not when ownership is diffuse.[6]

---

[6] When bank stability is regressed on a variable that indicates whether there is explicit deposit insurance or the complete guarantee of losses in the last banking crisis, its coefficient is significantly negative. However, when the regression also includes an interaction term between this deposit insurance variable and the cash-flow rights of the largest shareholder, the deposit insurance variable is not statistically significant, but the coefficient on the interaction term is negative and statistically significant.

2.4.2   Capital markets

The threat of a takeover provides management with a strong incentive to perform efficiently. As noted earlier in the chapter, the regulation of banking markets has historically limited takeovers. Notably, until the passage of the Interstate Banking and Branching Efficiency Act of 1994, US states limited entry of out-of-state banks into their banking markets. Schranz (1993) uses differences in entry restrictions across states to investigate how these differences in takeover threat, controlling for other sources of managerial discipline, are related to banking performance. Banks in states with an active takeover market are more profitable. On the cost side, Evanoff and Örs (2002) find that entry into local banking markets leads to improved cost efficiency among incumbent banks.

Brook, Hendershott, and Lee (1998) use the event-study method to consider how banking stock prices responded to the passage of the Interstate Banking and Branching Efficiency Act. They find a statistically significant positive cumulative abnormal return to the industry. Poorer performing banks, prior to the passage of the act, obtain the largest abnormal returns, limited only by evidence of entrenched management.

Hughes et al. (2003) find evidence of managerial entrenchment at US BHCs that have higher levels of managerial ownership, higher valued investment opportunities, poorer financial performance, and smaller asset holdings. At banks that do not exhibit signs of managerial entrenchment, asset sales and acquisitions are both associated with improved financial performance; however, at banks with entrenched management, only asset sales lead to improved performance. This suggests that entrenched managers may engage in empire building to feather their own nests at the expense of the shareholders.

2.4.3   Product markets

John Hicks (1935, p. 8) famously noted, 'The best of all monopoly profits is a quiet life'. The quiet life is not an automatic benefit of market power since the many other disciplining mechanisms of management are not short-circuited by market concentration. Berger and Hannan (1998) control for these other sources of discipline and find evidence that market concentration is negatively related to operating cost efficiency in US banking. As noted previously, Evanoff and Örs (2002) find that entry into local banking markets leads to improved cost efficiency among incumbent banks.

In contrast to these papers, Petersen and Rajan (1995) find a benefit to concentration in banking markets that flows from banks' special information about their depositors obtained from their deposit records, which allows banks to make information-intensive loans to relatively opaque borrowers more efficiently than nondepository lenders. With this information, banks are able to lend at a lower interest rate to young firms to reduce their probability of financial distress in the early years of their operation. However, banks must recover the cost of these subsidies in the later years of these lending relationships. In competitive markets, the ability of banks to price loans in this manner is limited. In the later years, competitors would take the lending business from banks that tried to recover their earlier subsidies.

We again use the data of Hughes et al. (2003) – publicly traded, top-tier banking holding companies in 1993 and 1994 – to investigate the relationship between performance and concentration. We measure a BHC's market power by a Herfindahl index constructed as

Table 2.5   Regression of market-value inefficiency on market concentration.

Variable	Parameter estimate	Heteroscedasticity-consistent		
		Standard error	t-Value	$Pr > |t|$
Dependent variable: market-value inefficiency				
Intercept	−0.88882*	0.21458	−4.14	<0.0001
Herfindahl index	− 0.28964*	0.09301	−3.11	0.0022
(Herfindahl index)^2	0.37171**	0.15097	2.46	0.0148
Investment opportunity ratio	1.68158*	0.13698	12.28	<0.0001
ln(Book value of assets)	−0.04593*	0.00526	−8.73	<0.0001
Number of observations = 169				
Adjusted R^2 = 0.9237				

The data are taken from Hughes et al. (2003) and consists of 169 publicly traded, top-tier BHCs in 1993 and 1994, and the regressions are used in the course Hughes (2011).
See Table 2.1 for data definitions.
The dependent variable is market-value inefficiency, and the performance equation is estimated by ordinary least squares.
* Significantly different from zero at the 1% level.
** Significantly different from zero at the 5% level.
*** Significantly different from zero at the 10% level.

the weighted sum of the BHC's squared share of deposits in each of its local markets, where the weights are the proportion of total deposits found in each market. Higher values imply the BHC has more market power in its markets. The performance measure, market-value inefficiency, is derived from a market-value frontier and is the difference between the best-practice potential (frontier) value and the noise-adjusted observed market value of assets, as a proportion of the potential value. Table 2.5 reports, as in previous regressions, a larger asset size is related to lower inefficiency, while having more valuable investment opportunities is related to greater inefficiency. The coefficients on the Herfindahl index, −0.28964, and on the squared index, 0.37171, imply that at degrees of market concentration below 0.39, concentration and inefficiency are negatively related, which is consistent with the Petersen–Rajan hypothesis. Above this value, the relationship switches signs so that higher concentration is positively related to higher inefficiency, which is consistent with the quiet-life hypothesis and the increasing ability of banks to extract surplus as concentration increases. While the regression reported in Table 2.5 controls for the value of investment opportunities, which results in part from market power, it does not control for prices charged on loans and paid on borrowed funds. Hence, we add these interest rates as control variables and report the results in Table 2.6.

The statistically significant negative coefficient on the Herfindhal index, −0.23524, and the positive coefficient on the squared index, 0.30791, again imply that the negative value of the derivative of inefficiency with respect to market concentration switches to a positive value – in this case, at 0.382. While controlling for prices reduces the sensitivity of inefficiency to concentration, it does not fundamentally change the nonmonotonic relationship.

Table 2.6 Regression of market-value inefficiency on market concentration controlling for bank input and output prices.

Variable	Parameter estimate	Heteroscedasticity-consistent		
		Standard error	t-Value	Pr > |t|
Dependent variable: market-value inefficiency				
Intercept	−0.79832*	0.20834	−3.83	0.0002
Herfindahl index	−0.23524**	0.10567	−2.23	0.0274
(Herfindahl index)^2	0.30791***	0.16708	1.84	0.0672
Average interest rate on loans	0.39197	0.48698	0.80	0.4221
Uninsured deposit interest rate	0.23088	0.35842	0.64	0.5204
Insured deposit interest rate	−1.37030**	0.59377	−2.31	0.0223
Other borrowed funds rate	0.02694	0.21220	0.13	0.8992
Investment opportunity ratio	1.61292*	0.13022	12.39	<0.0001
ln(Book value of assets)	−0.04826*	0.00523	−9.23	<0.0001
Number of observations = 169				
Adjusted R^2 = 0.9238				

The data are taken from Hughes et al. (2003) and consists of 169 publicly traded, top-tier BHCs in 1993 and 1994, and the regressions are used in the course Hughes (2011).
See Table 2.1 for data definitions.
The dependent variable is market-value inefficiency, and the performance equation is estimated by ordinary least squares.
* Significantly different from zero at the 1% level.
** Significantly different from zero at the 5% level.
*** Significantly different from zero at the 10% level.

2.4.4 Labor markets

The financial crisis that began in 2007 has called into question the performance of many managers. Does the labor market recognize bad managers and punish them? Cannella, Fraser, and Lee (1995) investigate the efficiency of labor markets in banking: they ask if labor markets can distinguish the efficiency of managers from the efficiency of their banks. To answer this question, they consider Texas banking during the troubled period 1985–1990 when many banks failed. In particular, they compare matching samples of banks that did not fail with banks that failed because they were insolvent and with banks that, while not insolvent, nevertheless failed because they were part of a multi-BHC that failed. The latter group is termed 'innocent bystanders' and the former 'equity insolvent'. They find that 67% of managers of nonfailed banks were still employed in 1993 in Texas banking. In contrast, only 22% of managers of equity insolvent banks were still employed, while 44% of the managers of the innocent bystander banks were employed. In addition, they find that managers of the top-tier bank within the holding company are less likely to find employment in banking after their banks fail than second-tier managers. Finally, they find that managers of nonfailed banks whose cost efficiency measured by the ratio of noninterest expense to total assets is lower are less likely to be employed in Texas banking in 1993. Thus, labor markets appear to distinguish good from bad managers and weed out the latter.

2.4.5 Boards of directors

Boards of directors are charged with oversight of management, hiring and firing the CEO, setting top executive compensation, and providing advice and strategic guidance. Hermalin and Weisbach (2003) and Adams, Hermalin, and Weisbach (2010) survey the theoretical and empirical literature on boards. In carrying out their oversight function, boards require a degree of independence from management. Gauging the effectiveness of oversight often considers the composition of the board in terms of the proportion of the board's members who are independent outsiders – not part of management and not related to management by family, by business relationships ('grey' directors), or by interlocking boards. In addition, oversight may be improved when the roles of board chair and CEO are separated – although this need not be the case: recall that Enron had such a separation. In contrast to these arguments for independence of board members and the chair, insiders serving on the board can provide important information for board oversight and strategic decision-making. Similar arguments can be made for combining the roles of CEO and board chair.

Hermalin and Weisbach (2003) note that empirical research on boards typically reaches only two conclusions: the composition of the board and firm performance do not appear related, while the size of the board and performance are usually negatively related. On the other hand, evidence provides a clearer picture of the relationship of board composition to board actions. For example, Core, Holthausen, and Larcker (1999) regress CEO compensation on three sets of variables: economic determinants of optimal compensation and characteristics of ownership structure and board structure. They find that most variables in all three sets are statistically significant, and, in the case of ownership and board structure variables, the signs of the coefficients suggest that higher compensation is correlated with characteristics associated with weaker governance structures. In terms of board structure, higher CEO compensation is associated with larger boards, boards chaired by the CEO, a higher proportion of inside directors, higher proportions of grey and interlocked outside directors, a higher proportion of 'busy' outside directors who serve on three or more boards, and outside directors over the age of 69. To determine whether the influence of board and ownership structure variables on compensation reflects agency problems or correlation with missing economic determinants, future performance is regressed on the portion of compensation predicted by board and ownership structure variables. Since future performance varies inversely with this 'excess' compensation, the authors conclude that the higher CEO compensation and poorer future performance reflect agency problems.

Examining data on US banks over the period 1964–1989, Adams and Mehran (2008, revised 2011) find that performance measured by Tobin's q is not related to board independence; however, performance is positively related to board size. They speculate that the growing complexity of banks and mergers and acquisitions during this period may account for the result. They test and reject these explanations. Instead, they find that larger boards are associated with increased value when they include a larger number of directors who also sit on subsidiary boards. These joint directors apparently facilitate communication among the subsidiaries and parent company.

Tanna, Pasiouras, and Nnadi (2011) examine the relationship between efficiency and board structure of 17 banks operating in the United Kingdom between 2001 and 2006 and find results that contrast with those of Adams and Mehran for US banks. Tanna, Pasiouras, and Nnadi measure efficiency based on data envelopment analysis techniques. When they control for bank asset size and capitalization, as measured by the equity-to-assets ratio, they find a positive and significant relationship between the share of nonexecutive members on the board (a proxy for board independence) and efficiency. The size of the board is not significantly related to efficiency, once the composition of the board in terms of share of nonexecutive members is controlled for.

2.4.6 Compensation

The board of directors sets top executive compensation and can structure it so that it ameliorates agency problems. However, the board itself, as noted earlier in the chapter, may suffer from misaligned incentives and fail to execute optimal pay arrangements. Thus, the consideration of compensation must account not just for optimal structures, but also for structures that result from agency problems. As noted earlier in the chapter, Core, Holthausen, and Larcker (1999) regress CEO compensation on three sets of variables, economic determinants of optimal compensation and characteristics of ownership structure and board structure, and find that most variables in all three sets are statistically significant and that the signs of the coefficients on the variables characterizing ownership and board structure suggest that higher compensation is correlated with weaker governance structures. The negative relationship of future performance with the portion of compensation predicted by the board and ownership structure variables suggests that the higher compensation results from agency problems rather than optimal contracting.

In banking, optimal contracting must account for the value of investment opportunities and the subsidy of risk-taking provided by explicit and implicit deposit insurance. Smith and Watts (1992) note that higher valued investment opportunities add complexity to a firm's decision-making and make monitoring of managerial actions and effort by the board more difficult. Using the stochastic frontier technique to estimate the difference between the best-practice market value and achieved market value of assets – firms' market-value inefficiency – Hughes et al. (2003) find that market-value inefficiency is significantly higher for banks with higher valued investment opportunities. As implied by Smith and Watts (1992), managers of these banks appear to waste much more potential value than those with lower valued opportunities. Consequently, Smith and Watts (1992) and Gaver and Gaver (1993) contend that incentive-based compensation such as stock and stock options is much more important to firms with valuable investment opportunities as a tool to reduce agency problems and to encourage managers to pursue these opportunities. On the other hand, banks with poorer investment opportunities maximize their expected value by exploiting the cost-of-funds subsidy of deposit insurance through additional risk-taking. Optimal compensation for these banks encourages risk-averse managers to pursue riskier investment strategies.

Houston and James (1995) examine data on US banks during the period 1981–1990 and find that banks with more valuable investment opportunities are more likely to rely on equity-based incentives, while banks classified as too big to fail are no more likely to rely on equity-based compensation. They conclude that compensation in banking does not appear to promote risk-taking at banks with low-valued investment opportunities. However, this result is contradicted by Cheng, Hong, and Scheinkman (2010) who regress top executive compensation in the financial industry during two periods, 1992–1994 and 1998–2000, on size and subindustry classification and use the residuals from the regression to analyze their relationship to risk-taking during the subsequent periods, 1995–2000 and 2001–2008, respectively. High residual compensation firms include Bear Stearns, Lehman, Citicorp, Countrywide, and AIG while low to moderate residuals characterize compensation at JP Morgan Chase, Goldman Sachs, Wells

Fargo, and Berkshire Hathaway. Firms with high residuals were associated with higher measures of risk during the subsequent periods. In addition, these firms were also more likely to perform extremely well in the earlier period when the economy prospered and extremely poorly during the financial crisis in the latter period. As the authors put it (p. 4), '...the aggressive firms that were yesterday's heroes when the stock market did well can easily be today's outcasts when fortunes reverse...'. Examining the structure of compensation, they find that bonuses and equity-based compensation (including options) are strongly correlated with risk-taking even controlling for insider ownership. Moreover, residual compensation and risk-taking are positively correlated with institutional ownership. They conclude that institutional investors appear to prefer higher risk strategies and compensation structures that reward risk-taking. In good times, these strategies result in extremely strong performance. Similarly, Fahlenbrach and Stulz (2011) find that banks whose CEOs enjoyed stronger equity-based performance incentives performed significantly worse during the crisis than banks whose CEOs had weaker incentives. Of course, the incentives were designed to create much higher than average *expected* performance, which would likely have been realized under good macroeconomic conditions.

2.5   Conclusions

In an interesting study, McConnell and Servaes (1995) allude to what they call the two faces of debt. This refers to a dichotomy in how debt influences performance in nonfinancial firms. For firms with valuable investment opportunities, debt creates underinvestment problems; thus, relatively low financial leverage maximizes value, while for firms with poorer investment opportunities, debt resolves overinvestment problems by imposing performance pressure on managers, so for these firms relatively high leverage maximizes value.

In our view, capital structure in banking also displays two faces, which result from entry restrictions and from the federal safety net. One face looks toward protecting the valuable charter and its associated investment opportunities by adopting relatively low leverage. The other face turns to risk-taking opportunities involving higher leverage that maximizes expected value by exploiting deposit insurance and the too-big-to-fail doctrine. One face encourages investment strategies that promote the stability of the financial system; the other face leans toward strategies that can undermine stability in difficult macroeconomic conditions.

The charter gives a bank located in markets with valuable growth opportunities a claim on them that is partially protected from entry by other banks. The charter also provides access to the federal safety net; however, such banks typically adopt low-risk investment strategies that include substantial capitalization to protect the charter from episodes of financial distress that could lead to its loss. These strategies maximize expected value and promote the stability of the financial system. These banks do not exploit the safety net, since the associated risk would erode their expected value. On the other hand, the balance between protecting the charter versus exploiting the safety net shifts for banks located in markets with poorer growth opportunities – perhaps because these markets are extremely competitive. For these banks, lower capitalization and investment strategies that are higher risk maximize expected value.

Thus, good governance defined by its goal of aligning the interests of management with outside owners also has two faces. In cases in which maximizing shareholder value results in the socially optimal level of risk-taking, it fosters financial stability. But in cases in which maximizing shareholder value results in excessive risk-taking, it requires attention from regulation and supervision.

Appendix 2.A: Measuring performance based on the highest potential market value of assets

The market value of a firm's assets captures several aspects of the firm's performance that cannot be evaluated by accounting measures based on profit and cost. First, the market value represents the market's expectation of current and future profits and costs. In contrast, accounting measures capture the firm's current or historical cash flow. Firms that take more risks expect higher profits. If the firm's additional risk enhances the market value of its assets, the additional expected cash flow more than compensates for the higher required return on assets occasioned by the extra risk. In other words, the *discounted* expected value of the cash flow increases; therefore, the firm's market value increases. Thus, market value offers two key advantages over accounting measures of performance: it captures a firm's current and future expected cash flow, and it incorporates the market's assessment of the discount rate required by the risks the firm takes.

Tobin's q ratio measures the ratio of the market value of assets to the replacement cost of the assets – the average market value of a dollar investment in assets – and is a commonly used measure of performance based on market value. For example, the classic study of the relationship of managerial ownership to performance by Morck, Shleifer, and Vishny (1988), discussed in Section 2.3.1, uses a proxy for Tobin's q to gauge performance. The authors find that the q ratio and managerial ownership are correlated even when controlling for firm size. In a firm owned entirely by its management, such a relationship would not be expected, since the agent (manager) is also the principal (owner). However, when some owners have no role in management, this separation of ownership and management can lead to agency problems that vary by the division of ownership between insiders and outsiders; this is consistent with the econometric correlation of managerial ownership and firm performance.

While the q ratio provides evidence consistent with agency problems, it does not gauge the extent of those problems. If it were possible to estimate the highest potential value of a firm's assets given its investment in those assets, the difference between the highest value and the achieved value might provide evidence on the magnitude of agency costs. A more accurate gauge of agency problems would need to eliminate the influence of luck or statistical noise on that difference. One approach for doing this, proposed by Hughes et al. (1997) and used by Hughes et al. (2003), is to fit a stochastic frontier (an upper envelope) of firms' market values to the firms' investment in assets, which is a statistical technique for separately estimating the firms' best-practice market value given investment in assets, the systematic shortfall of achieved market value from best-practice value, and the statistical noise. The difference between the firm's frontier value and the achieved market value is its market-value shortfall, that is, its lost market value. The ratio of this shortfall to the frontier value is a measure of the bank's market-value inefficiency.

In this chapter, we use the estimates of market-value inefficiency in Hughes et al. (2003) to study the relationship between bank performance and sources of managerial discipline; we report our findings in Table 2.3, Table 2.4, Table 2.5, and Table 2.6. Hughes et al. (2003) discuss the technique for estimating market-value shortfall as follows.

Tobin's q ratio is proxied by the ratio of the market value of the bank's assets to the book value of the bank's assets, where the market value of assets is measured as the sum of the market value of equity and the book value of liabilities and the book value of assets excludes goodwill.

Letting MVA_i denote the market value of the ith bank's assets and BVA_i their book value of assets less goodwill, the equation of the frontier is given by

$$MVA_i = \alpha + \beta BVA_i + \gamma BVA_i^2 + \varepsilon_i, \tag{2.A.1}$$

where $\varepsilon_i = v_i - \mu_i$ is a composite error term used to distinguish statistical noise, $v_i \sim iid(0, \sigma_v^2)$, from the systematic shortfall, $\mu_i(>0) \sim iid\ N(0, \sigma_\mu^2)$. The quadratic specification allows the fitted frontier to be nonlinear. This equation is estimated using maximum likelihood. The deterministic kernel defines the best-practice frontier:

$$FMVA_i = \alpha + \beta BVA_i + \gamma BVA_i^2. \tag{2.A.2}$$

The stochastic frontier is composed of the deterministic kernel and the two-sided error term:

$$SFMVA_i = \alpha + \beta BVA_i + \gamma BVA_i^2 + v_i \tag{2.A.3}$$
$$= FMVA_i + v_i.$$

The difference between the best-practice frontier market value and the achieved value actually observed for the firm defines the bank's *market-value shortfall*, μ_i, and is given by

$$\mu_i = SFMVA_i - MVA_i = FMVA_i - (MVA_i - v_i). \tag{2.A.4}$$

Note that $(MVA_i - v_i)$ gives the noise-adjusted market value of assets.

Since the inefficiency component of the composite error term cannot be directly estimated, it is computed as the expectation of μ_i conditional on ε_i:

$$E(\mu_i \mid \varepsilon_i) = FMVA_i - (MVA_i - E(v_i|\varepsilon_i)).^{[7]} \tag{2.A.5}$$

We measure a bank's *market-value inefficiency* by normalizing its inefficiency by its frontier value. This shortfall ratio gives lost market value (after correcting for luck) to its best-practice value and is a gauge of the degree of agency problems in the bank:

$$\text{Market-value inefficiency}_i = \frac{E(\mu_i|\varepsilon_i)}{FMVA_i}. \tag{2.A.6}$$

In this chapter, we present results of regressing market-value inefficiency on various measures of market discipline and governance.

References

Adams, R.B. and Mehran, H. (2008, revised 2011) Corporate performance, board structure, and their determinants in the banking industry. *Staff Reports no. 330*. Federal Reserve Bank of New York.

Adams, R.B., Hermalin, B.E., and Weisbach, M.S. (2010) The role of boards of directors in corporate governance: a conceptual framework and survey. *Journal of Economic Literature*, **48**, 58–107.

---

[7] See Jondrow et al. (1982), who first proposed computing the firm-specific inefficiency by the conditional expectation; more details of this procedure can also be found in the survey by Bauer (1990).

Ang, J.S., Cole, R.A., and Lin, J.W. (2000) Agency costs and ownership structure. *Journal of Finance*, **55**, 81–106.

Barth, J.R., Caprio, G., Jr, and Levine, R. (2004) Bank regulation and supervision: what works best? *Journal of Financial Intermediation*, **13**, 205–248.

Barth, J.R., Caprio, G., Jr, and Levine, R. (2006) *Rethinking Bank Regulation: Till Angels Govern.* Cambridge University Press, Cambridge, MA.

Bauer, P.W. (1990) Recent developments in the econometric estimation of frontiers. *Journal of Econometrics*, **46**, 39–56.

Beltratti, A. and Stulz, R.M. (2009) Why did some banks perform better during the credit crisis? A cross-country study of the impact of governance and regulation. Finance working paper 254/2009. European Corporate Governance Institute.

Berger, A.N. and Hannan, T.H. (1998) The efficiency cost of market power in the banking industry: a test of the 'quiet life' and related hypotheses. *Review of Economics and Statistics*, **80**, 454–465.

Berger, A.N. and Mester, L.J. (1997) Inside the black box: what explains differences in the efficiencies of financial institutions? *Journal of Banking and Finance*, **21**, 895–947.

Brook, Y., Hendershott, R., and Lee, D. (1998) The gains from takeover deregulation: evidence from the end of interstate banking restrictions. *Journal of Finance*, **53**, 2185–2204.

Cannella, A.A., Jr, Fraser, D.R., and Lee, D.S. (1995) Firm failure and managerial labor markets: evidence from Texas banking. *Journal of Financial Economics*, **38**, 185–210.

Cheng, I.-H., Hong, H., and Scheinkman, J.A. (2010) Yesterday's heroes: compensation and creative risk-taking. NBER working paper 16176.

Chortareas, G.E., Girardone, C., and Ventouri, A. (2012) Bank supervision, regulation, and efficiency: evidence from the European Union. *Journal of Financial Stability*, **8**, 292–302.

Core, J.E., Holthausen, R.W., and Larcker, D.F. (1999) Corporate governance, chief executive officer compensation, and firm performance. *Journal of Financial Economics*, **51**, 371–406.

Cronqvist, H. and Fahlenbrach, R. (2009) Large shareholders and corporate policies. *Review of Financial Studies*, **22**, 3941–3976.

Delis, M.D., Molyneux, P., and Pasiouras, F. (2011) Regulations and productivity growth in banking: evidence from transition economies. *Journal of Money, Credit, and Banking*, **43**, 735–764.

DeYoung, R.E., Hughes, J.P., and Moon, C.-G. (2001) Efficient risk-taking and regulatory covenant enforcement in a deregulated banking industry. *Journal of Economics and Business*, **53**, 255–282.

DeYoung, R., Spong, K., and Sullivan, R.J. (2001) Who's minding the store? Motivating and monitoring hired managers at small, closely held commercial banks. *Journal of Banking and Finance*, **25**, 1209–1243.

Ellul, A. and Yerramilli, V. (forthcoming) Stronger risk controls, lower risk: evidence from U.S. bank holding companies. *Journal of Finance*.

Evanoff, D.D. and Örs, E. (2002) Local market consolidation and bank productive efficiency. Federal Reserve Bank of Chicago working paper 2002-25.

Fahlenbrach, R. and Stulz, R.M. (2011) Bank CEO compensation and the credit crisis. *Journal of Financial Economics*, **99**, 11–26.

Fama, E.F. and Jensen, M.C. (1983) Separation of ownership and control. *Journal of Law and Economics*, **26**, 301–325.

Gaver, J.J. and Gaver, K.M. (1993) Additional evidence on the association between the investment opportunity set and corporate financing, dividend, and compensation policies. *Journal of Accounting and Economics*, **16**, 125–160.

Gorton, G. and Rosen, R. (1995) Corporate control, portfolio choice, and the decline of banking. *Journal of Finance*, **50**, 1377–1420.

Grossman, R.S. (1992) Deposit insurance, regulations, and moral hazard in the thrift industry: evidence from the 1930s. *American Economic Review*, **82**, 800–821.

Hermalin, B.E. and Weisbach, M.S. (2003) Boards of directors as an endogenously determined institution: a survey of the economic literature. Federal Reserve Bank of New York. *Economic Policy Review*, **9**, 7–26.

Hicks, J.R. (1935) Annual survey of economic theory: the theory of monopoly. *Econometrica*, **3**, 1–20.

Holderness, C.G. (2003) A survey of blockholders and corporate control. Federal Reserve Bank of New York. *Economic Policy Review*, **9**, 51–64.

Houston, J. and James, C. (1995) CEO compensation and bank risk: is compensation in banking structured to promote risk taking? *Journal of Monetary Economics*, **36**, 405–431.

Hughes, J.P. (2011) *Syllabus: Economics 408 Market Discipline.* http://economics.rutgers.edu/people/218-hughes-joseph-p (accessed on 12 December 2012).

Hughes, J.P. and Mester, L.J. (2010) Efficiency in banking: theory, practice, and evidence, in *The Oxford Handbook of Banking* (eds A.N. Berger, P. Molyneux, and J. Wilson), Oxford University Press, Oxford, pp. 463–485.

Hughes, J.P. and Mester, L.J. (2011) Who said large banks don't experience scale economies? Evidence from a risk-return-driven cost function. Wharton Financial Institutions working paper #11–47 and Federal Reserve Bank of Philadelphia working paper no. 11–27.

Hughes, J.P. and Moon, C.-G. (2003) Estimating managers' utility-maximizing demand for agency goods. Rutgers University, New Jersey, Department of Economics working paper 2003-24.

Hughes, J.P., Lang, W., Mester, L.J., and Moon, C.-G. (1996) Efficient banking under interstate Branching. *Journal of Money, Credit, and Banking*, **28**, 1045–1071.

Hughes, J.P., Lang, W., Moon, C.-G., and Pagano, M. (1997) Measuring the efficiency of capital allocation in commercial banking. Federal Reserve Bank of Philadelphia working paper no. 98-2 (revised as Rutgers University Economics Department working paper 2004-1).

Hughes, J.P., Lang, W., Mester, L.J., and Moon, C.-G. (2000) Recovering risky technologies using the almost ideal demand system: an application to U.S. banking. *Journal of Financial Services Research*, **18**, 5–27.

Hughes, J.P., Mester, L.J., and Moon, C.-G. (2001) Are scale economies in banking elusive or illusive? Evidence obtained by incorporating capital structure and risk-taking into models of bank production. *Journal of Banking and Finance*, **25**, 2169–2208.

Hughes, J.P., Lang, W., Mester, L.J., et al. (2003) Do bankers sacrifice value to build empires? Managerial incentives, industry consolidation, and financial performance. *Journal of Banking and Finance*, **27**, 417–447.

Jensen, M.C. and Meckling, W.H. (1976) Theory of the firm: managerial behavior, agency costs, and ownership structure. *Journal of Financial Economics*, **5**, 305–360.

Jondrow, J., Lovell, C.A.K., Materov, I.S., and Schmidt, P. (1982) On the estimation of technical efficiency in the stochastic frontier production function model. *Journal of Econometrics*, **19**, 233–238.

Keeley, M.C. (1990) Deposit insurance, risk, and market power in banking. *American Economic Review*, **80**, 1183–1200.

Koehn, M. and Santomero, A. (1980) Regulation of bank capital and portfolio risk. *Journal of Finance*, **35**, 1235–1244.

Laeven, L. and Levine, R. (2009) Bank governance, regulation and risk taking. *Journal of Financial Economics*, **93**, 259–275.

Marcus, A.J. (1984) Deregulation and bank financial policy. *Journal of Banking and Finance*, **8**, 557–565.

McConnell, J.J. and Servaes, H. (1995) Equity ownership and the two faces of debt. *Journal of Financial Economics*, **39**, 131–157.

Morck, R., Shleifer, R.M., and Vishny, R.W. (1988) Management ownership and market valuation: an empirical analysis. *Journal of Financial Economics*, **20**, 293–316.

Nakamoto, M. and Wighton, D. (2007) Citigroup chief stays bullish on buy-outs. Financial Times (July 9), p. 1.

Pasiouras, F. (2008) International evidence on the impact of regulations and supervision on banks' technical efficiency: an application of two-stage data envelopment analysis. *Review of Quantitative Finance and Accounting*, **30**, 187–223.

Pasiouras, F., Tanna, S., and Zopounidis, C. (2009) The impact of banking regulations on banks' cost and profit efficiency: cross-country evidence. *International Review of Financial Analysis*, **18**, 294–302.

Petersen, M.A. and Rajan, R.G. (1995) The effect of credit market competition on lending relationships. *Quarterly Journal of Economics*, **110**, 407–443.

Schranz, M.S. (1993) Takeovers improve firm performance: evidence from the banking Industry. *Journal of Political Economy*, **101**, 299–326.

Smith, C.W. and Watts, R.L. (1992) The investment opportunity set and corporate financing, dividend, and compensation policies. *Journal of Financial Economics*, **32**, 263–292.

Tanna, S., Pasiouras, F., and Nnadi, M. (2011) The effect of board size and composition on the efficiency of U.K. banks. *International Journal of the Economics of Business*, **18**, 441–462.

Weston, J.F. (1979) The tender takeover. *Mergers and Acquisitions*, **14**, 74–82.

3

Modeling economies of scale in banking: Simple versus complex models

Robert DeYoung

KU School of Business, University of Kansas, USA

3.1 Introduction

According to American folklore, Abraham Lincoln was once asked: 'Abe, how long should a man's legs be?' Lincoln, who was famously tall and long-legged for his time, is reported to have answered: 'Long enough to reach the ground'.

How large should a bank be? Unfortunately, we can no longer ask Abe for his wisdom. And although President Lincoln signed into law the National Banking Acts of 1863 and 1864 that created the national banking system in the United States, the optimal size of those newly chartered national banks was never an important concern in his time.

Today, things are different. The leading banks in most western economies have grown substantially larger over the past several decades – so large that their regulators and government legislatures take actions to prop them up when they approach insolvency, for fear that large bank failures could have contagious effects on other banks, financial markets, or the macroeconomy. While the realization that some banks that are too big to fail (TBTF) is not a new one, recent actions by western governments to prop up their large, insolvent banks have been heroic. During the global financial crisis in 2008, large banks in the United States (e.g., Citibank, Bank of America, Fannie Mae, Freddie Mac, AIG), the United Kingdom (Royal Bank of Scotland, Lloyd's, Northern Rock), Switzerland (UBS), The Netherlands (ING), Germany (CommerzBank, Hypo Real Estate), Belgium/France (Dexia), and elsewhere received huge government capital injections, extensive government financial guarantees, or were nationalized. Close on the heels of those bailouts, the European Central Bank took actions to stabilize large banks across Europe that had suffered losses during the sovereign debt crisis in 2011.

Bailing out some failed enterprises but not others raises questions of fairness in market economies, and losses incurred during government bailouts redound to the taxpayers. TBTF has become a very visible public policy problem, perhaps best summed up by Bernie Sanders on the floor of the US Senate on September 19, 2008: 'If a bank is too big to fail, then it is too big to exist'. So how big should we allow a *large bank* to get?

At the other end of the bank size spectrum, the best size for banks is less a public policy question than a business policy question. Small businesses are essential to innovation and job creation in most market economies, and these firms typically rely on small banks for credit. But small banks have been disappearing in recent years – in the United States, the number of so-called community banks has declined from around 14 000 in the 1990s to less than 7000 today – and this has engendered worries about small businesses' access to credit. Most of the banks that disappeared simply discovered that their small size and lack of expertise prevented them from being competitively viable in the deregulated banking environment of the 1990s and 2000s, and exited the market by selling themselves to other banks. So how big does a *small bank* have to be in order to be financially viable?

This chapter explores these two questions – how large is too large for banks, and how small is too small for banks – using two starkly different modeling approaches. First, we review and critique the findings of studies written over the past several decades by banking scholars and regulators. These studies use sophisticated econometric models to estimate bank cost functions, and then make inferences about the 'too big' and 'too small' questions based on the shapes of those estimated cost functions. The results of these studies vary drastically and can be very sensitive to small changes in data and method; after several decades of work, this body of research has not generated anything close to consensus answer regarding our two questions.

Second, we abandon complexity for more basic, commonsense approaches that lend themselves to graphical presentation. We use simple frameworks that allow us to draw inferences about bank scale economies simply by 'looking at' the data. These simple frameworks are based on first-principles economics; the frameworks use bottom-line data such as bank profits, bank asset size, and bank stock market value; and the methods we employ are familiar to first-year statistics students. By combining a series of these simple models, we can offer a definitive answer for the 'too small' question and a suggestive answer for the 'too big' question.

The real message of this chapter is not really about banks. Rather, it is about needless complexity in modeling, and harkens back to the lesson of Occam's Razor. Complex modeling is costly, and there is no guarantee that the answers generated by complex models are any better than those generated by simple models.

3.2   The increasing size of banks

As a rule, large banks have grown larger over the past several decades in most market economies. Figure 3.1 shows how the size distribution of commercial banks changed in the United States from 1976 through 2010. Although banks of all sizes have grown larger, the distribution of bank assets has skewed much more toward the largest banks. The real assets

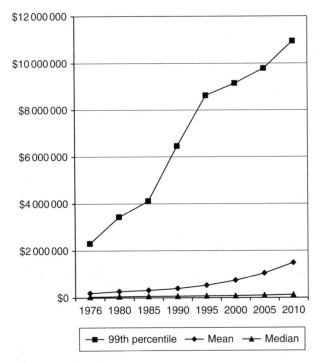

Figure 3.1   Changes in the distribution of assets at US commercial banks, 1976–2010. Assets in thousands of 2010 dollars. Data from FDIC and author's calculations.

(in 2010 dollars) of the median bank more than tripled from $37 million to $126 million – but the assets of the bank at the 99th percentile of the industry nearly quintupled from $2.3 billion to $10.9 billion. This upward skew is seen in the mean bank size, which increased by an order of magnitude from $192 million to $1.5 billion. Much of the increase in size was implemented by banks merging with each other, and those combinations are largely responsible for the reduction in the number of US commercial banks, as shown in Figure 3.2, from more than 15 000 banks to less than 7000 banks. (For a detailed description of the laws and regulations that have permitted geographic expansion of banks in Europe and the United States, and also some data showing the increased number of bank mergers in both places, see Berger et al. (2000).) Importantly, the increased skewedness in bank size has resulted in a large concentration of bank assets at a handful of very large banks. As shown in Figure 3.3, the asset share of the top 100 banks increased from around 25% to more than 50%, while the largest 10 banks – which traditionally have held a little less than half of all banking assets – now hold more than 80% of industry assets.

   A similar size transformation has occurred in European and Asian countries.[1] For example, the three largest banking companies by asset size in the United States, JPMorgan Chase, Bank of America, and Citigroup, grew their assets from $591 billion, $588 billion, and $591 billion, respectively, in 2001 to $1.8 trillion, $1.7 trillion, and $1.3 trillion by the end of 2010. The two largest banking groups in the United Kingdom, Barclays and Royal Bank of Scotland,

---

[1] All data in this paragraph is from Wikinvest.com.

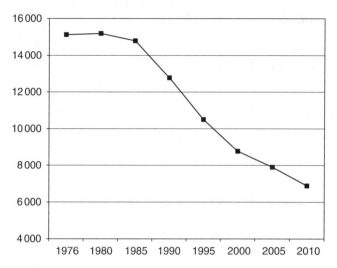

Figure 3.2   Number of US commercial banks, 1976–2010. Data from FDIC.

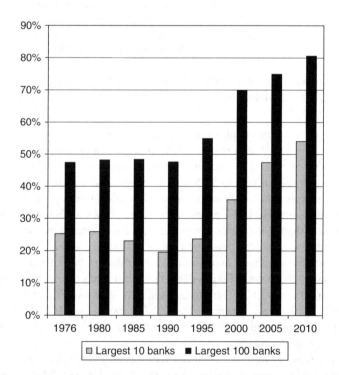

Figure 3.3   Percentage of industry assets held by the largest US commercial banks, 1976–2010. Data from FDIC and author's calculations.

grew their assets from $536 billion and $519 billion in 2001 to $2.2 trillion and $3.3 trillion in 2010. The largest bank in Germany, Deutsche Bank, grew its assets from $818 billion in 2001 to $2.4 trillion in 2010. The largest banks in other leading western countries experienced similar growth during the 1990s and 2000s. In all these countries, the fundamental drivers of larger bank size were the same: innovations in information and communication technologies, new and deeper financial markets, and banking industry deregulation.

3.3   What has allowed banks to grow larger?

At the beginning of the postwar period, most banks faced legal restrictions that governed the geography of their operations. For example, the McFadden Act of 1927 prohibited US banks from branching into bordering states, and within many of the 50 states banks were limited to operating within a city or county. The size of US banks was further restricted by the Glass-Steagall Act of 1933, which largely prohibited commercial banks from offering investment banking and securities brokerage services. In Europe, a cobweb of different currencies, tax laws, legal codes, and other trade restrictions largely prohibited banks from branching freely into neighboring countries. Some relief from these restrictions was available via limited cross-state ownership in the United States and limited cross-country ownership in Europe, but these arrangements made geographic expansion costly and effectively restricted the size of banking companies in both places. (For a discussion of laws and regulations that at first restricted, and then permitted, geographic expansion of banks in the United States and Europe, see DeYoung (2010) and Goddard, Molyneux, and Wilson (2010).)

The main rationale for geographic banking restrictions was at once political and economic. Limiting the entry of banks from outside the local market dampened competitive rivalry in those markets, which in turn (a) supported the profitability of local banks and (b) reduced the probability of bank failure. Competition was also held in check by restrictions on deposit pricing (especially in the United States) and difficulty securing a charter to start a new bank. Indeed, the number of commercial banks in the United States stayed around 14 000 from the mid-1930s through the mid-1980s, despite a near doubling of population and a sevenfold increase in real GDP. As inexorable progress in information, communications, and financial technology reduced the potential cost of operating banks with larger geographic footprints, pressure mounted to eliminate geographic restrictions on banks. In response to these new technological realities, banking markets in the United States, Europe, and elsewhere were deregulated, allowing banks to expand and take advantage of these efficiencies.

3.3.1   New banking technologies

Technological progress has diminished the importance of physical location by allowing banks to communicate cheaply with their customers at a distance. In banking, this started with the automated teller machine (ATM). By allowing branch transactions without branch buildings, ATMs provided greater convenience for retail customers, increased bank efficiency by substituting technology for expensive human tellers, and enhanced bank revenues by charging fees to customers of other banks. Internet banking and the increased use of electronic payments further diminished the importance of bank brick-and-mortar geography. Electronic payments allow depositors and borrowers to do their banking anywhere, unlike payments made with cash or checks, which require physical branches for distribution, collection, and clearing. Because the marginal cost of producing a basic Internet banking transaction is very low, electronic delivery channels exhibit larger scale economies than paper-based delivery channels (DeYoung, 2005). Moreover, there is evidence that small banks can enhance their profitability by offering Internet banking services (DeYoung, Lang, and Nolle, 2007). Today, most banks combine Internet banking with physical branches – the so-called click and mortar model – allowing banks to expand their deposit bases without adding a proportionate amount of branch office overhead. These innovations have made physical bank locations more productive over time: assets, operating income, and the number of transactions per banking office have all increased in the United States since the 1980s despite a doubling of the number of bank offices (DeYoung, Hunter, and Udell, 2004). And the increasing ability of bank managers to exchange information via electronic channels has steadily reduced the management problems associated with operating geographically far-flung banking organizations (Berger and DeYoung, 2006).

Technological progress has also introduced new scale-based efficiencies to bank lending. For example, the rapid expansion of syndicated loan markets has enabled banks to participate in very large commercial loans while maintaining both liquidity (syndicated loans can be sold in an over-the-counter secondary market) and a diversified loan portfolio (Berlin, 2007). But the most transformational technological advancement in bank lending is the combination of consumer credit scoring models and loan securitization known as 'transactions banking'. This business model has allowed banks to originate large amounts of consumer and real estate loans, made to borrowers who they have never met, and without having to raise any deposits to fund the loans.

A loan securitization is a trust that purchases existing home mortgage loans (or auto loans, or credit card receivables) from banks, using funds raised by selling mortgage-backed securities (MBSs) to third-party investors. The MBSs yield returns based on the performance of the mortgage loans held in the trust. This process allows banks to sell their otherwise illiquid loans to the securitization, and use the proceeds of these sales to fund additional loans or make other alternative investments. Many large retail banks have transformed themselves from traditional 'originate-and-hold' mortgage lenders to 'originate-and-securitize' mortgage bankers, relying less on traditional interest-based income and increasingly more on noninterest income from loan origination fees, loan securitization fees, and loan servicing fees.

Loan securitization would not be possible without credit scoring, a process that transforms quantitative information about individual borrowers (such as income, employment, or payment history) into a single numerical 'credit score'. Lending banks use credit scores to analyze loan applications quickly and inexpensively; investment banks use credit scores to group loans with similar risk characteristics into pools to be securitized; and bond-rating companies use credit scores to assign risk ratings to the mortgage-backed securities that finance those pools. Credit scoring has significantly reduced the unit cost of underwriting individual loans, because it economizes on expensive loan officer time in deciding whether to accept or reject loan applications. By substituting information technology for labor inputs, credit scoring has increased the minimum efficient scale of consumer loan underwriting operations and has expanded lenders' incentives to make additional credit available (Berger, Frame, and Miller, 2005; Frame, Srinivasan, and Woosley, 2001).

3.4   Why do banks choose to be large?

The nearly universal increase in the size of banks over the past two decades is, for some, prima facie evidence of scale economies in banking. But even in the absence of opportunities for scale economies, banks still have reasons to grow large. On one hand, the desire to rule a large corporate empire can drive bank managers to grow their firms beyond the size that maximizes shareholder value. On the other hand, gaining access to subsidies that the government bestows on very large banks is a powerful incentive for banks to grow inefficiently large at the taxpayers' expense.

3.4.1   Objectives of bank management

When a board of directors does not adequately monitor, direct, or discipline its top management, those managers are likely to take actions that enhance their own utility rather than shareholder value. A classic form of behavior in these situations is 'empire building', in which the manager makes negative net present value acquisitions and other investments that make the firm (aka, the manager's empire) larger. One way for the management to defend its actions is by claiming that growth is necessary to reduce costs, marketing synergies, or some other type of size-based efficiency that is unavailable to the original, smaller firm. Bliss and Rosen (2001) investigated bank acquisitions in the United States during the 1990s, and found evidence strongly consistent with empire building motives rather than efficiency motives.

Other managerial utility-maximizing behaviors can limit, rather than expand, the size of the firm. For example, some managers – perhaps those late in their careers who simply wish to hold on to their positions until retirement – may opt for a 'quiet life' in which the chance of experiencing a loss is small from year to year. These risk-averse managers will reject investment opportunities that expose the firm to risk, even if those investments have positive net present values and increase shareholder value. At publicly traded firms – where the shareholder can be characterized as the 'principal' and the manager can be characterized as the shareholder's 'agent' – such behavior qualifies as a principal–agent problem and results in an inefficiently small firm. But the majority of banks in the world are privately held, and many of these banks are owner-managed. In these cases, risk-averse investment practices that keep the bank relatively small do not constitute corporate governance problems, because the owner-manager represents himself or herself – he or she is both principal and agent. Operating these banks in a risk-averse fashion may be the value-maximizing strategy for owner-managers whose personal wealth is undiversified and dominated by their ownership stake in the bank. DeYoung, Spong, and Sullivan (2001) show that small owner-managed banks operate less efficiently than small banks that hire outside managers, a result that is consistent with risk-averse owner-manager behavior. Spong and Sullivan (2007) and DeYoung (2007) show that small owner-managed banks exhibit less risk when the manager's personal wealth is concentrated in bank stock.

Bank managers, like most other labor market participants, respond to the incentives placed in their employment contracts. Recent research (e.g., DeYoung, Peng, and Yan, 2012; Suntheim, 2011) shows that bank CEOs who get paid with stock options are more likely to pursue risky investment strategies relative to CEOs who get paid with stock shares. This is because an increase in earnings volatility enhances the value of a stock option, while an increase in earnings enhances the value of stock shares (and hence aligns the CEO's best interests with the shareholders' best interests). Hence, depending on the composition of his or her contract, a bank CEO might choose the bank size necessary to generate the amount of risk that maximizes his or her own wealth, rather than choosing the bank size necessary to maximize profit or minimize costs.

3.4.2   Government subsidies

Banking is more highly regulated than other industries in market economies for a variety of reasons – not the least important of which is that banks issue debt (deposits) that can be used as an efficient means of exchange (money). Because of this, insolvent banks are denied access to bankruptcy protection; allowing this would limit or delay depositors access to their deposits, which would reduce the money supply in the markets from which the insolvent bank collects its deposits. If the insolvent bank was large enough, this could generate economy-wide illiquidity. Instead, government regulators (e.g., the FDIC in the United States, the CFDIC in Canada, the FSA in the United Kingdom) seize and 'resolve' the insolvent institution, selling off the bank's assets in whole or piece by piece while providing depositors access to their funds. However, regulators sometimes lack the resources or legal powers to effectively seize and resolve large failed banks, and in these cases the failed banks have been bailed out; they are allowed to continue operating after receiving capital injections, loans, or other special considerations from the government treasuries or central banks. These banks are TBTF.

Because being bailed out is better than going belly up, such policies provide an incentive for banks to grow inefficiently large, so long as the benefits of TBTF status (e.g., lower cost of debt and equity capital) outweigh the costs associated with larger size (e.g., decentralization, reduced managerial control). In a study of large bank mergers, Penas and Unal (2004) found that banks' costs of debt funding declined significantly after acquisitions propelled them into the top size tier of banking companies. This size-related reduction in the credit risk premium can be thought of as a 'reward' for becoming TBTF. Molyneux, Schaeck, and Zhou (2010) look at large European banks, and Brewer and Jagtiani (2011) look at large US banks, and both find that banks pay higher merger premiums in acquisitions that push the acquiring bank close to or over the TBTF threshold. The belief that government regulators would not let the very largest banks fail was confirmed in recent years when large banks in Europe and the United States approached insolvency and were bailed out by their government treasuries.

3.4.3   Scale economies

For now, let us assume that banks give their managers contracts that penalize empire building, so that bank managers no longer have personal incentives to grow their banks inefficiently large. And let us also assume that the banking regulators have the ability to seize and resolve large complex banks, so that banks no longer have incentives to become TBTF.[2] In this optimal world, without the shareholder benefits of TBTF or the managerial benefits of empire-building, large banks might decide to shrink in size. Or perhaps not, because large size might still convey benefits to banks such as lower costs per unit, higher revenues per customer, or better returns per unit of risk. The question, then, is whether scale economies exist in bank production, marketing, and/or risk management. If so, banks would still have incentives to grow large.

Casual observation reveals that most banks are not large; indeed, quite the opposite is the case. For example, about 19 out of 20 commercial banks in the United States have less than

---

[2] In the United States, the Dodd-Frank 'Wall Street Reform and Consumer Protection Act' of 2010 provided enhanced resolution powers to the FDIC so that it will (arguably) be able to seize and resolve any size insolvent bank in the future.

$2 billion in assets – which makes them 'community banks' – and most of these small banks earn sustainable levels of profit (DeYoung, Hunter, and Udell, 2004). And over 1200 brand new, or 'de novo', banks have started up in the United States since 2000. These observations imply that scale economies in bank production, marketing, and risk management may exist, but they are not so great as to render small- and medium-sized banks financially nonviable. Conversely, there may be benefits to being small that partially or fully offset production, marketing, and risk-management diseconomies of scale. A large research literature, reaching back nearly 50 years, has attempted to gauge these very phenomena – with researchers often coming to very different conclusions. We now turn to a review of those studies.

3.5   Econometric modeling of bank scale economies

Can banks become more efficient by growing larger? This is an important question for the owners, boards, and managers of banks. Furthermore, if larger banks are more efficient than small banks, then are these private efficiencies (higher profits for banks, lower prices for bank customers) large enough to justify the potential social costs of large banks (the potential for macroeconomic disruption when a large bank fails)? These are important questions for society. Neither of these questions can be answered unless we have good measures of the cost savings from bank scale economies. Researchers in US universities and US regulatory agencies have been attempting to measure the size and extent of economies of scale in banking since the 1960s, and it is a rich and evolving literature.

Although scale economy research has been performed using data on banks in countries across the globe, I focus mainly on US studies in this chapter for two reasons. First, most of the technical innovations in bank scale economy measurement were developed and implemented by US researchers, and these techniques were only later applied to banks in other countries. Second, the sheer number of banks in the United States allows researchers to estimate these models with greater statistical precision.

Researchers typically test for the existence and magnitude of bank scale economies by estimating econometric cost functions. These are statistical estimates of the expenses necessary for banks to produce a given amount of banking products and services. A very simplified bank cost function, for a bank that produces loans but no other financial products or services, might look like the following:

$$\ln(\text{cost}) = \alpha + \beta \times \ln(\text{loans}) + \gamma \times \ln(\text{deposit rate}) + \delta \times \ln(\text{wage rate}). \qquad (3.1)$$

The cost of production has a fixed component α (overhead costs); costs increase as the bank makes more loans ($\beta > 0$); and costs increase if the price the bank has to pay to get deposits or hire labor increases ($\gamma > 0$, $\delta > 0$). Note that all of the variables are expressed in terms of natural logs – for example, the left-hand side of the equation has ln(cost) instead of just cost – and this is referred to as a 'functional form'. This particular functional form is the Cobb–Douglas functional form, used in many of the earliest research studies. To perform the analysis, the researcher would collect data on costs, loans, deposit rate, and wage rate for a large sample of banks, and then estimate the Cobb–Douglas cost function using multiple regression analysis. The regression would generate values for the coefficients α, β, γ, and δ, and these coefficient estimates would tell us, on average, by how much bank costs increase when loans, deposit rates, or wage rates increase. In this simple example, banks would be revealed to have scale economies if the estimated value for β was less than 1. That is, $\beta < 1$ indicates that increasing the size of the bank by 1% would increase bank costs by less than 1%. Conversely, banks would be revealed to have scale diseconomies if $\beta > 1$, indicating that increasing the size of the bank by 1% would increase bank costs by more than 1%.

Over time, bank researchers have attempted to increase the realism and the accuracy of their cost function estimations; in the process, these econometric exercises have grown increasingly complex. Some of the main improvements include: using more than one loan output (e.g., business loans, consumer loans, real estate loans), adding nonloan outputs (e.g., depositor services, off-balance sheet activities), adding variables to capture the cost effects of financial leverage and loan quality, using functional forms that are more 'flexible' than the Cobb–Douglas form (and hence better able to closely fit the data), adjusting the cost function to reveal the 'best practices' cost frontier, and estimating revenue scale economies and profit scale economies. Clark (1988) provides a good early explanation of bank cost function estimation.

3.5.1   Findings from 50 years of studies

While our interest today in bank scale economies is due to our concern that our largest banks may be too large, earlier interest in bank scale economies was due to concerns that our largest banks were *too small*. In the United States, during the 1960s, federal law (the McFadden Act of 1927) prohibited banks from expanding naturally across state lines, and laws in many of the individual states prohibited banks from expanding naturally across county lines. Bankers and others argued that these rules made the banking system inefficient. Removing these constraints, they said, would enable banks to expand their geographic footprints and by growing larger capture scale economies. Because banking services are sold in competitive markets – as stated earlier, there were about 14 000 commercial banks in the United States at the time – much of the resulting cost savings would be passed along to customers and not simply accrue to bank shareholders as increased profits.

The unknown element in this argument was whether scale economies actually existed in banks. Did the cost of producing loans and deposits decline as banks grew larger? If so, by how much? And for what size bank did costs stop declining, that is, what was the 'minimum efficient scale' for a bank? Researchers began studying banking data in search of answers to these questions. At first, the research suggested that larger banks were no more efficient than small banks; but as time passed and researchers' methodologies improved, estimates of bank scale economies increased. Gilbert (1984) and Clark (1988) review the first 25 years of bank scale economy research: the former reports on 18 studies published between 1963 and 1983, while the latter reports on 13 studies published between 1983 and 1987. These studies were nearly uniform in finding scale economies for banks with less than $100 million in assets (approximately $200–$400 million in today's dollars), but only a handful of these studies found scale economies for larger banks. Thus, the research provided justification for some expansion of bank size – say, the merging of two small banks from adjacent states to create a medium-sized bank with a more efficient cost structure.

The results of these early studies were constrained by the availability of data and (by today's standards) the relative crudeness of the available statistical methods for analyzing this data. At the time, only a small portion of banks had more than $100 million in assets, and those that did were operating under regulatory constraints (i.e., geographic limitations) that required them to use suboptimal production processes. One of the popular data sets of the day – the Federal Reserve's Functional Cost Analysis data – included only a subset of banks and only a very few had more than $1 billion (Gilbert, 1984). And the earliest of these studies used the Cobb–Douglas functional form, which is not capable of fitting a cost function that is 'U-shaped', the usual shape predicted by economic theory. (In a U-shaped cost function, a bank just large enough to be at the bottom of the U-shape is considered to be operating at minimum efficient scale.)

During the 1980s and 1990s, a number of US states relaxed or eliminated their geographic banking limits. Banks quickly took advantage of this to grow larger (usually via merger with other banks), and this provided important new data for researchers to include in their studies. Researchers also developed better analytical methods – for example, the widespread adoption of the 'translog' functional form for estimating cost functions. For the very simple bank we discussed earlier that produces only loans, the translog cost function would look like the following:

$$
\begin{aligned}
\ln(\text{cost}) = {} & \alpha + \beta \times \ln(\text{loans}) + \gamma \times \ln(\text{deposit rate}) + \delta \times \ln(\text{wage rate}) \\
& + \eta \times \ln(\text{loans})^2 + \theta \times \ln(\text{deposit rate})^2 + \lambda \times \ln(\text{wage rate})^2 \\
& + \rho \times \ln(\text{loans}) \times \ln(\text{deposit rate}) + \phi \times \ln(\text{loans}) \times \ln(\text{wage rate}) \\
& + \omega \times \ln(\text{deposit rate}) \times \ln(\text{wage rate}).
\end{aligned}
\tag{3.2}
$$

Unlike the Cobb–Douglas functional form, the translog functional form is capable of finding a U-shaped cost function. Berger, Demsetz, and Strahan (1999) review econometric bank scale economy studies published between 1986 and 1997, nearly all of which used this functional form. These studies found a substantially higher range for minimum efficient scale. Depending on the study in question, the estimated bottom of the U-shape occurred for banks with as little as $100 million in assets or for banks with as much as $25 billion in assets, with the resulting cost reductions anywhere between 5% and 20%. Thus, these studies provided justification for further expansion of bank size – say, the merging of two medium-sized banks from adjacent states to create a region-wide bank with a more efficient cost structure.

These scale economy studies provided an objective, empirical argument with which to support the Riegle-Neal Act of 1997, which by rescinding the last major restrictions to geographic expansion in the United States, paved the way for truly nationwide banking franchises. This ushered in the final phase of industry consolidation, with region-wide banks purchasing other region-wide banks, creating companies with a national presence. These combinations resulted in companies that were far larger than the banks examined in existing studies. For example, today the three largest US banking firms (Bank of America, JPMorgan Chase, and Citigroup) all exceed $2 trillion in assets, while the three next largest (Wells Fargo, Goldman Sachs, and Morgan Stanley) all have assets in excess of $800 billion, well above the range covered previously by academic researchers. Did these new banking behemoths also benefit from scale economies?

Compared to the frequency with which new studies on bank scale economies emerged during the 1970s, 1980s, and 1990s, fewer studies on bank scale economies have been performed since 2000. But those that have surfaced are important, because they are the first to include data on the huge nationwide banks that emerged after interstate deregulation in the United States and intercountry deregulation in Europe. These studies are also important for their new and startling findings that even the very largest of today's incredibly large banks may have access to scale economies. For example, Hughes and Mester (2011) estimate the cost function for US commercial banking companies in 2007, the last full year of data prior to the financial crisis. Among other innovations, these authors recognize an important point that may have been causing earlier studies to understate the scale economies enjoyed by large banks. Because large banks can diversify their risk across larger portfolios of loans, they can operate with a smaller precautionary cushion of equity capital, all else equal. But the savings generated by this shift in funding mix (i.e., reduced opportunity costs for bank owners) do not show up in the accounting cost figures typically used in scale economy studies. Quite to the contrary, this funding shift will result in *higher* accounting costs, because interest-bearing debt that generates positive accounting expenses is replacing equity funding that generates no accounting expenses.

After controlling for this shift in funding mix at large banks (as well as other phenomena), Hughes and Mester find scale economies for banks of all sizes. Moreover, they estimate that scale savings per dollar of bank output actually increase with bank size. For banks with more than $100 billion in assets, their model indicates that a 10% increase in bank output results in only a 5.4% increase in costs, all else equal. An earlier study by the same authors (Hughes and Mester, 1998) that used US banking data from 1994 also found increasing scale economies. And these findings are largely confirmed by Wheelock and Wilson (2012), who use a large panel of data on US banks between 1984 and 2006 and a different econometric approach, but also find scale economies or banks of all sizes. Using their model, reducing the size of the four largest US banks by half would increase costs at these banks by about 19%. Considering these results in a different context indicates that these estimated annual cost increases are huge – approximately equal to four years of accumulated profits at these banks.

This is certainly not the end of the 50-year debate over bank scale economies. Research will continue so long as growth and innovation in the banking industry continues to change the costs and benefits of bank size. Whether or not the findings of these new studies are influential in affecting bank regulatory policy is perhaps a separate question. Consider the following: In June 2011, Daniel Tarullo, then a member of the Federal Reserve Board of Governors, stated that 'There is little evidence that the size, complexity and reach of some of today's strategically important financial institutions is necessary to realize achievable economies of scale (Tarullo, 2011)'. The year before, Alan Greenspan, former Chairman of the Federal Reserve Board, wrote the following: 'For years the Federal Reserve was concerned about the ever-growing size of our largest financial institutions. Federal Reserve research had been unable to find economies of scale in banking beyond a modest size (Greenspan, 2010)'. Why would Federal Reserve officials – with access to some of the best banking economists and most innovative bankers in the world – so explicitly disregard the notion that large banks have access to scale economies, at a time when some econometric studies were suggesting that they do?

I suggest two reasons: First, the methods used in these studies have weaknesses that make industry regulators doubtful about their conclusions. Second, industry regulators are looking at other evidence. I entertain these two possibilities in the next two sections.

3.6   Weaknesses in econometric modeling of bank scale economies

Econometric estimates of bank scale economies have hardly been constant over time. As discussed earlier, studies published during the mid-1960s through the mid-1980s tended to find that scale economies diminished or disappeared once a bank had accumulated several hundred million dollars in assets; studies during the mid-1980s through the mid-1990s find scale economies continued for banks up to $25 billion in assets; and more recent studies have found scale economies for banks of all sizes. The increasing estimates of bank minimum efficient scale reflect at least three phenomena. First, advances in communications, information, and financial technologies have drastically altered the nature of financial services and the processes used to produce those services. Second, industry deregulation permitted banking companies to grow large enough to efficiently exploit those new production processes, and also fostered a nationwide competitive environment so that large banks could no longer get by with inefficient operations. And third, the techniques and methods used by bank researchers have improved, becoming better able to accurately describe bank production and costs.

But still, these techniques and methods are imperfect. One weakness is the inability of these models to separate large bank cost advantages derived from private production processes from large bank cost advantages derived from TBTF subsidies. Given the huge government bailouts of large banks in the United States and Europe in 2008–2011, there is no longer any doubt that the largest financial companies in the world are considered TBTF by their governments. If capital markets also consider these banks to be TBTF, then bank creditors are much less likely to incur losses when these banks perform poorly and should be willing to provide funding to these banks for a lower return. Thus, the largest banking companies should have a cost-of-capital advantage over their smaller rivals, and this advantage will create the appearance of private scale economies where none really exists. Davies and Tracey (2012) estimate a relatively standard translog bank cost function model with one difference: they adjust the cost of bank debt funding (the market yield of bonds issued by banks) upward to what it might be if TBTF policies no longer existed. After estimating their model for large banks in the United States and Europe, they show that after this adjustment most of the scale economies found by Hughes and Mester (1998, 2011) disappear. They conclude that large bank scale economies found in other studies are not a result of the production functions to which these banks have access, but simply reflect investor acknowledgment of TBTF subsidies, subsidies that reduce downside risk for investors and create incentives for banks to grow ever larger.

It is important to note that Davies and Tracey (2012) is not the only recent study that rejects large bank scale economies based on a traditional econometric cost function approach. Inanoglu et al. (2012) find constant returns to scale for the largest 50 US banks after adjusting their estimated cost function to reflect 'best-practices' cost performance. The larger point is this: A small adjustment to these models – e.g., measuring input prices differently, using a different specification for the error term – can lead to dramatically different results. Let us explore three other potential challenges facing researchers who use these classes of models to measure bank scale economies.

3.6.1   Few and far between

It is well known that the statistical techniques employed in most scale economy studies deliver very accurate estimates for the average-sized banks in the data, but deliver less precise estimates for banks that are substantially larger than average. One reason for this is that the distribution of bank size is highly skewed, so that the size gap between banks grows increasingly larger as banks grow in size. For example, as of March 2010, the three largest US banking companies each had assets of over $2 trillion, 10 times larger than the 13th largest banking company, Bank of New York Mellon ($220 billion), and 100 times larger than the 43rd largest bank, BOK Financial of Tulsa, Oklahoma ($23 billion). Threading an estimated bank cost function through the gaping spaces in between these large banks leaves lots of room for error.

A visualization of this problem can be had by considering the space between planets in our solar system. The Earth is 1 astronomical unit (AU) away from the Sun, or about 8 light minutes. The inner planets (Mercury, Venus, Earth, and Mars) are located relatively close to the center of the solar system, with Mercury located approximately 0.4 AU from the Sun, and Mars approximately 1.5 AU from the Sun. But the outer planets (Jupiter, Saturn, Uranus, Neptune, and Pluto) are located between 5.2 and 39.5 AU from the Sun. In other words, Pluto is on average about 40 times further away from the Sun than is Earth (1 AU). Hence, charting a course between Mercury and Venus, or between Earth and Mars, is relatively easy because these planets are close together. But the huge gaps between the outer planets make navigation much more difficult, and there is a bigger chance that our spaceship will miss its destination. By comparison, the huge asset size discontinuities between the largest US banks are even more dramatic than the huge distances between the outer planets – and unlike the planets, whose positions follow the laws of physics, bank costs can increase or decrease due to unpredictable events and sometimes are not even accurately observable. Hence, statistical estimates of scale economies among large banks can be quite sensitive to the changing financial fortunes of just one or two of these largest banks. Indeed, recall how dramatically the Davies and Tracey (2011) estimates of large bank scale economies changed when they adjusted banks' cost of debt financing for TBTF subsidies.

This does not mean that innovative researchers cannot eventually conquer these measurement challenges. For instance, the Wheelock and Wilson (2012) study cited earlier employs techniques explicitly designed to deal with the size gaps between large banks. Nevertheless, the outsized magnitudes of the cost savings they derive with this method have to be taken with a large grain of salt, an indication of how difficult a task this is.

3.6.2   Strategic groups

The number of small banks in the United States dwarfs the number of large banks. As of 2010, there were 19 banks with more than $100 billion in assets; 66 banks with between $10 billion and $100 billion; 473 banks with between $1 billion and $10 billion; and 6331 banks with less than $1 billion in assets. Hence, because the bulk of the banking data comes from very small banks, models that use this data are typically constructed based on the business processes most often used by small banks. This segment of the industry relies predominantly on traditional banking approaches: they make illiquid loans and hold them in portfolio, finance those loans by issuing liquid deposits, and earn most of their profits from the resulting interest margin. Small banks often make business decisions based on their stores of 'soft' (nonquantifiable) information collected over time from person-to-person communication with their borrowers, depositors, and the local community. This 'relationship-based' approach to banking allows small banks to serve local businesses that are unable to access public capital markets and households who require in-person financial services. Because these small banks differentiate themselves from their larger rivals by offering personalized products and services (e.g., small business loans, financial planning), and because they have a store of customer information to which their small local competitors may not have access, they can charge higher prices.

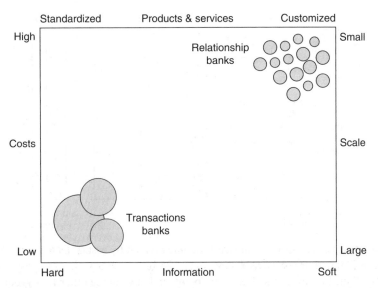

Figure 3.4   A strategic map of the banking industry. Relationship banks earn interest income by selling personalized products and services using a low-volume, high-value-added approach. Transactions banks earn noninterest income by selling commodity-like products and services using a high-volume, low-value-added approach.

In contrast, the very largest banking companies rely less on deposits and more on short-term market financing; they sell or securitize many of their loans rather than hold them; and they earn a substantial portion of their profits from customer fees rather than interest margins. Large banks favor low-cost, impersonal retail delivery channels (e.g., ATM networks, online banking, and automated payments) over customer interactions with expensive human tellers in physical branch offices. Their retail and small business lending decisions are based on 'hard' (quantifiable) information supplied to them by credit bureaus, and their large business loans are often syndicated arrangements that can be sold in the secondary market. Large banks compete against other large banks with access to the same hard information, so they face intense price competition for the nondifferentiated financial products they sell (e.g., credit cards, mortgage loans, and transactions services). Given these differences, it is not hard to believe that cost models built around small bank production processes may be incapable of accurately describing the production processes and costs at large banking companies.

DeYoung, Hunter, and Udell (2004) provide a simple but powerful framework for thinking about the differences between large and small banks. The strategic map in Figure 3.4 is an illustration of that framework, showing an industry equilibrium with two different bank clusters or strategic groups. Each bank is represented by a circle – the size of the circle represents bank size, and the position of the circle on the map indicates banks' strategic characteristics. The vertical dimension measures bank size, with large banks at the bottom and small banks at the top. The vertical dimension also measures unit costs, with low unit costs at the bottom and high unit costs at the top. Large banks have low unit costs in this framework not necessarily because of scale economies, but because large banks and small banks produce and sell fundamentally different types of banking services, and the production function used by the large banks to do this simply has lower unit costs.

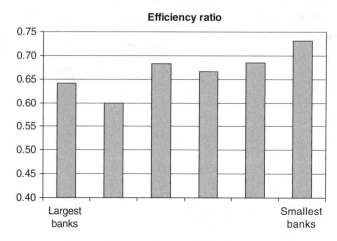

Figure 3.5   Efficiency ratio = noninterest income divided by assets. Data from US commercial banks in 2007. 'Largest Banks' have over $25 billion in assets. 'Smallest Banks' have less than $500 million in assets. The break points in between the other four groups of banks occur at $10 billion, $5 billion, and $1 billion in assets, respectively.

The horizontal dimension measures product differentiation. Banks that produce mostly differentiated or customized products and services (e.g., human tellers, customized loan contracts, personalized private banking) are located on the right, and banks that produce mostly nondifferentiated or standardized products and services (e.g., online banking, securitized mortgage loans, discount brokerage) are located on the left. The distinction between standardized products and customized products on the horizontal dimension is also consistent with the distinction between hard information (an important input for standardized products) and soft information (an important input for customized products).

The framework predicts that only two places on the strategic map are profitable – banks using a 'relationship banking' strategy will locate in the upper right corner, and banks using a 'transactions banking' strategy will locate in the lower left corner. Banks have to be large to operate the low-cost transactions banking model, while banks have to be small to maintain the local focus necessary for the relationship banking strategy. Relationship banks sacrifice sales volume and low costs in exchange for loyal customers willing to pay high prices. Transactions banks sacrifice high prices in exchange for high sales volume and low unit costs.

If this highly stylized framework was found to be consistent with the actual differences between large and small banks, it would suggest that small banks and very large banks use different production functions, and that researchers should be estimating separate cost functions for transactions banks and relationship banks. Both of these cost functions might exhibit within-strategy scale economies, but observing cost differences across strategies would not indicate scale economies.

Figure 3.5, Figure 3.6, Figure 3.7, Figure 3.8, Figure 3.9, and Figure 3.10 display various financial ratios for US banking companies in 2007, the last full year before the onset of the financial crisis. Banks are grouped by asset size. The largest banks have over $25 billion in assets; the smallest banks have less than $500 million in assets, with intermediate break points at $1 billion, $5 billion, and $10 billion. The data in the figures shows that large banks

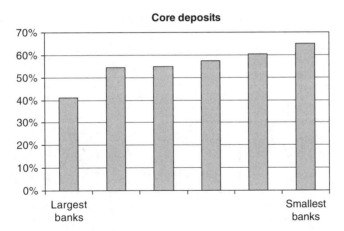

Figure 3.6 Core deposits=transactions accounts and small CDs, divided by assets. Data from US commercial banks in 2007. 'Largest Banks' have over $25 billion in assets. 'Smallest Banks' have less than $500 million in assets. The break points in between the other four groups of banks occur at $10 billion, $5 billion, and $1 billion in assets, respectively.

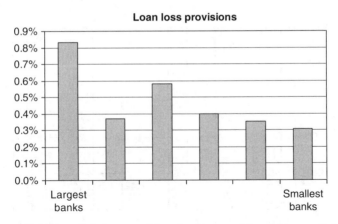

Figure 3.7 Loan loss provisions=provision for loan and lease losses divided by assets. Data from US commercial banks in 2007. 'Largest Banks' have over $25 billion in assets. 'Smallest Banks' have less than $500 million in assets. The break points in between the other four groups of banks occur at $10 billion, $5 billion, and $1 billion in assets, respectively.

use a fundamentally different approach to banking than the small banks, and these differences are consistent with the strategic framework.

The efficiency ratio (noninterest expenses-to-operating income) in Figure 3.5 roughly declines with bank size. This is consistent with scale economies, but it is also consistent with large banks using a different, low-unit-cost production function. The evidence in Figure 3.6, Figure 3.7, Figure 3.8, Figure 3.9, and Figure 3.10 supports the latter.

On average, small banks funded over 60% of their assets with core deposits, compared with only about 40% for the largest banks (Figure 3.6). Core deposit funding is based on long-term relationships with customers (transactions deposits, small savings deposits, small

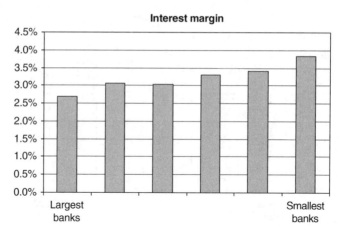

Figure 3.8   Interest margin=interest income minus interest expense, divided by assets. Data from US commercial banks in 2007. 'Largest Banks' have over $25 billion in assets. 'Smallest Banks' have less than $500 million in assets. The break points in between the other four groups of banks occur at $10 billion, $5 billion, and $1 billion in assets, respectively.

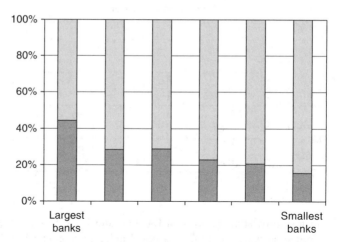

Figure 3.9   The bars show the percentage of bank operating income derived from net interest income (top portion of each bar) and the percentage derived from noninterest income (bottom portion of each bar). Data from US commercial banks in 2007. 'Largest Banks' have over $25 billion in assets. 'Smallest Banks' have less than $500 million in assets. The break points in between the other four groups of banks occur at $10 billion, $5 billion, and $1 billion in assets, respectively.

certificates of deposit (CDs)). The dramatic difference in core deposit funding across the $25 billion threshold indicates the use of nonrelationship funding (brokered deposits, negotiable CDs, interbank loans) at the very largest banks. Loan loss provisions (banks' expectations of the percentage of their loans that will default) also provide evidence of relationship banking (Figure 3.7). Provisions are relatively low at small banks – banks that know their borrowers

Return on equity

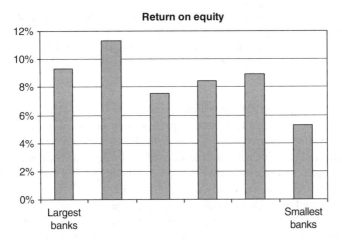

Figure 3.10 Return on equity=net income divided by equity capital. Data from US commercial banks in 2007. 'Largest Banks' have over $25 billion in assets. 'Smallest Banks' have less than $500 million in assets. The break points in between the other four groups of banks occur at $10 billion, $5 billion, and $1 billion in assets, respectively.

(i.e., have soft information) will write fewer bad loans, and borrowers that rely on their banks for funding will not want to default and lose that relationship. Provisions are substantially higher at the very largest banks, consistent with a strategy of making arm's-length transactions loans based on hard information.

These differences in lending and funding strategies are reflected in the interest margins earned by the two sets of banks (Figure 3.8). The net interest margin is about 2.7% for the very largest banks, compared with over 3% for all of the other groups and over 3.5% for the very smallest banks. Consistent with the strategic framework, relationship loans (deposits) will carry high (low) contractual interest rates, while nonrelationship or transactions loans (deposits) will carry low (high) contractual rates. In contrast, noninterest income increases with bank size (Figure 3.9). This is consistent with differences in relationship banking (e.g., charging low fees on core deposit accounts) versus transactions banking (e.g., fees associated with loan origination and securitization), and also reflects fee income from the nontraditional lines of business that have become central to the business models of the very largest banks (e.g., investment banking, brokerage, insurance).

Conceptually, bank profits are equivalent to the net interest margin (Figure 3.8) plus noninterest income (Figure 3.9) minus noninterest expenses (Figure 3.5). How do these elements, which differ substantially across banks of different sizes, balance out in terms of profitability? Figure 3.10 shows that bank profitability, as measured by return on equity (ROE), does not vary systematically with bank size. The very smallest banks (assets less than $500 million) are substantially less profitable than all the other banks, but this is not surprising given that nearly all of the research studies discussed earlier indicated that these very small banks are operating below minimum efficient scale. The two largest groups of banks (assets over $10 billion) are substantially more profitable than the smaller banks – but these ROE measures are not adjusted for risk. As seen in Figure 3.7, larger banks make riskier loans, and there is strong evidence (discussed in the next section) that fee-based banking activities generate riskier earnings than do margin-based banking activities. Hence, large banks need to generate higher equity returns to reward their owners for higher risk exposure. In the end, it appears that the ROE results are also consistent with the strategic map framework – both relationship banking and transactions banking are profitable and viable business strategies.

3.6.3   External costs

Most academic and regulatory studies of bank scale economies are based on banks' accounting costs, that is, the expenses that banks report on their annual income statements. These expenses are sometimes referred to as internal costs, because they measure the costs borne by the bank (salaries and benefits, interest expense, rent, materials and services, etc.) in the process of delivering financial products and services to its customers. But banks can also generate external costs that are borne by society (taxpayers) rather than the banks' shareholders. For example, the financial instability of many US and European banks during the financial crisis required various forms of government subsidies to keep these banks afloat and the financial sector from collapsing. In these cases, taxpayers bore some of the costs of bank risk-taking that, in a perfectly competitive industry, would have been borne by bank shareholders. Because the very largest banks received the bulk of the capital injections, loan guarantees, and other subsidies, any study of bank costs that does not include both internal and external costs will overstate scale economies for the very largest banks.

Only a few studies have attempted to include, or at least control for, these external costs. As discussed earlier, Davies and Tracey (2012) impute the cost of debt finance that TBTF banks would have to pay in the absence of their TBTF protections. When they use the imputed cost of debt finance in their costs function – rather than banks' actual and substantially lower cost of debt finance, which reflects TBTF protections – they no longer find cost scale economies for large banks. Additional work in this area is crucially important for public policy.

3.7   Other evidence on bank scale economies

For the most part, academic and regulatory economists searching for evidence of bank scale economies have used the econometric cost function approach discussed earlier. But there are other approaches that, while less rigorous or less formal, also allow us to test for the existence of bank scale economies. This section provides some examples of simple, transparent analyses that simply 'let the data speak'.

3.7.1   Survivor analysis

Rather than estimating complex models of bank scale economies, why could not we simply depend on the market to reveal the best size for banks? The argument goes like this: If a large bank exists, then scale diseconomies probably do not exist for banks of that size. If this were not the case, managers of large banks would be operating inefficiently large firms and their ill-served shareholders would sell their shares. Investors would purchase these shares, gain control of the bank, pull it apart, and reallocate its assets to more efficient uses.[3] The same logic would hold for small banks: if a small bank exists, then it must not have access to scale

---

[3] Because changes in ownership of banks require regulatory approvals, this 'market for corporate control' mechanism would likely work more slowly in the banking industry than in other industries.

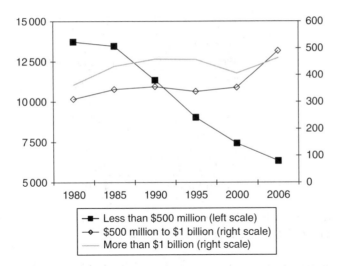

Figure 3.11   Number of US commercial banks in various size categories, 1980–2006. Data from the FDIC and author's calculations.

economies. If this were not the case, investors would have purchased it and increased its value by growing it to a larger, more efficient size. This ongoing process of market discipline would reveal for what size bank economies end (banks below this size would eventually disappear) and for what size bank scale diseconomies begin (banks would not exist above this size). Simply sitting back and watching this process happen is known as 'survival analysis' and is attributable to Stigler (1958).

The data in Figure 3.11 is a kind of survival analysis for US banks. Beginning in 1985 and ending in 2006 (just before the financial crisis), the number of banks with assets less than $500 million declined by about 7000. This is approximately equal to the total number of banks that disappeared from the industry during this time period (see Figure 3.2). By contrast, the number of banks with more than $500 million in assets increased. This analysis is crude, but it delivers a powerful conclusion: commercial banks with assets less than $500 million have access to substantial scale economies.

The analysis is not perfect; for instance, it does not tell us anything about the existence of scale economies or diseconomies for banks larger than $500 million. The increase in the number of these larger banks indicates that being large (more than $500 million) is better than being small (less than $500 million), but this does allow us to discern whether these larger banks experience additional economies of scale or merely constant returns to scale. Breaking up the large banks into additional size categories might give us some answers, although that exercise would be problematic for at least two reasons. First, the industry has not yet fully adjusted to deregulation, so the large bank portion of the industry has not yet settled into a final equilibrium structure. Second, as discussed earlier, these larger banks might continue growing for reasons unrelated to efficient size, for example, bank managers may be empire building and/or trying to capture TBTF subsidies.

Nevertheless, we might learn something further from Figure 3.12, which lists the 10 largest financial holding companies operating in the United States as of June 2007. Of these 10 firms, 4 failed during the financial crisis. In September 2008, Washington Mutual was seized by the Office of Thrift Supervision, placed into receivership at the FDIC, and sold to

1	Citigroup	$2 220 866
2	Bank of America	$1 535 684
3	JPMorgan Chase	$1 458 042
4	Wachovia	$719 922
5	Deutsche Bank	$579 062
6	MetLife	$552 564
7	Wells Fargo	$539 865
8	Washington Mutual	$349 140
9	U.S. Bancorp	$222 530
10	SunTrust Banks	$180 314

Figure 3.12   Ten largest bank holding companies in the United States as of June 2007. Assets in thousands of US dollars. Data from the FDIC.

JPMorgan Chase. Wachovia was illiquid and nearing insolvency when it was purchased by Wells Fargo in October 2008.[4] Bank of America and Citigroup remained solvent only because of government equity injections of $45 billion each under the TARP program, plus hundreds of billions of dollars in government loan guarantees. Thus, 40% of the banks in this 'very large' size category failed. In comparison, only about 4% of US banks smaller than this failed during the crisis. Professor Stigler would claim that this is compelling evidence of scale diseconomies for very large banks.

3.7.2   The market price of banks

The market might also reveal useful information about optimal bank size during bank mergers and acquisitions. If the acquisition price varies systematically with the size of the purchased bank, then differences in the price paid per dollar of target bank assets might be an indicator of bank scale economies or diseconomies.

Figure 3.13 shows a scatter plot of the purchase price for all 796 acquisitions of US commercial banks during the years 2003 through 2007. The purchase price is expressed on the vertical axis as the ratio of market value to book value. The horizontal axis measures the size of the acquired bank in natural logs. The scatter plot itself reveals nothing; to make sense of the data, I regressed the market-to-book purchase price on the log of target bank size, the squared log of target bank size, and included year fixed effects to control for changes in the value of banks from year to year:

$$\text{Market-to-book} = -4.1898 + 0.8863 \times (\ln \text{assets}) - 0.0293 \times (\ln \text{assets})^2 - 0.1319 \times D2003$$
$$+ 0.0033 \times D2004 + 0.2028 \times D2005 + 0.2937 \times D2006.$$

The estimated relationship is the solid parabolic line laid over the top of the scatter diagram. Acquiring banks paid higher prices for larger target banks, but only up to about $3.7 billion in assets, after which acquiring banks paid lower prices. While this is once again a crude test,

---

[4] Days before the Wells Fargo deal, the Office of the Comptroller of the Currency and the FDIC had made plans to close Wachovia, seize its assets, sell the bank to Citigroup, and absorb up to $42 billion in Citigroup's losses on Wachovia assets.

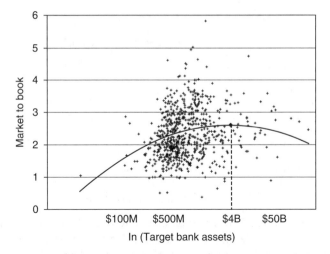

Figure 3.13 Data for all 796 acquisitions of US commercial banks during the years 2003 through 2007. Horizontal axis shows the natural log of target bank assets. Vertical axis shows the ratio of market value (per share acquisition price times number of target bank shares) to target book value. The solid line is the fitted values of a quadratic ordinary least squares regression with year fixed effects. Merger data from The American Banker data archives.

it implies that scale economies of a sort exist that make banks more valuable up to about $3.7 billion in assets, and less valuable above that.

Clearly, there are weaknesses with this type of analysis. First, the quadratic shape imposed on the regression guarantees that we will find either a maximum or a minimum somewhere within the range of the data – but of course, the fact that this simple test found a maximum rather than a minimum is telling. Adding a cubic term to the model resulted in a worse statistical fit, but other more flexible functional forms might be tried. Second, acquiring banks made these purchases in order to expand their own size, so these results may simply reveal the most sought-after target bank size for banks that are expanding via acquisition. In other words, this analysis reveals the most efficient size for a bank that wants to sell itself to another bank.

3.7.3   Bank size and bank risk-return trade-offs

Scale economy analysis focuses on finding minimum efficient scale, that is, the smallest firm for which unit costs are at a minimum. Economists like to focus on minimizing costs (rather than, say, maximizing profits) because they are interested in putting society's scarce resources to their most efficient use. In this case, the scarce resources are deposit funding, bank workers, and the real estate, buildings, and machines used by the banks. Banks that operate below minimum efficient scale will have higher than necessary costs of production: they are too small to use society's resources efficiently. And if banks operate so far above minimum efficient scale that they encounter diseconomies of scale, they will also have higher than necessary costs of production: they are too large to use society's resources efficiently.

The trouble with this line of logic is that it only holds under special environmental conditions, namely, when all of the markets in which banks purchase inputs are perfectly efficient

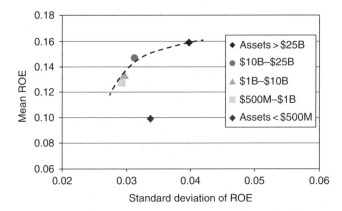

Figure 3.14 Mean and standard deviation of quarterly ROE for all US banks and bank holding companies that operated for 40 consecutive quarters between 1998 and 2007. Data is separated into five asset size classes and then averaged. Dotted line is suggestive, not estimated.

and competitive, and when all of the markets in which banks sell their services are perfectly efficient and competitive. But this is not always the case. For example, banks have private information about borrowers that can give them market power in loan markets. And as discussed earlier, when banks are TBTF, credit markets lend banks money at inefficient (from society's viewpoint) prices that do not fully reflect bank risk. Also discussed earlier, bank managers are unlikely to be pursuing cost minimization as an objective, but rather may be running their banks in order to maximize their own well-being and preferences. Some managers may be willing to sacrifice some amount of profits in exchange for a quiet life, while other managers may be willing to expose the bank to high levels of risk in order to build their empire.

Given the undeniable role of bank profits and bank risk in determining bank size, we may reveal additional information about bank scale economies by looking at the relationships among these three variables. As a simple exercise, I collected quarterly data on ROE for all US banks and bank holding companies that operated for 40 consecutive quarters between 1998 and 2007. (This exercise is similar to the analysis in DeYoung and Rice (2004).) Newly chartered banks and tax-advantaged banks organized as subchapter S corporations are excluded from the analysis. I then calculated the mean quarterly ROE for each bank, as well as the standard deviation of ROE for each bank, over the entire 40 quarters, which provides a risk-return profile for each bank. Finally, I separated the banks into five size groups and calculated the average risk-return profile in each of the size groups.

The results of this exercise are plotted in Figure 3.14. The dotted line is hand-drawn, meant to indicate the standard risk-return trade-off that this data strongly suggests. Three points are worth mentioning. First, the only size group that is nowhere close to the risk-return frontier is small banks with assets less than $500 million. Once again, this evidence suggests substantial scale economies for banks below this size threshold. In this analysis, the evidence indicates that small banks not only can increase their earnings by growing larger, but they can also reduce their risk (earnings volatility) by growing larger. Second, the other four groups of banks located on the risk-return line are arrayed from left-to-right in order of bank size. On average, banks with more than $500 million worth of assets have already exploited the risk-free earnings scale economies; further earnings improvements via scale economies can be achieved only by taking more risk. Third, this risk-return trade-off becomes increasingly expensive as banks grow larger, with greater increases in risk necessary for a given improvement in earnings. Without properly controlling for the costs of risk-taking – including controls for the costs to society when risk-taking results in large bank insolvencies – academic-style estimates of cost-based bank scale economies will overstate the efficiency of large banks. Hughes and Mester (2010) provides a thorough review of the internal costs of risk-taking in bank scale economy research.

3.8   Conclusions

The purpose of this chapter was not to provide a complete overview of research on bank scale economies – for those interested in a more complete review of the literature, Berger, Demsetz, and Strahan (1999) and Hughes and Mester (2010) are good places to start. Rather, the purpose of this chapter is to discuss why it is important for society to measure bank scale economies, to illustrate the challenges of doing so, to offer some alternative perspectives, and to posit some intriguing questions.

In the end, there are three reasons for doing research in bank scale economies. First, pure academics are mainly interested in finding out how the world works. These researchers provide us with important basic learning, although this learning does not always line up well with the questions that business people and policymakers need to address. Moreover, the studies produced by academic researchers are often too technical and densely written to be understood outside of the profession! Second, banking industry analysts are mainly interested in finding out how bank size affects bank costs, revenues, profits, risk, and competitive advantage – ultimately, how bank size affects the price of bank shares. The work of these researchers is generally accessible to industry decision makers, but leaves unaddressed the larger societal questions concerning the external costs of large banks (systemic risk, TBTF subsidies) and the potential external benefits of small banks (small business finance, local community development). Third, today's bank regulators and policymakers are mainly interested in how large bank size affects the risk of bank failure, systemic risk, and macroeconomic disruptions. But a set of policies designed to limit bank size based solely on concerns about external costs would be similar to the deregulatory policies of the 1980s and 1990s that encouraged bank size based solely on concerns about internal costs; such single-minded public objectives often result in incomplete and inefficient policies. Good research on bank scale economies should inform all three of these constituencies.

So what do we currently know for sure about bank scale economies? Let us start with the bad news: There is currently no consensus about whether large banks truly exhibit economies of scale. This is truly unfortunate, given that we are on the cusp of making banking policy decisions that would restrict and/or discourage large bank size. Thomas Hoenig, a former Federal Reserve Bank President and a current FDIC Board Member, has long advocated simply breaking up the largest US banks. New international bank regulations promulgated by the Bank for International Settlements would impose capital surcharges on the largest banks. Unfortunately, we are making such decisions without good information about relative internal and external costs of bank size. But there is also good news: Although existing studies provide no definitive answers about scale economies at the largest banks, they provide very consistent readings at the other end of the size spectrum. Whether one uses sophisticated models of bank cost functions or simply looks at the data on bank size and performance, it is clear that banks with less than about $500 million in assets have access to substantial scale economies. And policy decisions are not necessary to guide the efficiency of this part of the banking sector: the market for bank ownership is gradually eliminating most of the banks that operate well below this size threshold.

References

Berger, A.N. and DeYoung, R. (2006) Technological progress and the geographic expansion of commercial banks. *Journal of Money, Credit and Banking*, **38** (6), 1483–1513.

Berger, A.N., Demsetz, R.S., and Strahan, P.E. (1999) The consolidation of the financial services industry: causes, consequences, and implications for the future. *Journal of Banking and Finance*, **23** (2–4), 135–194.

Berger, A.N., DeYoung, R., Genay, H., and Udell, G.F. (2000) Globalization of financial institutions: evidence from cross-border banking performance, in *Brookings-Wharton Papers on Financial Services*, vol. 3 (eds R. Litan and A. Santomero), Brookings Institution Press, Washington, DC, pp. 23–125.

Berger, A.N., Frame, W.S., and Miller, N.H. (2005) Credit scoring and the availability, price, and risk of small business credit. *Journal of Money, Credit and Banking*, **37** (2), 191–222.

Berlin, M. (2007) Dancing with wolves: syndicated loans and the economics of multiple lenders. Federal Reserve Bank of Philadelphia. Business Review (Quarter 3), pp. 1–8.

Bliss, R.T. and Rosen, R.J. (2001) CEO compensation and bank mergers. *Journal of Financial Economics*, **61**, 107–138.

Brewer, E. and Jagtiani, J. (2011) How much did banks pay to become too-big-to-fail and to become systemically important? Federal Reserve Bank of Philadelphia working paper 11–37.

Clark, J.A. (1988) Economies of scale and scope at depository financial institutions: a review of the literature. Federal Reserve Bank of Kansas City. Economic Review (Sep/Oct), pp. 16–33.

Davies, R. and Tracey, B. (2012) Too big to be efficient? The impact of implicit funding subsidies on scale economies in banking. Bank of England working paper.

DeYoung, R. (2005) The performance of Internet-based business models: evidence from the banking industry. *Journal of Business*, **78** (3), 893–947.

DeYoung, R. (2007) Corporate governance at community banks: one size does not fit all, in *Corporate Governance in Banking: An International Perspective* (ed. B. Gup), Quorom Books, Westport, pp. 62–76.

DeYoung, R. (2010) Banking in the United States, in *The Oxford Handbook of Banking* (eds A.N. Berger, P. Molyneux, and J.O.S. Wilson), Oxford University Press, Oxford, pp. 777–806.

DeYoung, R. and Rice, T. (2004) How do banks make money? A variety of business strategies. Federal Reserve Bank of Chicago. *Economic Perspectives*, **28** (4), 52–67.

DeYoung, R., Spong, K., and Sullivan, R.J. (2001) Who's minding the store? Motivating and monitoring hired managers at small closely held commercial banks. *Journal of Banking and Finance*, **25**, 1209–1244.

DeYoung, R., Hunter, W.C., and Udell, G.F. (2004) The past, present, and probable future for community banks. *Journal of Financial Services Research*, **25** (2–3), 85–133.

DeYoung, R., Lang, W.W., and Nolle, D.L. (2007) How the Internet affects output and performance at community banks. *Journal of Banking and Finance*, **31**, 1033–1060.

DeYoung, R., Peng, E.Y., and Yan, M. (2013) Executive compensation and business policy choices at U.S. commercial banks. *Journal of Financial and Quantitative Analysis*, in press.

Frame, W.S., Srinivasan, A., and Woosley, L. (2001) The effect of credit scoring on small business lending. *Journal of Money, Credit and Banking*, **33**, 813–825.

Gilbert, A.R. (1984) Bank market structure and competition. *Journal of Money, Credit and Banking*, **16**, 617–645.

Goddard, J., Molyneux, P., and Wilson, J.O.S. (2010) Banking in the European Union, in *The Oxford Handbook of Banking* (eds A.N. Berger, P. Molyneux, and J.O.S. Wilson), Oxford University Press, Oxford, pp. 807–843.

Greenspan, A. (2010) The crisis. *Brookings Papers on Economic Activity*, **41** (Spring), 201–261.

Hughes, J.R. and Mester, L.J. (1998) Bank capitalization and cost: evidence of scale economies in risk management and signaling. *The Review of Economics and Statistics*, **80**, 314–325.

Hughes, J.R. and Mester, L.J. (2010) Efficiency in banking: theory, practice and evidence, in *The Oxford Handbook of Banking* (eds A.N. Berger, P. Molyneux, and J.O.S. Wilson), Oxford University Press, Oxford, pp. 463–485.

Hughes, J.R. and Mester, L.J. (2011) Who said banks don't experience scale economies? Evidence form a risk-return-driven cost function. Federal Reserve Bank of Philadelphia working paper.

Inanoglu, H., Jacobs, M., Jr, Junrong, L., and Sickles, R.C. (2012) Analyzing bank efficiency: are 'too-big-to-fail' banks efficient? Rice University working paper.

Molyneux, P., Schaeck, K., and Zhou, T.M. (2010) Too-big-to-fail and its impact on safety net subsidies and systemic risk. CAREFIN research paper no. 09.

Penas, M.F. and Unal, H. (2004) Gains in bank mergers: evidence from the bond markets. *Journal of Financial Economics*, **74**, 149–179.

Spong, K. and Sullivan, R.J. (2007) Managerial wealth concentration, ownership structure, and risk in commercial banks. *Journal of Financial Intermediation*, **16** (2), 229–248.

Stigler, G.J. (1958) The economies of scale. *Journal of Law and Economics*, **1**, 54–71.

Suntheim, F. (2011) Managerial compensation in the financial service industry. Bocconi University working paper.

Tarullo, D.K. (2011) Regulating systemically important financial firms. Speech at the Peter G. Peterson Institute for International Economics, Washington, DC, June 3, 2011.

Wheelock, D.C. and Wilson, P.W. (2012) Do large banks have lower costs? New estimates of returns to scale for U.S. banks. *Journal of Money, Credit and Banking*, **44**, 171–200.

4

Optimal size in banking: The role of off-balance sheet operations

Jaap W.B. Bos[1] and James W. Kolari[2]

[1]*Finance Department, Maastricht University, The Netherlands*
[2]*Finance Department, Texas A&M University, USA*

The last two decades have witnessed a dramatic shift in banks' product mix from traditional on-balance sheet activities (e.g., loan and deposit services) to nontraditional off-balance sheet (OBS) activities (e.g., loan commitments, credit risk guarantees of various kinds, and derivative securities). In Europe, banks dramatically increased OBS activities from less than 50% of total outstanding loans in the early 1990s to 150% of loans in the early 2000s. While previous studies have shown that X-efficiency improves with the inclusion of OBS activities and that economies of scale and scope are not available to large banks with OBS activities as measured across all bank outputs, most of the literature overlooks *product-specific* economies of scale and scope. A likely explanation is the well-known finding by Berger and Humphrey (1991) and Berger, Hunter, and Timme (1993) that banks reap larger cost gains from improving their efficiency relative to the cost frontier, or X-efficiency, than from expanding their size to reach an optimal scale or jointly producing outputs to obtain scope economies. Numerous studies have confirmed this result for both cost and profit functions (e.g., see Berg, Førsund, and Jansen,1992; Berger and DeYoung, 1997; Berger and Mester, 1997; Bos and Kolari, 2005; Lovell, 1993; McAllister and McManus, 1993; Mester, 1996b, c, and others). Since this efficiency literature generally finds that scale economies are exhausted rapidly as output is expanded and scope economies are relatively small (if any), efficiency studies have not explored potential scale and scope economies unique to OBS activities per se.

In this chapter, we examine the possibility that there are product-specific cost economies of scale as well as benefits derived from the nonseparability of bank outputs available to banks that encourage the expansion of OBS activities to relatively high output levels. For this purpose, we collect data for European banks for the period 1989–2005. Also, we revisit the nonseparability

Efficiency and Productivity Growth: Modelling in the Financial Services Industry, First Edition. Edited by Fotios Pasiouras.
© 2013 John Wiley & Sons, Ltd. Published 2013 by John Wiley & Sons, Ltd.

of bank outputs in a manner that is based on within-sample data, rather than out-of-sample data extrapolated beyond the range of actual output levels. Both direct and indirect cost and profit economies with respect to OBS activities are investigated. In brief, our empirical analyses reveal that direct scale economies are important, as OBS activities have a downward sloping average cost curve, as opposed to the U-shaped curve for bank loans and investments. Hence, while some banks currently produce loans and investments at increasing average costs, increases in OBS operations are rewarded with decreases in average costs. Additionally, we find that banks indirectly benefit from the fact that there is nonseparability in the production of bank outputs; more specifically, banks that increase their OBS operations benefit from increasing scale economies in their other outputs. Particularly, in the case of OBS operations and investments, size increases lead to important cost advantages as scope and scale economies reinforce each other. We conclude that size matters in OBS operations, and its impact on average costs constitutes an important reason for the rise of OBS operations. By implication, banks will continue to expand their OBS activities and further shift their product mix in the years ahead.

This chapter continues as follows. Section 4.1 overviews relevant OBS literature. Section 4.2 provides background discussion on OBS activities among European banks in the period 1985–2005. Section 4.3 describes the empirical methods employed to estimate banks' cost functions. In Section 4.4, we introduce our data. Section 4.5 reports the results. Section 4.6 contains a summary and conclusion.

4.1   Literature review

A growing body of literature has examined the risk, regulatory, and tax motivations for OBS activities. Both theoretical studies (Benveniste and Berger, 1987; Boot and Thakor, 1991; Hull, 1989; James, 1988; Thakor and Udell, 1987) and empirical analyses (Angbazo, 1997; Avery and Berger, 1991a, b; Benveniste and Berger, 1987; DeYoung and Roland, 2001; Hassan, 1993; Hassan, Karels, and Peterson, 1994; Jagtiani, Nathan, and Sick, 1995a; Jagtiani, Saunders, and Udell, 1995b; James, 1988; Koppenhaver and Stover, 1991; Pavel, 1988; Saunders and Walter, 1994; Stiroh, 2004; Templeton and Serveriens, 1992) have focused on the potential risk effects of OBS activities on banks. In this regard, while theory posits that OBS activities should increase diversification and thereby reduce bank risk, the empirical evidence is mixed with the greater weight favoring higher bank risk. For example, Stiroh (2004) found that noninterest income from fees, service charges, trading revenue, etc. is relatively volatile compared to interest income and increasingly correlated with interest income over time. He concluded that as banks increase their OBS activities, profit per unit risk declines. Another branch of literature examines the regulatory motivations for OBS activities. Studies by Jagtiani, Nathan, and Sick (1995a) and Pavel and Phillis (1987) revealed that banks facing binding capital constraints use more swaps and loan securitization than other banks. Baer and Pavel (1987) similarly found that banks avoid reserve requirements and deposit insurance premiums via standby letters of credit and loan securitization. However, other papers by Benveniste and Berger (1987) and Koppenhaver (1989) do not find that binding capital constraints are related to OBS activities. Related papers by Pennachi (1988) and Flannery (1989) have argued that there are tax advantages of OBS activities due to on-balance sheet deposit financing and required reserves as well as loan loss and capital gains rules.

An alternative explanation for the growth of OBS activities is that economies of scale and scope and X-efficiencies are associated with nontraditional banking services. In this regard, Jagtiani, Nathan, and Sick (1995a) provided some early evidence on OBS activities using

1988 data for 134 large banks. They found that neither economies nor diseconomies of scale existed for large banks with OBS activities. Also, little or no cost complementarities of loans, deposits, and OBS activities outputs were available to large banks. Another paper by Jagtiani and Khanthavit (1996) updated the analyses to the period 1984–1994 for 91 large banks. Expansion path subadditivity results based on in-sample data indicated that scope economies exist between loans, securities, and OBS activities. However, other results led the authors to conclude that money center and super-regional banks are too large to be efficient. Rogers (1998) examined X-efficiency among all US banks in the period 1991–1995 and found that cost, revenue, and profit X-efficiency increases when nontraditional OBS activities are included in standard models with only traditional bank outputs. DeYoung (1994) obtained a similar result for cost efficiency.

Another paper by Clark and Siems (2002) confirms that cost X-efficiency increases upon including OBS activities, but profit X-efficiency does not. Other results revealed that bank size and off-to-on-balance sheet mix of banking activities are not related to cost or profit X-efficiency. Tortosa-Ausina (2003) also reports mixed evidence, with OBS activities improving the cost efficiency of some groups of banks but lessening efficiency among other groups. However, Rao (2005) examines banks in the United Arab Emirates and finds OBS activities tend to reduce bank operating costs. More recently, Pasiouras (2008) finds that OBS items do not have a significant impact on efficiency scores of Greek commercial banks in the period 2000–2004. Finally, numerous efficiency studies (e.g., see Berger and Mester, 1997; Bos and Kolari, 2005; Saunders and Walter, 1994, and others) have incorporated OBS activities in cost and profit efficiency analyses, but their inclusion is peripheral to their main purposes. In general, these studies establish that OBS activities should be counted as an output to accurately measure bank efficiency. However, no studies investigate potential scale and scope economies of OBS activities themselves. In the forthcoming analyses, we seek to fill this gap in the literature.

4.2   Off-balance sheet activities of European banks

OBS activities are comprised of loan commitments, credit risk guarantees (e.g., standby letters of credit, revolving underwriting facilities, and credit derivative swaps), and various derivatives contracts (e.g., swap, futures, forwards, and options contracts). Banks utilize OBS operations to manage on-balance sheet risks as well as provide client hedging risk management services. With respect to on-balance sheet risk management, securitization of home and other loans has allowed banks to create vehicles that move these loans OBS. While the notional value of derivative instruments is held OBS, their fair value is carried on the balance sheet. OBS activities generate revenues but have little or no effect on total assets (see Ronen, Anthony, and Sondhi, 1990). Consequently, OBS can boost profit ratios greater than traditional on-balance sheet activities per unit of labor. Since little or no investment in assets is required, operating costs are reduced per unit aggregate output. Hence, there are practical reasons to believe that OBS cost and profit efficiencies exist.

Panel cost of Figure 4.1 graphically displays the growth of bank outputs for European banks in 15 countries during the period 1985–2005 for loans, investments, and OBS operations, in addition to associated total operating costs.[1] Two observations relevant to the

---

[1] Outputs and total cost are measured as asset-weighted averages. See Section 4.4 for a detailed description of the data.

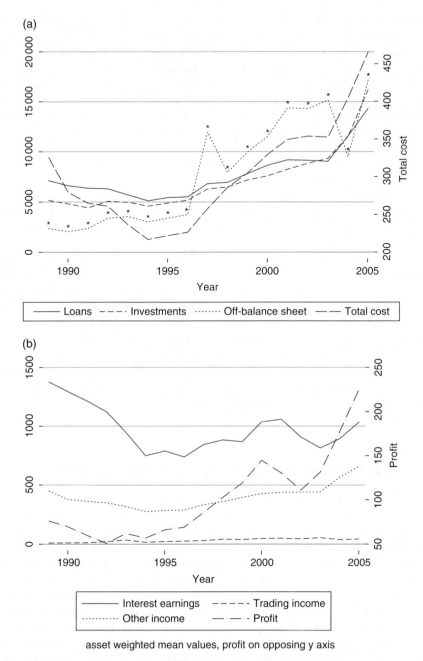

Figure 4.1   Trends in (a) outputs and total cost and (b) earnings and profit.

present study arise from casual inspection. First, we observe a dramatic rise of OBS operations after 1995. Second, we observe a concomitant rise in bank operating costs. Of course, the latter largely reflects the output expansion of the European banking industry in this period. Panel earnings of Figure 4.1 show that other income associated with OBS activities tends to mimic the pattern of interest income but is more stable over time. Also, bank profits have grown in step with costs and output growth.

Table 4.1 provides some evidence on the characteristics of banks that have higher proportions of loans, investments, or OBS activities compared to other banks. In particular, we are interested in distinguishing characteristics of banks specializing in OBS activities. We consider six different bank characteristics, including bank profitability (net income/ total assets, or ROA), size (total assets), risk (ROA/σ_{ROA} similar to the Sharpe ratio, where ROA and σ_{ROA} are measured per year and for each of 20 size classes based on a k-means cluster analysis of total assets), turnover (total revenues/total assets), cost efficiency (X-efficiency estimated from forthcoming translog cost function analyses), and capitalization (total equity capital/total assets). For each output variable, we compare the 10% of most specialized banks to other banks by computing the ratio y_k/y_{total}, for $k = 1, 2, 3$ (=loans, investment, and OBS operations, respectively). For each output, the resulting two groups are compared using a nonparametric Kruskall–Wallis (KW) rank test and an independent sample t-test. In addition, elasticities are reported from a three-stage estimation of a multiple regress system of three simultaneous equations, where the output shares of each of the k outputs are regressed on the six bank characteristics as well as on the levels of the three outputs.

Focusing primarily on the characteristics of OBS specializing banks, the results in panels A to G in Table 4.1 can be summarized as follows:

Panel (a) – Banks that specialize in OBS activities have higher average profit rates (ROAs) than banks that specialize in other outputs. However, the elasticities for all three outputs are negative and significant, which indicates that a higher profitability is expected in banks with lower shares of a specific output. Hence, specialization in any of the three outputs lowers profitability, *ceteris paribus*.

Panel (b) – Banks that specialize in OBS activities are larger than other banks in terms of mean total assets but not in terms of median asset size due to skewness in the distribution of banks' total assets. Indeed, the elasticities suggest that all three types of specialization are expected to occur among relatively small banks.

Panel (c) – Consistent with the intuition that OBS activities are high-risk, high-return business services, banks specializing in OBS tend to have a relatively lower Sharpe ratio than other banks, as do banks that specialize in investments. A decrease in the Sharpe ratio (i.e., an increase in risk) is significantly associated with higher shares of OBS activities but not loans or investments.

Panel (d) – Banks that specialize in OBS activities and loans have significantly higher turnover. However, the estimated elasticity is only positive and significant for banks that specialize in OBS. These results may be related to the Basel I capital accord amendment in 1995 that lowered capital requirements on securitized assets (e.g., home loans), thereby freeing up capital and increasing turnover.

Panel (e) – Specialization in any of the three outputs appears to lower efficiency. However, the elasticity for OBS activities is not significant, which suggests that efficiency does not necessarily diminish as OBS services increase.

Panel (f) – Banks with high levels of OBS activities do not require as much investment in fixed assets as other banks. However, specializing in loans and investments tends to demand higher fixed asset investment than other banks. The elasticities confirm the

Table 4.1 How important are other explanations for the rise in OBS operations?

Explanation	Variable	Loans (y_1)	Investments (y_2)	Off-balance (y_3)
	(a) *Are banks that specialize in OBS more profitable?*			
Profitability	Δ means	1.05–0.89	1.16–0.88	1.25–0.86
	KW (*p* values)	0.0001***	0.0001***	0.0542*
	t-Test (*p* values)	0.1952	0.0057*	0.0001***
	Elasticity	−0.0011	−0.0018***	−0.0013***
	(b) *Are banks that specialize in OBS larger?*			
Size	Δ means	3 500–17 032	3 434–17 495	18 257–15 847
	KW (*p* values)	0.0001***	0.0001***	0.0001***
	t-Test (*p* values)	0.0000***	0.0000***	0.1854
	Elasticity	−0.0031***	−0.0025***	−0.0046***
	(c) *Are banks that specialize in OBS more risky?*			
Risk	Δ means	0.43–0.36	0.31–0.39	0.33–0.37
	KW (*p* values)	0.0001***	0.0001***	0.0001***
	t-Test (*p* values	0.0000***	0.0000***	0.0000***
	Elasticity	−0.0001	−0.0001	−0.0005***
	(d) *Do banks that specialize in OBS have a higher turnover?*			
Turnover	Δ means	0.06–0.04	0.05–0.04	0.05–0.04
	KW (*p* values)	0.0001***	0.0001***	0.0001***
	t-Test (*p* values)	0.0000***	0.0064*	0.0000***
	Elasticity	0.0020	0.0017	0.0170***
	(e) *Are banks that specialize in OBS more efficient?*			
Efficiency	Δ means	0.63–0.75	0.67–0.75	0.71–0.75
	KW (*p* values)	0.0001***	0.0001***	0.0001***
	t-Test (*p* values)	0.0000***	0.0000***	0.0000***
	Elasticity	0.0111**	−0.0131**	−0.0066
	(f) *Do banks that specialize in OBS have less fixed assets?*			
Fixed assets	Δ means	0.015–0.013	0.014–0.008	0.013–0.014
	KW (*p* values)	0.0001***	0.0001***	0.0001***
	t-Test (*p* values)	0.0574*	0.0000***	0.2710
	Elasticity	0.0056***	−0.0049***	0.0002
	(g) *Are banks that specialize in OBS better capitalized?*			
Capitalization	Δ means	13.25–17.54	37.73–15.99	18.88–17.28
	KW (*p* values)	0.0001***	0.0001***	0.0619*
	t-Test (*p* values)	0.0140**	0.0000***	0.1915
	Elasticity	−0.0076***	0.0167***	−0.0003

Mean difference between OBS-specialized (i.e., banks with OBS shares of total output in the top 10%) and other banks for each bank characteristic. Bank characteristics are defined as follows: profitability (net income/total assets, or ROA), size (total assets), risk (ROA/σ_{ROA}, similar to the Sharpe ratio, where ROA and σ_{ROA} are measured per year and for each of 20 size classes based on a k-means cluster analysis of total assets), turnover (total revenues/total assets), cost efficiency (X-efficiency estimated from forthcoming translog cost function analyses), and capitalization (total equity capital/total assets). The p-values are based on the nonparametric Kruskall–Wallis (KW) rank test and independent sample t-test. Elasticity is the estimated coefficient between each bank characteristic and specific output in a system regression of three simultaneous equations of output shares of loans, investments, and OBS activities on the above bank characteristics and the levels of the three outputs. The pseudo-R^2 values for these three equations are 0.0997, 0.1206, and 0.4784, respectively. Unrestricted linear predictions for shares are 0.3431, 0.3413, and 0.3156, respectively. The total number of observations is equal to the sample size used for estimating the cost frontier, or 16 882 observations, except for capitalization with 3651 observations.
*, **, and *** indicate significance at the 1%, 5%, and 10% levels, respectively.

latter findings but are not significant for OBS activities. These results are consistent with the fact that OBS operations do not normally require brick-and-mortar bank production (e.g., a large branch network).

Panel (g) – Banks that specialize in lending (investments) hold relatively less (more) capital than other banks, which is consistent with the risk results in Table 4.1. However, despite the higher risk of OBS activities, banks with higher proportions of OBS activities have capital ratios that are close to the average for the whole sample. As such, Basel I capital accord rules may have motivated banks to expand OBS operations.

In sum, we find that banks specializing in OBS activities have higher profits, median assets, risk, and turnover but lower fixed assets than other banks. Also, unlike loans and investments, OBS operations are not necessarily associated with lower cost efficiency or a change in capitalization.

4.3 Methodology

In this section, we begin by discussing banks' production set and stochastic cost frontier analysis. We then overview banks' production technology. Finally, we discuss the measurement of scale economies.

4.3.1 Stochastic frontier analysis

Following standard practice in bank efficiency studies, we employ the intermediation approach to model bank production (Sealey and Lindley, 1977). Banks' production set consists of output vector Y, input vector X, and a control variable Z. The output vector Y consists of three outputs: loans, investments, and OBS operations. Banks are price takers in factor markets and, consequently, face the vector X of exogenous prices for their inputs W. Input price vector W consists of the prices of labor, financial capital, and physical capital. Finally, the control variable equity (Z) is included as a risk measure (Hughes and Mester, 1993).

Banks convert their inputs X into outputs Y while maintaining equity Z via a transformation function $T(Y, X, Z)$. Hence, to produce outputs Y, banks minimize cost by choosing optimal input quantities $X^*(Y, W, Z)$ conditional on the available level of equity Z, given input prices W, and subject to the technology constraint $T( )$. We can utilize duality to model the resulting optimization as a cost minimization problem, where the minimum cost level is obtained by substituting the optimal input demand functions into the total cost function to obtain $TC^* = W' \times X(Y,W,Z)^* = TC^*(Y,W,Z)$.

Deviations from optimal cost in year t can be due to either random noise or suboptimal employment of inputs at given prices. We therefore write a baseline stochastic cost frontier for a bank k in logs and add a composed error term ε to the deterministic kernel $f(y_{kt}, w_{kt}, z_{kt}; \beta)$ as[2]

$$tc_{kt} = f\left(y_{kt}, w_{kt}, z_{kt}, t; \beta\right) + v_{kt} + \upsilon_{kt}, \tag{4.1}$$

where β is a vector of parameters to be estimated, time trend t captures technological change (Altunbas, Goddard, and Molyneux, 1999), and the total error $\varepsilon_{kt} = v_{kt} + \upsilon_{kt}$, with the term v_{kt}

---

[2] Lowercase letters denote natural logs.

denoting random noise and υ_{kt} representing deviations due to inefficiency. The random error term v_{kt} is assumed i.i.d. with $v_{kt} \sim N\left(0, \sigma_v^2\right)$ and independent of the explanatory variables. The inefficiency term υ_{kt} is distributed i.i.d.$N \mid \left(\mu, \sigma_\upsilon^2\right) \mid$ and independent of v_{kt} (Stevenson, 1980). For a cost frontier, inefficient input use entails higher than optimal cost, such that υ_{kt} is strictly positive.[3]

Following Kumbhakar and Lovell (2000), and imposing homogeneity of degree one in input prices and symmetry, we estimate Equation (4.1) with ordinary least squares (OLS).[4] We then use OLS parameter estimates as starting values and reparametrize using $\lambda = \sigma_\upsilon / \sigma_v$ and $\sigma = \sigma_v + \sigma_\upsilon$, where λ is the variance accruing to inefficiency relative to the random variance, and σ is the total variance. A point estimator of technical efficiency is then given by $E(\upsilon_{kt} | \varepsilon_{kt})$, the conditional distribution of υ given ε (Jondrow et al., 1982). Estimates of bank-specific cost efficiency are obtained by calculating $CE_{kt} = [exp(-\upsilon_{kt})]$. Cost and profit efficiencies equal 1 for a fully efficient bank that operates on the efficient stochastic frontier.

4.3.2 Functional form

We next specify a functional form for the deterministic kernel of Equation (4.1) so that the parameter vector β can be estimated to measure scale economies. There are three types of functional forms from which to choose. First, we can choose a nonflexible form and, for example, estimate a Cobb–Douglas specification. Second, we can choose a semiflexible form, such as the translog functional form. Third, we can choose a fully flexible form by estimating, for example, a Fourier specification.[5] We follow most of the bank efficiency literature and estimate the following translog cost specification[6]:

$$
\begin{aligned}
tc = {} & \beta_0 + \beta_1 w_1 + \beta_2 w_2 + \beta_3 y_1 + \beta_4 y_2 + \beta_5 y_3 + \beta_6 z + \frac{1}{2}\beta_7 w_1^2 \\
& + \beta_8 w_1 w_2 + \frac{1}{2}\beta_9 w_2^2 + \frac{1}{2}\beta_{10} y_1^2 + \beta_{11} y_1 y_2 + \beta_{12} y_1 y_3 + \frac{1}{2}\beta_{13} y_2^2 \\
& + \beta_{14} y_2 y_3 + \frac{1}{2}\beta_{15} y_3^2 + \frac{1}{2}\beta_{16} z^2 + \beta_{17} y_1 w_1 + \beta_{18} y_1 w_2 + \beta_{19} y_2 w_1 \\
& + \beta_{20} y_2 w_2 + \beta_{21} y_3 w_1 + \beta_{22} y_3 w_2 + \beta_{23} y_1 z + \beta_{24} y_2 z + \beta_{25} y_3 z \\
& + \beta_{26} w_1 z + \beta_{27} w_2 z + \beta_{28} t + \frac{1}{2}\beta_{29} t^2 + \beta_{30} y_1 t + \beta_{31} y_2 t + \beta_{32} y_3 t \\
& + \beta_{33} w_1 t + \beta_{34} w_2 t + \beta_{35} z t + v + \upsilon.
\end{aligned} \tag{4.2}
$$

---

[3] Stochastic frontier analysis (SFA) is the most common parametric frontier approach (Kumbhakar and Lovell 2000). Alternative parametric methods include the thick frontier approach (Berger and Humphrey 1991, 1992) and the distribution-free approach (Berger, Hunter, and Timme, 1993).

[4] We also check if the expected value of the OLS residuals is significantly different from zero and exhibits positive skewness (Waldman 1982; Carree 2002).

[5] Each of these suggested specifications is nested in the latter. Hence, we can test for the joint significance of additional parameters and thereby find a preferred specification. In this regard, Swank (1996) has demonstrated that the choice between the translog specification and the Fourier does not significantly affect efficiency measurement. In addition, the full Fourier specification often suffers from multicollinearity problems, and estimation of a partial Fourier leads to biased scale economies estimates (see Brambor, Clark, and Golder, 2006, on the inclusion of all interaction terms).

[6] We test whether this specification is preferred to the Cobb–Douglas (not reported here).

For the present purposes, the translog specification has a crucial advantage compared to the Cobb–Douglas specification. Since for multi-output, multi-input banks the production of each output is *nonseparable* from the production of the other outputs, the significance of the interaction terms in the translog function enables a test of the production technology, where the cost of producing one output may depend on the amount of other outputs that are produced.

4.3.3   Scale economies

We can use Equation (4.2) to calculate output-specific scale economies. For example, for output Y_1, loans, we take the partial derivative and calculate in logs:

$$\frac{\partial tc}{\partial y_1} = \beta_3 + \beta_{10}y_1 + \beta_{11}y_2 + \beta_{12}y_3 + \beta_{17}w_1 + \beta_{18}w_2 + \beta_{30}t. \tag{4.3}$$

Equation (4.3) shows that scale economies for output Y_1 are nonseparable from the level of the other outputs and input prices.[7] In addition, the level of scale economies changes with the time trend t. Further information on nonseparability can be gleaned from the variance of the output-specific scale economies[8]:

$$\begin{aligned}
\sigma^2_{\partial tc/\partial y_1} = {} & \mathrm{var}\left(\beta_3\right) + 2\,\mathrm{var}\left(\beta_{10}\right)y_1^2 + \mathrm{var}\left(\beta_{11}\right)y_2^2 + \mathrm{var}\left(\beta_{12}\right)y_3^2 \\
& + \mathrm{var}\left(\beta_{17}\right)w_1^2 + \mathrm{var}\left(\beta_{18}\right)w_2^2 + \mathrm{var}\left(\beta_{30}\right)t^2 + 2\,\mathrm{cov}(\beta_3\beta_{10})y_1 \\
& + 2\,\mathrm{cov}(\beta_3\beta_{18})w_2 + 2\,\mathrm{cov}(\beta_3\beta_{11})y_2 + 2\,\mathrm{cov}(\beta_3\beta_{12})y_3 + 2\,\mathrm{cov}(\beta_3\beta_{17})w_1 \\
& + 2\,\mathrm{cov}(\beta_3\beta_{30})t + 2\,\mathrm{cov}\left(\beta_{10}\beta_{11}\right)y_1y_2 + 2\,\mathrm{cov}\left(\beta_{10}\beta_{12}\right)y_1y_3 \\
& + 2\,\mathrm{cov}\left(\beta_{10}\beta_{17}\right)y_1w_1 + 2\,\mathrm{cov}\left(\beta_{10}\beta_{18}\right)y_1w_2 + 2\,\mathrm{cov}\left(\beta_{10}\beta_{30}\right)y_1t \\
& + 2\,\mathrm{cov}\left(\beta_{11}\beta_{12}\right)y_2y_3 + 2\,\mathrm{cov}\left(\beta_{11}\beta_{17}\right)y_2w_1 + 2\,\mathrm{cov}\left(\beta_{11}\beta_{18}\right)y_2w_2 \\
& + 2\,\mathrm{cov}\left(\beta_{11}\beta_{30}\right)y_2t + 2\,\mathrm{cov}\left(\beta_{12}\beta_{17}\right)y_3w_1 + 2\,\mathrm{cov}\left(\beta_{12}\beta_{18}\right)y_3w_2 \\
& + 2\,\mathrm{cov}\left(\beta_{12}\beta_{30}\right)y_3t + 2\,\mathrm{cov}\left(\beta_{17}\beta_{18}\right)w_1w_2 + 2\,\mathrm{cov}\left(\beta_{17}\beta_{30}\right)w_1t \\
& + 2\,\mathrm{cov}\left(\beta_{18}\beta_{30}\right)w_2t. \tag{4.4}
\end{aligned}$$

Combining Equations (4.3) and (4.4), we can calculate the sign and significance of the contribution of each of the elements of the production set to the output-specific scale economies.

Before we turn to our results, consider the consequences of nonseparability in output production to the measurement of scope economies. Berger, Hanweck, and Humphrey (1987) observe that for translog functions complementarities cannot exist at all levels of output. Berger and Mester (1997) note that an additional problem with the estimation of scope economies is the possible existence of zero outputs. Another potential pitfall is that there often is an extrapolation problem. Given a sample containing both universal banks and other banks, only the former banks typically offer the full range of financial services. Consequently, the economies of scope derived from the cost function tend to overestimate the true economies of scope among most sample banks.

---

[7] To further separate out the individual effects of each output on the other outputs, one can also follow the approach presented in Ozer-Balli and Sørensen (2010).

[8] See Brambor, Clark, and Golder (2006).

4.4   Data

Table 4.2 gives descriptive statistics for our variables. Data is collected from Bureau van Dijk's BankScope. We include independent banks from 15 European countries in the period 1989–2005 for a total of 16 882 observations. Our variables are typical for bank efficiency and productivity studies. We identify a production set consisting of three outputs, three inputs, and a control variable. The three outputs are loans, investments, and OBS activities. Loans capture all lending activities and are measured by the total volume of loans. Investments are also taken from the balance sheet and represent securities held by the bank. OBS products and services are included as reported by banks, that is, the notional value of loan commitments, credit risk guarantees, and derivatives activities.

As shown in Table 4.2, OBS is the only output for which the within standard deviation is as high as the between standard deviation, where the former is bank-specific and the latter is for the whole sample. Hence, it is the only output that has a large bank-specific standard deviation. We attribute this finding to the rapid rise of OBS activities in our sample period as well as their volatile nature. Notice also that the strong consolidation in the banking industry is reflected in the relatively large size of the between standard deviation for total assets compared to the within standard deviation.

Input prices of financial capital, labor, and physical capital are measured as interest expenses/total interest-paying liabilities, total personnel expenses/total assets,[9] and total write-offs (i.e., depreciation)/fixed assets, respectively.

Lastly, as already mentioned, we include the equity ratio to control for differences in risk-taking (see Mester, 1996a).

4.5   Results

In this section, we provide empirical evidence on the cost efficiency effects of OBS activities. We test two research hypotheses: (a) banks can realize more scale economies by increasing OBS operations than by increasing other outputs, and (b) increases in OBS operations contribute to scale economies for other outputs due to product mix (scope) economies.

4.5.1   Increasing OBS operations

We begin by examining 'pure' scale economies via the marginal effect of an increase in an output on total cost as that output increases. In Figure 4.2a–c, we plot the conditional marginal effect of each output as the respective output is increased over the range that is available in our sample. These findings are based on the partial derivative of Equation (4.2) with respect to each output, for example, for loans y_1, see Equation (4.3).

It may seem appropriate to measure scale economies for y_1 using β_{10} from Equation (4.3), which is associated with the conditioning effect of y_1 on the marginal effect. However, in so doing we would implicitly assume that all other elements of the production set, including the other outputs, remain constant at their mean values. Figure 4.1a and b show that this

---

[9] A well-known problem with BankScope is the fact that the number of employees is missing for most observations.

Table 4.2 Descriptives.

Variable		Mean	Std. dev.	Minimum	Maximum
TC	Total cost	429.37			
	Overall		1460.56	0.04	20086.5
	Between		1352.79	0.14	18645.8
	Within		441.31	−5899.02	13853.3
Y_1	Loans	11311.87			
	Overall		42056.35	0.10	722700.0
	Between		40625.54	0.10	722700.0
	Within		12662.44	−123166.30	413346.7
Y_2	Investments	9661.87			
	Overall		34177.65	0.04	952406.4
	Between		40547.71	0.10	952406.4
	Within		12488.85	−90598.32	679911.6
Y_3	OBS operations	13664.03			
	Overall		151156.90	0.02	5358907
	Between		97619.43	0.03	2545398
	Within		81375.34	−2470752.00	4813238
W_1	Price of financial capital	0.08			
	Overall		0.80	0.00	66.88461
	Between		1.20	0.00	52.61964
	Within		0.58	−17.03	58.55614
W_2	Price of labor	0.02			
	Overall		0.02	0.00	0.46977
	Between		0.02	0.00	0.20499
	Within		0.01	−0.16	0.40376
W_3	Price of physical capital	1.81			
	Overall		2.37	0.06	29.16918
	Between		2.85	0.06	24.30973
	Within		1.75	−11.14	22.52992
Z	Equity ratio	9.17			
	Overall		9.25	0.00	95.8031
	Between		10.25	0.26	90.3867
	Within		4.43	−40.76	80.1772
TA	Total assets	22807.75			
	Overall		79609.77	0.75	1591339
	Between		84907.22	1.58	1591339
	Within		25623.99	−218295.40	1138607

N (between), 16882; n (within), 2413; period, 1989–2005; countries included, Austria, Belgium, Denmark, France, Germany, Greece, Italy, Luxembourg, the Netherlands, Norway, Portugal, Spain, Sweden, Switzerland, and the United Kingdom.

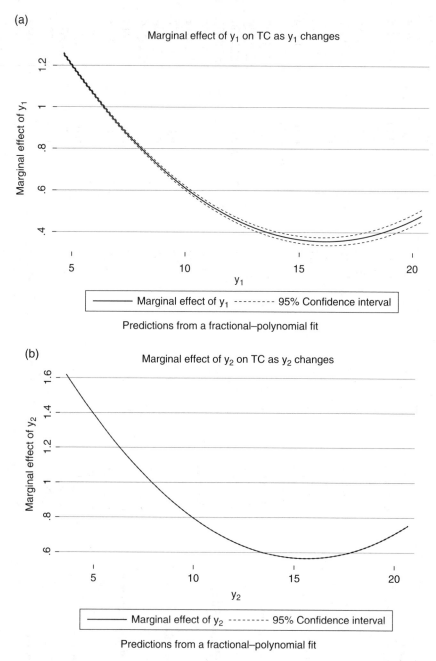

(a)

Marginal effect of y₁ on TC as y₁ changes

Figure 4.2 Scale economies. Partial derivative of cost function for (a) loans, (b) investments, and (c) off-balance sheet operations.

assumption is clearly violated. Hence, we fit a polynomial over the entire partial derivative of
Equation (4.3) using the output itself as the explanatory variable.

For loans, the resulting fit is depicted in Figure 4.2a as well as a 95% confidence interval.
The resulting scale economies for loans and investments reflect the U-shaped average cost

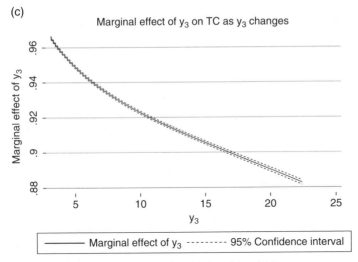

(c)

Marginal effect of y_3 on TC as y_3 changes

Predictions from a fractional–polynomial fit

Figure 4.2   (*Continued*)

curves that previous cost studies typically found. For OBS operations, however, the results are different in two ways. First, scale economies are always positive for OBS operations, as the marginal effect is always smaller than 1. Second, the continuously downward sloping line indicates that scale economies keep increasing with the size of OBS operations. We infer that these findings support our first research hypothesis.

4.5.2   Nonseparability effects of OBS operations

Turning to the effect of changing product mix on scale economies, Figures 4.3a and b summarize the effects of the other outputs on the marginal effect (i.e., scale economies) of each output.[10] For example, for loans y_1 we consider all elements from Equation (4.3) except y_1. Hence, in Figure 4.3a in the top left graphs, we consider the effect of increasing investments y_2 on the scale economies for loans as y_2 increases (and where the remainder of the production set is held constant). We are mainly interested in the magnitude of the resulting partial conditional marginal effects, which reflect cost complementarities if they are less than 1. Also, we are interested in the effects of increases in the outputs.

For loans in panel (a) of Figure 4.3, we observe strong negative complementarities with investments. OBS operations y_3, however, create positive complementarities, and the effect increases somewhat as OBS activities increase. As such, joint cost benefits of loans and OBS activities exist.

For investments in panel (b) of Figure 4.3, the same results are obtained as for loans. Negative complementarities exist with respect to loans, but complementarities with OBS activities are positive and even increasing. Hence, banks can reap scope economies between investments and OBS activities.

For OBS operations themselves in panel (a) of Figure 4.3, complementarities created by the other outputs are significantly negative, except for very high values of loans y_1. Thus, the

---

[10] The effects of the input prices, time t, and the equity ratio z on each output are available upon request.

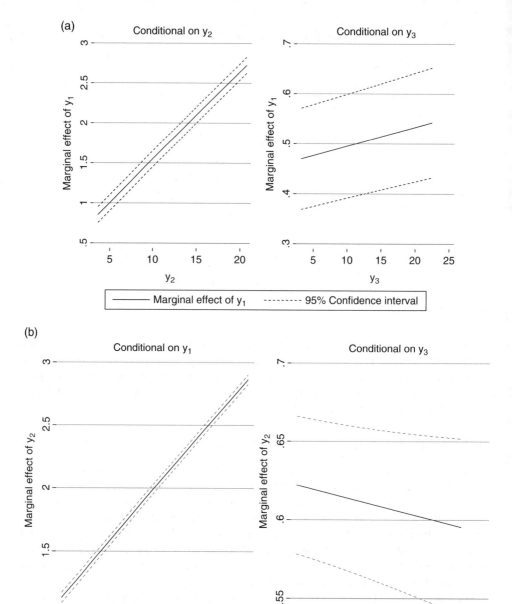

Figure 4.3 Product mix effects. Partial conditional marginal effects for (a) loans, (b) investments, and (c) off-balance sheet operations.

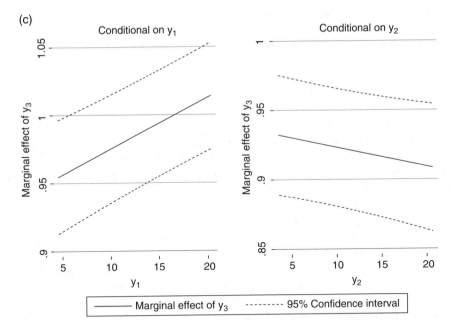

Figure 4.3   (*Continued*)

evidence supports our second research hypothesis that scope economies exist in the joint production of OBS services with loans and investments.

4.6   Conclusion

In this chapter, we provide empirical evidence on cost efficiency motivations for the rapid expansion of OBS activities among European banks in 15 countries between 1989 and 2005. Initial comparative analyses reveal that banks specializing in OBS activities are different from other banks in terms of higher profits, median assets, risk, and turnover as well as lower fixed assets. Further cost efficiency analyses show that OBS activities have a downward sloping average cost curve, rather than the U-shaped cost curves as in the cases of bank loans and investments. Also, based on a new approach to estimating nonseparability of bank outputs using within-sample data, we find that scope economies between OBS activities and both loans and investments exist but not between loans and investments. Together, these results lead us to conclude that scale and scope economies are important potential reasons for the recent growth of OBS financial services in the banking industry. Given these cost benefits, we expect that banks will continue to expand their OBS activities in the future.

References

Altunbas, Y., Goddard, J., and Molyneux, P. (1999) Technical change in European banking. *Economics Letters*, **64** (2), 215–221.

Angbazo, L. (1997) Commercial bank net interest margins, default risk, interest-rate risk, and off-balance sheet banking. *Journal of Banking and Finance*, **21**, 55–87.

Avery, R. and Berger, A.N. (1991a) Loan commitments and bank risk exposure. *Journal of Banking and Finance*, **15**, 173–193.

Avery, R.B. and Berger, A.N. (1991b) Risk-based capital and deposit insurance reform. *Journal of Banking and Finance*, **15**, 147–174.

Baer, H.L. and Pavel, C.A. (1987) Does regulation drive innovation? Federal Reserve Bank of Chicago. Economic Perspectives (March), pp. 3–16.

Benveniste, L.M. and Berger, A.N. (1987) Securitization with recourse: an instrument that offers uninsured bank depositors sequential claims. *Journal of Banking and Finance*, **11** (3), 403–424.

Berg, S.A., Førsund, F.R., and Jansen, E.S. (1992) Malmquist indices of productivity growth during the deregulation of Norwegian banking, 1980–89. *Scandinavian Journal of Economics*, **94**, 211–228.

Berger, A.N. and DeYoung, R. (1997) Problem loans and cost efficiency in commercial banks. *Journal of Banking and Finance*, **21** (6), 849–870.

Berger, A.N. and Humphrey, D. (1991) The dominance of inefficiencies over scale and product mix economies in banking. *Journal of Monetary Economics*, **28**, 117–148.

Berger, A.N. and Humphrey, D.B. (1992) Megamergers in banking and the use of cost efficiency as an antitrust defense. *Antitrust Bulletin*, **37**, 541–600.

Berger, A.N. and Mester, L.J. (1997) Inside the black box: what explains differences in the efficiencies of financial institutions? *Journal of Banking and Finance*, **21** (7), 895–947.

Berger, A.N., Hanweck, G., and Humphrey, D. (1987) Competitive viability in banking: scale, scope and product mix economies. *Journal of Monetary Economics*, **20**, 501–520.

Berger, A.N., Hunter, W.C., and Timme, S.G. (1993) The efficiency of financial institutions: a review and preview of research past, present and future. *Journal of Banking and Finance*, **17** (2–3), 221–249.

Boot, A.W.A. and Thakor, A.V. (1991) Off-balance sheet liabilities, deposit insurance and capital regulation. *Journal of Banking and Finance*, **15** (4–5), 825–846.

Bos, J.W.B. and Kolari, J.W. (2005) Large bank efficiency in Europe and the united states: are there economic motivations for geographic expansion in financial services? *Journal of Business*, **78** (4), 1555–1592.

Brambor, T., Clark, W., and Golder, M. (2006) Understanding interaction models: improving empirical analyses. *Political Analysis*, **14** (1), 63–82.

Carree, M.A. (2002) Technological inefficiency and the skewness of the error component in stochastic frontier analysis. *Economics Letters*, **77** (1), 101–107.

Clark, J.A. and Siems, T.F. (2002) X-efficiency in banking: looking beyond the balance sheet. *Journal of Money, Credit and Banking*, **34**, 987–1013.

DeYoung, R. (1994) *Fee-Based Services and Cost Efficiency in Commercial Banks.* . Proceedings of a Conference on Bank Structure and Competition. Federal Reserve Bank of Chicago, Chicago, pp. 501–520.

DeYoung, R. and Roland, K.P. (2001) Product mix and earnings volatility at commercial banks: evidence from a degree of total leverage model. *Journal of Financial Intermediation*, **10**, 221–252.

Flannery, M. (1989) Capital regulation and insured banks' choice of individual loan default rates. *Journal of Monetary Economics*, **24**, 235–258.

Hassan, M.K. (1993) The off-balance sheet bank risk of large U.S. commercial banks. *The Quarterly Review of Economics and Finance*, **33**, 51–69.

Hassan, M.K., Karels, G.V., and Peterson, M.O. (1994) Deposit insurance, market discipline and off-balance sheet banking risk of large U.S. commercial banks. *Journal of Banking and Finance*, **18** (3), 575–593.

Hughes, J. and Mester, L.J. (1993) A quality and risk adjusted cost function for banks: evidence on the 'too-big-to-fail-doctrine'. *Journal of Productivity Analysis*, **4**, 292–315.

Hull, J. (1989) Assessing credit risk in a financial institutions' off-balance sheet commitments. *Journal of Financial and Quantitative Analysis*, **24**, 489–501.

Jagtiani, J. and Khanthavit, A. (1996) Scale and scope economies at large banks: including off-balance sheet products and regulatory effects (1984–1991). *Journal of Banking and Finance*, **20** (7), 1271–1287.

Jagtiani, J., Nathan, A., and Sick, G. (1995a) Scale economies and cost complementarities in commercial banks: on-and off-balance-sheet activities. *Journal of Banking and Finance*, **19** (7), 1175–1189.

Jagtiani, J., Saunders, A., and Udell, G. (1995b) The effect of bank capital requirements on bank off-balance sheet financial innovations. *Journal of Banking and Finance*, **19** (3–4), 647–658.

James, C. (1988) The use of loan sales and standby letters of credit by commercial banks. *Journal of Monetary Economics*, **22**, 395–422.

Jondrow, J., Lovell, C.A.K., Van Materov, S., and Schmidt, P. (1982) On the estimation of technical inefficiency in the stochastic frontier production function model. *Journal of Econometrics*, **19** (2–3), 233–238.

Koppenhaver, G.D. (1989) The effects of regulation of bank participation in the guarantee market. *Research in Financial Services*, **4**, 165–180.

Koppenhaver, G.D. and Stover, R.D. (1991) Stand-by letters of credit and large bank capital: an empirical analysis. *Journal of Banking and Finance*, **15**, 315–327.

Kumbhakar, S.C. and Lovell, C.A.K. (2000) *Stochastic Frontier Analysis*, Cambridge University Press, Cambridge.

Lovell, C.A.K. (1993) Production frontiers and productivity, in: *The Measurement of Productivity Efficiency: Techniques and Applications* (eds H.O. Fried, C.A.K. Lovell, and S.S. Schmidt), Oxford University Press, Oxford.

McAllister, P.H. and McManus, D. (1993) Resolving the scale efficiency puzzle in banking. *Journal of Banking and Finance*, **17** (2–3), 389–405.

Mester, L.J. (1996a) Measuring efficiency at US banks: accounting for heterogeneity is important. Federal Reserve Bank of Philadelphia Research working paper 96/11, 14.

Mester, L.J. (1996b) Measuring efficiency at U.S. banks: accounting for heterogeneity is important. Federal Reserve Bank of Philadelphia research working paper.

Mester, L.J. (1996c) A study of bank efficiency taking into account risk-preferences. *Journal of Banking and Finance*, **20** (6), 1025–1045.

Ozer-Balli, H. and Sørensen, B. (2010) Interaction effects in econometrics. CEPR discussion paper 7929. Centre for Economic Policy Research, London.

Pasiouras, F. (2008) Estimating the technical and scale efficiency of Greek commercial banks: the impact of credit risk, off-balance sheet activities, and international operations. *Research in International Business and Finance*, **22** (3), 301–318.

Pavel, C. (1988) Loan sales have little effect on bank risk. *Economic Perspectives*, **12**, 23–31.

Pavel, C.A. and Phillis, D. (1987) Why commercial banks sell loans: an empirical analysis. Federal Reserve Bank of Chicago. Economic Perspectives (May/June), pp. 3–14.

Pennachi, G.G. (1988) Loan sales and the cost of bank capital. *Journal of Finance*, **43**, 375–396.

Rao, A. (2005) Cost frontier efficiency and risk-return analysis in an emerging market. *International Review of Financial Analysis*, **14** (3), 283–303.

Rogers, K.E. (1998) Nontraditional activities and the efficiency of us commercial banks. *Journal of Banking and Finance*, **22** (4), 467–482.

Ronen, J., Anthony, S., and Sondhi, A.C. (1990) *Off-Balance Sheet Activities*, Quorum Books, Westport.

Saunders, A.W. and Walter, I. (1994) *Universal Banking in the United States: What Could We Gain?* Oxford University Press, New York.

Sealey, C.W. and Lindley, J.T. (1977) Inputs, outputs, and a theory of production and cost and depository financial institutions. *The Journal of Finance*, **32** (4), 1251–1265.

Stevenson, R. (1980) Likelihood functions of generalized stochastic frontier estimation. *Journal of Econometrics*, **13** (1), 57–66.

Stiroh, K.J. (2004) Diversification in banking: is noninterest income the answer? *Journal of Money, Credit and Banking*, **36**, 853–881.

Swank, J. (1996) How stable is the multi-product translog cost function? Evidence from the Dutch banking industry. *Kredit und Kapital*, **29**, 153–172.

Templeton, W.K. and Serveriens, J.T. (1992) The effect of nonbank diversification on bank holding companies. *Quarterly Journal of Business and Economics*, **31**, 3–16.

Thakor, A.V. and Udell, G.F. (1987) An economic rationale for the pricing structure of bank loan commitments. *Journal of Banking and Finance*, **11** (2), 271–289.

Tortosa-Ausina, E. (2003) Nontraditional activities and bank efficiency revisited: a distributional analysis for Spanish financial institutions. *Journal of Economics and Business*, **55** (4), 371–395.

Waldman, D.M. (1982) A stationary point for the stochastic frontier likelihood. *Journal of Econometrics*, **18** (2), 275–279.

5

Productivity of foreign banks: Evidence from a financial center*

Claudia Curi[1] and Ana Lozano-Vivas[2]

[1] School of Economics and Management, Free University of Bolzano, Italy
[2] Department of Economic Theory and Economic History, University of Malaga, Spain

5.1 Introduction

Over the last few decades, the waves of foreign banks entering European (and non-European) countries substantially altered the geography and governance of the banking industry. This process, often referred to as internationalization process, has certainly affected the financial sector of those countries which host foreign banks, facilitating the local access to financial services and enhancing the local economic performance. As a consequence, this process has increasingly led the local banking sector to largely depend on foreign banks' activities. In Europe, there are several financial systems dominated by foreign-owned assets: on the one side, we can find new member states (ECB, 2004, 2006, 2010); on the other side, we have Luxembourg, the European country with the highest per capita GDP, which hosts a financial center.

Today, policy makers, financial institution managers, and academics devote significant attention to understand the functioning of foreign banks operating in the host country, patterns of behavior and configurations of activities, in the belief that the success of a foreign bank in financial intermediation can result in gains for the bank group and benefits for the host countries.

Operating as a foreign bank in a host country raises a number of performance challenges linked to differences in languages, laws, social practices, regulations, and customer

---

*The opinions expressed in this paper are those of the authors and do not necessarily reflect the view of their institutions.

Efficiency and Productivity Growth: Modelling in the Financial Services Industry, First Edition. Edited by Fotios Pasiouras.
© 2013 John Wiley & Sons, Ltd. Published 2013 by John Wiley & Sons, Ltd.

expectations, as well as the sheer geographic distance between the home and host countries (Berger et al., 2004). Moreover, foreign banks run different intermediation activities, reflected in various balance sheet characteristics (e.g., ECB, 2010; Curi, Guarda, and Zelenyuk, 2011; Curi et al., 2012, 2013; Claessens and Van Horen, 2012a, b; Chen and Liao, 2011), compared to their relative counterparts operating in the domestic country. A distinguishing feature of foreign banks is the organizational form. In fact, a foreign bank could operate as either a subsidiary bank or a branch bank. Depending on the organizational form, different degrees of legal responsibility are ascribed to foreign banks in case of their failure (e.g., Dermine, 2006). In addition, recent empirical research found that the organizational form may matter also on the operating behavior of foreign banks along several dimensions. For instance, it may matter on the performance and business model run (e.g., Curi, Guarda, and Zelenyuk, 2011; Curi et al., 2012, 2013), on the competitive structure of the local banking systems (e.g., Cerruti, Dell'Ariccia, and Peria, 2007), and it may have stability implications not only for the parent bank but also for local regulators, who care about the stability of the host country, and for local depositors, who care about the safety of their savings (e.g., Cerruti, Dell'Ariccia, and Peria, 2007; Fiechter et al., 2011).

As well-documented in the international banking literature, multinational banks use foreign banks to reap benefits of superior efficiency, profitability, and risk diversification across nations and regions of the world at the group banking level. So far, studies focused either on the impact of foreign banks on competition and performance in their host country or on the comparison of banking performance across countries (see Berger, 2007 for an overview). There is also a research area of interest to academics, regulators, and policy makers (Park and Essayyad, 1989) related to multinational banks which establish their presence in a financial center in the pursuit of 'going where other banks are' (Tschoegl, 2000). This rubric includes three main aspects. First, multinational banks consider financial center as the place where they could easily interact with other banks for doing their business, in particular some specialized activities such as wholesale. Second, in a financial center, banks could behave strategically, causing different geographic patterns, such as 'oligopolistic reaction'. Third, banks located in a financial center could benefit from agglomeration economies.

In contrast to the case of foreign banks operating in 'normal' banking sectors, the existing literature on the performance of foreign banks in financial centers is still limited. Rime and Stiroh (2003) analyzed data from Switzerland, Kwan (2006) studied banks in Hong Kong, Sufian and Majid (2007) worked with banks in Singapore. Curi et al. (2012) and Curi et al. (2013) analyze the interesting topic on the determinants of foreign bank performance, linking the operational efficiency to both the home–host country characteristics and the business model run. Guarda and Rouabah (2006) and Guarda and Rouabah (2007), instead, focus on the analysis of productivity growth and constitute the first investigation of productivity evolution of foreign banks. Both sets of studies are important contributions to the literature on foreign banks and constitute the first attempts to study the performance of a financial center. However, they show some drawbacks. The first set of papers, based on efficiency measurement, do not capture shifts in the technology, due to technical progress (or regress), of the industry over time, yielding to possible misleading measure of the well-being of a bank (Wheelock and Wilson, 1999). This aspect could not be neglected in the analysis of financial centers' performance as financial centers are, more than other banking sectors, shaped by ongoing centrifugal and centripetal forces with the effects of high innovation (Tschoegl, 2000). For example, given a bank, the efficiency assessment could register a decrease in efficiency, while the bank could have experienced productivity gain due to progress in technology. Studying productivity growth allows measuring the overall performance of a bank by capturing simultaneously the efficiency gain (or loss) and the possible innovation effect. On the other hand, the second set of papers neglects the foreign nature of banks operating in a financial center. This could lead to misleading results due to different performance patterns followed by the different bank categories coexisting in a financial center.

Today, it seems urgent to approach the analysis of productivity growth of banks in a financial center disentangling the heterogeneity across banks and the effects of possible innovation. Moreover, both regulators and bankers are interested in understanding how foreign banks behave and how they responded to the recent financial crisis.

The goal of this chapter is to fill this void by examining the evolution of productivity of foreign banks located in the Luxembourg financial center throughout different periods: normal times and during financial crises. In doing so, we use a detailed and original database of Luxembourg banks operating throughout 1995 and 2010, provided by the Central Bank of Luxembourg (BCL). Based on previous findings, the productivity growth is examined by controlling for high level of heterogeneity among foreign banks, ascribed to different organizational forms (subsidiary vs branch banks), size (big, medium, and small banks), and nationality (domestic vs foreign banks). The productivity analysis based on multiple break-downs allows to control for possible different effects of technology and regulatory changes not only on different-sized banks, as emphasized by Berger et al. (2005), but also on different organizational forms and nationality, as shown for the case of efficiency analysis (see Sturm and Williams, 2008; Curi et al., 2012, 2013).

The bank technology, against which we benchmark each bank, is modeled through a global best-practice frontier (Berger, 2007) and built by using the nonparametric method based on DEA, as in several previous studies (e.g., Wheelock and Wilson, 1999; Casu, Girardone, and Molyneux, 2004). The productivity measurement is, in turn, based on Malmquist index (M) and its decomposition (Färe et al., 1994), which enables us to gauge the degree to which changes in productivity depend on technical progress (innovation) and/or catching-up effect (or falling behind). We provide with bias-corrected productivity estimates, a more accurate version of the traditional DEA-based measures (Simar and Wilson's, (1999) procedure).

The contribution of this chapter is threefold. First, we assess the evolution of productivity growth of foreign banks over the period 1995–2010. This is interesting as it allows us to better understand how foreign banks fulfill their intermediation role in a financial center during different structural changes that occur within the sector over time, financial crisis included. Moreover, we empirically investigate the performance behavior of subsidiary and branch banks and possibly identify the most desirable organizational form for a highly productive and stable financial center. We also address the same question with regards to bank size and nationality.

We can briefly summarize our findings as follows: First, looking at the evolution of productivity over the entire period 1995–2010 and considering the whole sample without controlling for heterogeneity among foreign banks, it seems that Luxembourg banks registered an overall, although slow, improvement in productivity, driven by technological change. However, the reverse trend is found over the period 1995–2006, financial crisis excluded. In fact, over the period 1995–2006, the productivity growth was negative, although close to 0, and driven by efficiency gains. This means that the high gain in productivity registered during the period of the financial crisis contributed to reverse the pattern of performance behavior in the Luxembourg financial center. Regarding the organizational form, subsidiary and branch banks followed the same patterns of productivity change, although branch banks exhibited a more volatile behavior. However, branch banks approached better the financial crisis.

In terms of size, big banks seem to have experienced positive productivity change, driven by technical change throughout the entire period, while efficiency improvements are registered during the period 2003–2006. Medium and small banks are found to be less innovative. During the financial crisis, all banks responded with technical changes, and small banks also responded with efficiency improvement. Lastly, our results show Swiss foreign banks as those banks with the highest rate of productivity growth, driven by important improvements in technology during the financial crisis. However, prior to the financial crisis, banks from Belgium were the most productive.

The rest of the chapter is structured as follows: Section 5.2 reviews the prior literature. Sections 5.3 and 5.4 present the methodology and data used, respectively. Section 5.5 discusses the results and Section 5.6 concludes.

5.2   Literature overview

In this section, we focus on two strands of empirical literature related to our analysis of productivity evolution of foreign banks. The first strand relates to the analysis of the effects of deregulation on productivity change of banks operating within the boundary of their home countries, either in the EU, in the United States, or in other countries around the world. The second strand relates to the performance of foreign banks operating in financial centers. Although our analysis is in the same spirit of the second strand, the first strand is presented to assess the possible similarities in performance patterns between the productivity evolution of the Luxembourg financial center and other banking sectors (made up of domestic banks).

In the context of the first strand, starting from the seminal work of Berg, Forsund, and Jansen (1992), several papers have investigated the productivity growth at both country and cross-country levels. For US banks, findings show that US commercial banks account for a slow productivity growth during much of the 20th century (e.g., Humphrey, 1992; Bauer, Berger, and Humphrey, 1993; Wheelock and Wilson, 1999; Alam, 2001; Berger and Mester, 2003; Tirtiroglu, Daniels, and Tirtiroglu, 2005). Findings also show pronounced differences in productivity change among banks of different sizes throughout the period analyzed, supporting the conclusions argued by Berger and Mester (2003) according to which deregulation and technical change probably had differential effects on banks of different sizes. For EU banks, the picture is unclear and more puzzling due to controversial results and lack of evidence for some countries. Overall, during the period of deregulation, European banks faced a decline in productivity. For instance, decreases in productivity growth have been found for the case of Spanish banks during the deregulation period (e.g., Grifell-Tatjé and Lovell, 1996, 1997; Lozano-Vivas, 1998; Kumbhakar and Lozano-Vivas, 2005), while positive over the years 1992–1998, the post-deregulation period (Tortosa-Ausina et al., 2008). Evidence of productivity decline, driven by technological regress, is found for Portuguese banks over the period 1990–1995 (e.g., Mendes and Rebelo, 1999; Canhoto and Dermine, 2003). Moreover, the authors found very similar values in the aggregated productivity index set over the period analyzed as well as heterogeneity across banks. The authors also point out lack of any relationship between technological change (and efficiency change) to size. Fiorentino et al. (2010) analyzed whether consolidation and privatization fostered productivity growth among Italian and German banks during the period 1994–2004. The authors find improvements in productivity in both countries (albeit that there was faster growth in Italy). Battese, Heshmati, and Hjalmarsson (2000) study of Swedish banks finds that technical change became exhausted with 'average' banks catching up with industry best practice. As far as the Luxembourg case is concerned, two papers study the evolution of productivity change. These will be discussed in the following text related to financial center.

Unlike the studies mentioned, empirical evidence on countries outside Europe and the United States finds improvement in productivity. For the case of Indian banks, a study by Rezvanian, Rao, and Mehdian (2008) finds improvement in productivity over the period of post-liberalization (1998–2003) for three groups of banks, with foreign banks experiencing the highest productivity growth compared to private and public-owned banks. Empirical evidence on Australian retail banks (Avkiran, 2000) shows an overall rise in total productivity driven more by technological progress than technical efficiency in the deregulated period (1986–1995). The authors found the performance of major trading banks on technical efficiency similar to that of regional banks but higher on technological progress. This observation implies that technological progress in banks has historically been expensive and that major banks with a larger capital base have been better positioned to undertake such investments. These findings are confirmed by a recent work (Abbott, Wu, and Wang, 2011). Empirical evidence of Turkish commercial banks (Isik and Hassan, 2003) sheds some light on the productivity gains over the period 1981–1990. The authors found that all forms of Turkish banks (private, public, foreign banks), although in different magnitudes, have recorded significant productivity gains driven mostly by efficiency increases rather than technical progress. Gilbert and Wilson (1998) also found large changes in productivity of Korean banks after the deregulation. Empirical evidence on listed Indonesian banks (Hadad et al., 2011) shows that productivity evolves through unstable patterns, associated very often to the volatility in technical efficiency.

There is also a cross-country analysis (Casu, Girardone, and Molyneux, 2004). This study estimates the productivity growth of four European countries (France, Germany, Italy, and Spain) over the period 1994–2000 and finds the total factor productivity (TFP) growth for all the countries. However, the analysis of the decomposition of the TFP index into technological change and technical efficiency change components highlights different trends. Productivity growth is brought by a positive technological change for French, Italian, and Spanish banks, while for German and British banks by efficiency change, although very modest.

The second strand of the literature analyzes the performance of banks in international financial centers. Most of them are oriented to study operational, cost, or profit efficiency. For instance, Rime and Stiroh (2003) examined production structure of Swiss banks using the distribution-free approach. They found relatively large cost and profit inefficiencies across all types of banks, with cost-scale economies for small- and medium-sized banks and little gains from diversification economies for the largest, universal banks in Switzerland. Recently, Burgstaller and Cocca (2011) studied the technical efficiency of banks operating in the private banking and wealth management industry in Switzerland and Liechtenstein over the period 2003–2007. They found private banks in Liechtenstein perform better than their Swiss counterparts, and the level of inefficiency depends on the degree of specialization in private banking activity. For the case of Singapore financial center, Sufian and Majid (2007) focused on the analysis of efficiency gain (or loss) on the performance of banks during the consolidation process. By using DEA and Tobit regression, the results suggest that mergers increased banks' efficiency; bank profitability had a positive impact on bank efficiency, whereas poor loan quality had a negative influence.

For the case of Luxembourg, a growing number of papers document both the efficiency, as in the mentioned cases of financial center, and the productivity evolution of this banking sector. Rouabah (2002) offered the first study of the efficiency of Luxembourg banks over the period 1995–2000 within the context of the comparative advantage theory. The analysis reveals positive efficiency of the socioeconomic variables (effects of the home country in terms of GDP, business cycle, and industrial composition) and significant technological progress, but no evidence of scale economies. Two more recent papers analyze the performance of banks in Luxembourg within the context of multinational banks and cross-border activities, considering a time span including the financial crisis (from 1995 to 2009). The first work by Curi et al. (2012) provided a detailed analysis of the efficiency of foreign banks in Luxembourg over the period 1995–2009. The authors find that the organizational form, business model, and geographical origin matter for the performance of foreign banks. The second work by Curi et al. (2013) investigated the effects of business model on the performance of foreign banks in Luxembourg's financial center accounting for three dimensions: asset mix, funding mix, and income mix. They found that asset diversification is associated with higher efficiency only during normal financial times, whereas diversification in funding and income has a negative impact on efficiency both during normal and distress financial times. Finally, only two papers were oriented to analyze the productivity of the banking sector on financial centers, Guarda and Rouabah (2007) and Guarda and Rouabah (2009). In these papers, the authors analyzed the productivity growth of Luxembourg banks prior to the financial crisis (1994–2007). By using the Malmquist productivity indices and the Tornquist indices, respectively, on quarterly data, they found productivity growth in Luxembourg positive since the mid-1990s, with persistent and procyclical dynamics. Moreover, productivity seems to vary across banks, and larger banks are found to be more productive. Lastly, the M analysis suggested that efficiency change dominated technical change.

Overall, previous literature suggests that banks operating in their home country followed different productivity patterns. Unclear evidence is found on the impact of size and ownership on productivity growth. Moreover, lack of empirical evidence is found regarding the impact of foreign ownership nationality and organization form on productivity growth. Lastly, lack of a long time period has not allowed capturing changes that take time to happen.

This chapter attempts to fill these gaps, having in mind the following questions: (a) Since foreign banks are subject to different regulations and business orientation, given their organizational form (Cerruti, Dell'Ariccia, and Peria, 2007), does organizational form matter in foreign bank productivity? (b) Since foreign banks' performance could depend on home parent banks' characteristics and differences between domestic and foreign banks can arise, does foreign bank nationality matter? (c) Does size of foreign banks matter? (d) Does the performance behavior change over different periods?

5.3   Methodology

5.3.1   TFP growth measures

We measure the performance of banks operating in a financial center through the TFP change. Compared to the concept of X-efficiency change, the concept of TFP change is broader: it captures the performance change of the entire industry as a whole over time, incorporating both changes in managerial best practices in the industry and changes in the cross-section X-efficiency (Berger et al., 2000).

In this respect, recent literature (e.g., Caves, Christensen, and Diewert, 1982; Färe et al., 1994; Simar and Wilson, 1999; Zofio, 2007) has proposed the use of DEA-based M as a comprehensive tool to estimate the TFP growth when economy and management need to be accounted properly. The most appealing features of DEA-based M are briefly described hereby. First, it does not require input or output prices, which makes it particularly useful in situations where prices are misrepresented or nonexistent as in the case of commercial banking industry. In the banking sector, in fact, the measurement of price is particularly difficult, because the services are jointly produced and prices are typically assigned to a bundle of financial services (Fixler, 1993). Second, it does not require the profit maximization (or cost minimization) assumption, which makes it useful in situations where the firms' objectives are unknown or not achieved, as in our special case of foreign banks operating in a host country. Foreign banks, in fact, may pursue specific goals associated to their home parent bank as a whole rather than goals associated to cost minimization or profit maximization, as typically assumed with banks operating in their home countries (Berger, 2007). For instance, a multinational bank could establish a foreign bank in a host country to diversify its risk profile or to serve multinational corporate customers, rather than to pursue profit maximization objectives. Third, the use of M allows disentangling the sources of productivity change, due to either *technical efficiency change* or *technological change*, or both. In particular, these two sources capture simultaneously the individual bank shifting relative to the best practice (*catch-up effect*) and the individual bank shifting relative to the frontier technology (*innovation effect*). Lastly, M provides with an easy economic interpretability (increase, decrease, or no change depending on the value).

In this study, we derive productivity change measures by estimating M in the spirit of Färe et al. (1994). This index involves comparisons of the observed performance of a particular bank to the efficient frontier (or *technology*) at two consecutive time periods, t_1 and t_2. Although banks' ownership differs in nationality within a financial center, banks of our sample operate within the same country. For this reason, we assume a global (i.e., common) frontier so that banks from different countries are compared against the same benchmark, as typically assumed in cross-border literature (e.g., Berger, 2007). To provide insights into whether foreign banks from some countries are more efficient than foreign banks from other countries, we then distinguish the results by their nationality.

To formalize these concepts, we briefly introduce a notation. Consider N banks, which employ P inputs to produce Q outputs over T time period. Let $x \in \Re_+^P$ and $y \in \Re_+^Q$ denote a nonnegative vector of P inputs and a nonnegative vector of Q outputs, respectively. The production possibilities set at time t can be represented by the closed set:

$$\Psi^t = \left\{ (x,y) \in \Re^{P+Q} \,\middle|\, x \text{ can produce } y \text{ at time } t \right\}, \tag{5.1}$$

and its boundary denotes the efficient frontier (or technology). This technology may change over time due to innovation, regulatory changes, or perhaps other factors. Since we are interested in the TFP change at bank individual level, let $\left(x_i^{t_1}, y_i^{t_1} \right)$ denote the input and output vectors of bank i at time t. The Farrell–Debreu (Farrell, 1957) output distance function for bank i at time t_1, relative to the technology existing at time t_2 is defined as

$$D_i^{t_1 t_2}\left(x_i^{t_1}, y_i^{t_1} \right) = \sup\left\{ \lambda \geq 0 \,\middle|\, \left(x_i^{t_1}, \lambda y_i^{t_1} \right) \in \Psi^{t_2} \right\}. \tag{5.2}$$

This measure is the radial distance from the ith bank in the input/output space at time t_1 to the boundary of the production set at time t_2 in the hyperplane, where inputs remain constant. If $t_1 = t_2$, then we have a measure of efficiency relative to the contemporaneous technology.

The Färe et al. (1994) version of M combines the different Farrell–Debreu output distance functions and gives the following definition of the index for two periods of time t_1 and t_2, with $t_2 > t_1$, as

$$M_i\left(t_1, t_2\right) = \frac{D_i^{t_2, t_2}}{D_i^{t_1, t_1}} \times \left(\frac{D_i^{t_2, t_1}}{D_i^{t_2, t_2}} \times \frac{D_i^{t_1, t_1}}{D_i^{t_1, t_2}}\right)^{0.5}. \tag{5.3}$$

The first term measures the change in output technical efficiency between periods t_1 and t_2 defining an output-based index of *Efficiency Change* (EC$_i(t_1,t_2)$). The second term measures the change in technology between periods t_1 and t_2 and defines an output-based index of *Technical Change* (TC$_i(t_1,t_2)$). The easy economic interpretation of these indexes is given by the fact that values greater (less) than unity for both $M_i(t_1,t_2)$ and its components indicate productivity, efficiency, and technical regress (progress) between time t_1 and t_2. Values equal to unity indicate no change.

5.3.2 Estimation of the TFP growth measures

As seen in the previous section, M is based on the measurement of a set of Farrell–Debreu (Farrell, 1957) output distance functions $\left\{D_i^{t_1|t_1}, D_i^{t_2|t_2}, D_i^{t_1|t_2}, D_i^{t_2|t_1}\right\}$. However, this set is not observed in reality as the frontier is unobserved. Several estimators of the frontier are possible. In this study, we estimate the frontier from the sample $\chi_n^t = \left\{\left(X_i^t, Y_i^t\right)\right\}_{i=1}^n$ of data on firms' input and output quantities for this period by using DEA, which is a popular nonparametric envelopment estimator. This relies on very few assumptions of the Data Generating Process[1] (DGP), which allows handling multiple inputs and outputs and does not require any parametric assumptions in the form of the production relationship (or the distribution of the inefficiency term). DEA estimator provides with a frontier built as a piecewise-linear convex hull of observed input–output bundles, given by

$$\hat{\Psi}_{\text{DEA}} = \left\{(x,y) \in \Re_+^P \times \Re_+^Q \middle| \sum_k^N z_k y_k^q \geq y^q, \; q = 1,\dots,Q\right.$$

$$\left.\sum_{k=1}^N z_k x_k^p \leq x^p \;\; p = 1,\dots,P \;\; \text{for}\left(z_1,\dots,z_N\right), \; \text{such that} \; \sum_{k=1}^N z_k = 1; \, z_k \geq 0 \;\; k = 1,\dots N\right\}. \tag{5.4}$$

The properties of DEA estimator are well known today (Kneip, Simar, and Wilson, 2008) and show that DEA is a downward-biased estimator (for the case of output orientation). This implies that the technology in Equation (5.1) is estimated with some error and, then, estimated set of distances to derive M and its components are too optimistic.

This study uses the bias-corrected version of DEA estimator. Bootstrapping technique is the only statistical tool, which could retrieve the bias-correction term, and the algorithm by Simar and Wilson (1999) is the only algorithm suited to the case of productivity inference,

---

[1] See Korostelev, Simar, and Tsybakov (1995) and Park, Jeong, and Simar (2010) for proof of consistency and rates of convergence of the DEA estimator under constant returns to scale. The consistency property is required then for the bootstrap approximation, as described in the next paragraph.

which provides with a consistent bootstrap. The rationale behind this procedure is to simulate the true sampling distribution of the estimators of interest by mimicking the DGP. Considering the $M_i(t_1,t_2)$ index, this procedure mimics the unknown distribution, $\hat{M}_i(t_1,t_2) - M_i(t_1,t_2)$, where $\hat{M}_i(t_1,t_2)$ and $M_i(t_1,t_2)$ are respectively the estimates and real values of interest, with the known bootstrap distribution $\hat{M}_i^*(t_1,t_2) - \hat{M}_i(t_1,t_2)$, where $\hat{M}_i^*(t_1,t_2)$ is the bootstrapped estimate for bank i, completely known once $\hat{M}_i(t_1,t_2)$ is supposed true. This approximation held as the DGP is consistently replicated by using a bivariate kernel density estimate (smooth bootstrap procedure), which allows to preserving the possibility of temporal correlation due to the panel data structure (Simar and Wilson, 1999), via the covariance matrix of the data of adjacent years. Practically, the procedure consists of replicating the DGP by generating an appropriate large number of B pseudosamples, and then for each bootstrap pseudosample generated, estimating the frontier and the set of distances $\left\{ \hat{D}_i^{t_1|t_1*}(b), \hat{D}_i^{t_2|t_2*}(b), \hat{D}_i^{t_1|t_2*}(b), \hat{D}_i^{t_2|t_1*}(b) \right\}_{b=1}^{B}$, for each bank $i=1,\ldots,N$, by reapplying the DEA estimator. These estimates are then used to obtain $\hat{M}_i^*(t_1,t_2)(b)$, $T\hat{C}_i^*(t_1,t_2)(b)$, and $T\hat{E}C_i^*(t_1,t_2)(b)$, where $i=1,\ldots,N$ and $b=1,\ldots,B$. The bootstrap bias estimate for the original estimator $\hat{M}_i(t_1,t_2)^2$ is $bi\hat{a}s\left(\hat{M}_i(t_1,t_2)\right) = (1/B)\sum_{b=1}^{B}\hat{M}_i^*(t_1,t_2) - \hat{M}_i(t_1,t_2)$, which is the empirical analog of $E\left[\hat{M}_i(t_1,t_2)\right] - \hat{M}_i(t_1,t_2)$.

Therefore, we correct the estimate of $M_i(t_1,t_2)$ and derive the bias-corrected estimates by shifting the original estimates of the bootstrapped bias estimated as

$$\hat{M}_i^{bc} = \hat{M}_i(t_1,t_2) - bi\hat{a}s\left(\hat{M}_i(t_1,t_2)\right) = 2\hat{M}_i(t_1,t_2) - \frac{1}{B}\sum_{b=1}^{B}\hat{M}_i^*(t_1,t_2), \tag{5.5}$$

Where $i=1,\ldots,N$ and $b=1,\ldots,B$ are the number of replications. We use Silverman's (1986) suggestion for bivariate data by setting $h=(4/5N)^{1/6}$ since we are using a bivariate normal kernel scaled to have the same shape as the data, as suggested by Simar and Wilson (1999), though acknowledging other optimization criteria are possible (Curi, 2007).

5.4   Data and sources

We use individual bank data derived from nonconsolidated banks' financial statements, provided by BCL. We exclude from the sample the cooperative banks (which are domestic banks) and a central securities depository with bank status. We also excluded banks that reported nonpositive values for any of the input or output measures, as such values are frequently indicative of reporting error or anomalous operation. The final sample consists of domestic and foreign banks which provide the full range of financial services in the financial center, namely, (a) customer, (b) interbank, (c) financial market, and (d) private banking activities. We carry out our analysis over a 15-year sample period (1995–2010). The choice of such a long period is related to the fact that it allows to analyze the dynamics, or the lack of it, of performance indexes over time, highlighting possible effects of bank survivals in a financial center in the long run and quantify the impact of structural changes on performance over time and across banks. We consider four subperiods: the first period characterized by strong consolidation process and high rate of banks' entry and closure (1995–2001); the

---

[2] Efficiency and technology change indices can be analyzed similarly.

second by the stock market crisis (2002–2003); the third by recovery (2004–2006); and the last one by the financial crisis (2007–2010). Moreover, the long length of the time period allows capturing the effects due to investment in new capital and technology and possible management changes. Therefore, we analyze the productivity change (and relative components) on the following time intervals: 1995–2001; 2001–2003; 2003–2006; and 2006–2010. Our sample has from 126 observations in 1995 to 77 observations in 2010 and is a good representative of the Luxembourg banking sector as it constitutes 78% of the system's total assets on average.

Our choice of bank inputs and outputs is based on the intermediation approach (Sealey and Lindley, 1977), commonly used in the bank efficiency literature (Berger and Humphrey, 1997) that views banks as financial intermediaries with labor, capital, and other funding sources treated as inputs to obtain bank outputs related with assets. Since most institutions in financial centers are commercial banks, the intermediation approach should be suitable for the definition of banking output and input. Specifically, on the output side, we select interbank loans, customer loans, securities, and off-balance sheet. In particular, interbank activities include those within the parent banking group as well as with other banks. Customer activities include those with households and with nonfinancial corporations, and securities include government securities, fixed-income securities, shares, participations, and other variable-income securities. Moreover, as discussed in Curi et al. (2012) and Curi et al. (2013), Luxembourg banks are also concentrated on private banking, relying on off-balance sheet items. Due to lack of this information and acknowledging the importance of the inclusion of a proxy in the frontier estimation (Lozano-Vivas and Pasiouras, 2010), we use the noninterest income as a measure of the off-balance sheet fee services (Clark and Siems, 2002).

On the input side, we include (a) labor, measured by total labor expenses; (b) capital, measured as the sum of tangible and intangible assets; (c) interbank deposits, including other liabilities, such as debt certificates and subordinated debts; and (d) customer deposits. Moreover, we follow Fixler and Zieschang (1992) in identifying additional inputs as purchased materials and services, including nonwage administrative costs and commissions paid. It is crucial to include this commission flow as the net commission income is of the same order of magnitude as the income from interest rate margin.[3] Since any efficiency and productivity analysis involved comparisons across time, data in nominal values is converted to real terms using the GDP deflator with base year 1995.

Table 5.1 presents the summary statistics of input and output variables for the pooled sample. Given the high presence of foreign ownership in the Luxembourg financial center, we suspect heterogeneity across banks in terms of output and input mix, depending on organizational form, origin, and size. This suspicion suggests that, as we pointed out in Section 5.2, when productivity of foreign banks is analyzed, then some specific aspects of banks need to be controlled for. Thus, Table 5.1 also reports the summary statistics of input and output variables according to three breakdowns: *organization form* (branch vs subsidiary), *nationality* (Luxembourg, Belgium, France, Germany, Switzerland, Italy, and other countries), and *size* (big, medium, and small). In terms of organizational form, we have distinguished according to the corporate governance, namely, branch versus subsidiaries. In terms of nationality, we group the banks according to home parent bank nationality, considering domestic banks

---

[3] As it has been noted by Guarda and Rouabah (2009).

Table 5.1 Descriptive statistics of bank inputs and outputs (pooled data 1995–2010).

Group		Labor	Capital	Deposits by banks	Deposits by customers	Other costs	Loans to banks	Securities	Loans to customers	Noninterest income
All	Mean	3.72	27.18	2970.37	1816.53	4.33	2420.70	1311.71	1139.20	7.03
	Median	1.20	4.82	448.01	481.58	1.51	743.95	121.87	195.98	2.44
	s.d.	7.46	71.19	5582.41	3214.70	7.63	4528.50	2650.34	2455.33	11.03
Subsidiary	Mean	4.05	30.01	3090.24	1968.21	4.65	2529.28	1389.13	1230.87	7.61
	Median	1.38	5.83	458.71	542.31	1.71	797.92	129.23	202.00	2.89
	s.d.	7.77	74.57	5780.31	3346.15	7.91	4712.34	2729.77	2564.01	11.37
Branch	Mean	0.80	2.16	1908.51	472.97	1.49	1458.82	625.88	327.18	1.87
	Median	0.24	0.76	385.88	214.41	0.36	498.06	101.42	151.55	0.22
	s.d.	2.23	3.61	3174.43	834.72	3.41	2114.85	1654.63	685.42	4.94
Big	Mean	13.04	114.05	12718.47	6819.47	13.88	9321.39	5495.85	4946.17	20.87
	Median	7.91	58.01	10841.98	5554.40	9.57	7812.77	4441.81	4013.99	19.05
	s.d.	14.10	142.67	7818.29	5084.49	13.33	7656.46	4072.43	4137.96	17.03
Medium	Mean	3.16	17.44	2033.51	1505.91	4.08	1952.40	957.38	724.85	7.28
	Median	1.96	9.61	1089.91	1086.99	2.46	1401.86	354.37	393.53	4.01
	s.d.	3.43	23.09	2106.79	1447.73	4.63	1551.36	1331.06	860.06	8.38
Small	Mean	0.72	3.46	165.40	221.36	0.97	256.85	64.05	88.10	1.59
	Median	0.51	2.21	85.73	181.33	0.51	207.08	28.08	52.19	0.85
	s.d.	0.71	3.73	193.06	181.52	1.31	195.54	84.37	105.86	2.04
Domestic	Mean	11.88	100.13	4842.52	4914.21	7.94	3677.82	2922.99	2461.34	9.61
	Median	2.68	13.86	833.49	647.17	1.25	550.63	271.16	135.69	2.52
	s.d.	13.74	143.06	7068.50	6209.06	9.52	4738.92	4122.76	3504.91	11.35
Belgium	Mean	12.25	105.36	5883.33	4611.30	10.82	5190.95	3571.85	1613.57	15.42
	Median	3.75	44.96	588.67	1021.54	6.78	600.31	492.70	127.09	10.15
	s.d.	14.77	140.61	9114.07	5808.50	11.57	7392.93	4953.73	2660.02	16.37

(continued overleaf)

Table 5.1 (continued)

Group		Labor	Capital	Deposits by banks	Deposits by customers	Other costs	Loans to banks	Securities	Loans to customers	Noninterest income
Germany	Mean	2.24	19.36	5796.62	1700.36	3.32	3739.11	2084.64	1923.58	5.58
	Median	1.27	4.91	3854.11	623.53	1.93	1801.93	1272.08	565.46	2.90
	s.d.	2.37	30.89	7192.17	2427.94	4.31	6238.18	2501.01	3314.14	7.62
France	Mean	7.50	61.14	3344.85	3582.86	10.74	3400.09	2064.48	1549.50	15.75
	Median	4.61	12.15	509.68	1770.19	5.67	1372.09	163.98	316.66	8.72
	s.d.	9.80	129.69	4965.33	4807.88	13.04	4472.70	3706.92	2671.17	18.23
Switzerland	Mean	2.87	13.26	617.04	1274.05	2.74	1481.94	62.56	410.64	7.49
	Median	0.79	4.77	254.31	306.51	1.28	494.12	18.67	80.13	2.73
	s.d.	3.87	16.92	1025.71	2030.68	3.43	2228.14	154.01	777.80	9.82
Italy	Mean	1.03	7.13	1123.79	773.21	1.70	1213.48	515.12	225.71	2.87
	Median	0.56	4.17	375.76	410.79	0.52	733.53	123.71	103.99	0.48
	s.d.	1.10	7.99	1790.75	833.20	2.44	1372.35	950.01	378.23	4.55
Others	Mean	2.35	9.41	697.11	875.81	2.54	799.94	307.02	536.38	4.46
	Median	1.02	3.85	256.10	383.55	1.22	391.86	49.51	163.53	2.22
	s.d.	6.51	16.85	2040.47	1478.48	5.87	1060.13	1157.51	1429.73	7.52

The table reports mean and median for the variables used as inputs and outputs in the estimation of TFP growth rate. All variables are measured in million euro in constant (2000) terms.

(Luxembourg banks), banks from neighboring countries (Belgium, France, Germany), and the other two countries (Italy and Switzerland), countries that show the highest presence in the Luxembourg financial center, plus banks from the rest of foreign countries, called as 'others'. Finally, we control for three bank size categories. These categories are defined according to the total assets held by the bank, and we follow the categories used by BCL in reporting bank data. In particular, big banks are those with total assets greater than €10 billion, medium banks hold total assets between 10 billion and 1 billion, and small banks hold less than 1 billion in total assets.

The statistics suggest heterogeneity across groups in terms of input and output compositions, justifying our discussion. Heterogeneity is also found within each group, as the high value of standard deviation shows. Regarding heterogeneity between branch and subsidiary banks, branch banks have, on average, the smallest banking activity as well as employ less capital and labor in their activities than subsidiary banks do. It follows that they face a lower level of other cost. The notable difference between the two organizational forms might be explained by the different involvement in customer-related activity of lending and funding as well as activities related to noninterest income. Regarding heterogeneity of banks depending on the nationality, we can see interesting aspects. It seems that both domestic and Belgian banks invest the most in labor and investment in capital. On overage, the highest share of interbanking activity (both borrowing and lending) is concentrated in Belgian banks. However, looking at the median values, it is evident that the distribution is skewed and, therefore, there are possibly few big banks where this activity is concentrated. On the contrary, German banks seem to be more involved in borrowing interbanking activities. On average, domestic banks seem to be the largest banks in terms of customer activity (both borrowing and lending), followed by Belgian banks on the borrowing side and German banks on the lending side. Here again, it is clear that there are few domestic banks concentrated in this activity, and the average statistics could be misleading. Belgian and German banks show a less skewed distribution.

Both Belgian and German banks are the most involved banks in the securities activity. However, as before for the case of interbanking activity, only few Belgian banks are involved in this activity while a larger number of German banks are.

Regarding activity related to noninterest income, Belgian and French banks are the most involved. Concerning Swiss banks, it is evident that compared to others, they are less involved in interbanking but much more involved in customer-related activity and non-interest-related activity. On the contrary, Italian banks are more involved in interbanking and securities activity.

Heterogeneity is also given by the size effects of banks. In addition to the difference between groups of different sizes, differences within each group exist. This is notable both within big and medium banks.

For a deep analysis of the sample used, we show in Table 5.2 the distribution in each year by organization form, size, and origin. It is evident that most of the banks are subsidiaries. In terms of size, the number of small banks is always the highest, but since 2007, the number of medium banks is the highest one. The uppermost percentage of foreign banks in Luxembourg is from Germany, followed by banks from the neighboring countries, France and Belgium. However, among the three neighboring countries, the number of Belgian banks constitutes the minority. Nonetheless, as we have seen from Table 5.1, they control the largest portion of banking activity in Luxembourg. This implies that there is a high concentration of banks with Belgian ownership, although the highest number consist of German banks.

Table 5.2 Distribution of banks by type, size, and origin.

Year	Total banks	Subsidiary	Branch	Big	Medium	Small	LUX	BE	DE	FR	SW	IT	Others
1995	126	109	17	10	50	66	8	7	33	17	13	14	34
1996	133	116	17	10	48	75	8	8	34	17	13	15	38
1997	123	107	16	12	50	61	7	7	33	15	11	16	34
1998	111	97	14	14	39	58	4	7	33	12	11	16	28
1999	99	89	10	13	39	47	4	5	29	12	8	14	27
2000	99	89	10	15	39	45	4	6	28	11	8	15	27
2001	99	92	7	18	33	48	3	6	26	11	8	14	31
2002	94	86	8	18	35	41	3	6	25	10	8	14	28
2003	92	82	10	14	36	42	1	5	27	12	8	12	27
2004	84	75	9	15	34	35	1	4	25	12	7	8	27
2005	79	71	8	17	30	32	1	4	25	10	7	8	24
2006	79	71	8	19	29	31	2	4	25	10	7	9	22
2007	80	74	6	18	34	28	3	4	25	9	7	7	25
2008	66	60	6	16	33	17	2	3	17	9	6	7	22
2009	75	69	6	14	38	23	3	4	24	10	7	7	20
2010	77	71	6	15	38	24	3	4	25	10	6	7	22

5.5   Empirical results

The discussion of the empirical results on the productivity evolution of banks operating in the Luxembourg financial center will be structured in two parts. In the first part, we evaluate the efficiency of the full sample of banks and study their evolution over time. In the second, we focus on the specific issues of the relative productivity evolution of subsidiary versus branch banks, big versus medium and small banks, and in relation to the nationality.

Consistent with the literature, we estimate bias-corrected Malmquist productivity (and relative components) for each bank in each year. For the sake of space and discussion, we summarize our results in Table 5.3, Table 5.4, Table 5.5, Table 5.6, Table 5.7, and Table 5.8, structured as follows: in Panel A, we report the annual geometric mean, in Panel B the means of the annual geometric means over the four subperiods, namely, (a) phase of growth of the financial center given to consolidation and entry and exit of banks, controlling the changes over the period 1995–2001; (b) collapse of the stock market, controlling the changes over the period 2001–2003; (c) phase of recovery, controlling the changes over the period 2003–2006; and (d) financial crisis, where the changes are controlled over the period 2006–2010. Lastly, in Panel C we report the mean of the annual geometric means over the period prior to the financial crisis and the period during the financial crisis. Since we correct the Malmquist estimates for the bias term, our estimates might be either greater than 1 implying a decline or less than 1 implying a growth.

5.5.1   Productivity growth over time

Table 5.3 (Panel A) shows the evolution of average year-on-year efficiency as well as technical and productivity changes for the entire sample. Over the 15-year period, efficiency gains are found in only 5-year periods, with the greatest gain of 4.59% between 2005 and 2006 and technology progress in 9-year periods, with the highest value of 10% between 2007 and 2008. It is interesting to note that the highest loss in technology, i.e., −7.56%, is reached in 2005–2006, the same time interval when banks reached the highest efficiency gain. Similar cases of opposite patterns of behavior between efficiency and technological change are shown for most of the years. The picture that emerges is a continuous adjustment in banking performance of the banking sector as a whole, driven by a trade-off between technological improvements and slight efficiency deterioration. The result is a positive, although slow, rate of productivity growth. On average, in fact, the Luxembourg financial center experienced a slight improvement in productivity (0.84%), driven by annual average of technology change of about 1.45% and efficiency change of −0.46%. These first results are in line with the results from Abbott, Wu, and Wang (2011), the only study analyzing a long and recent time period. As Luxembourg banks, the four biggest Australian banks were characterized by slow productivity growth, mainly driven by technology change.

Table 5.3 (Panel B) shows the evolution of productivity change (and relative components by time intervals) and allows disentangling the effects of possible structural change in the financial center. This closer analysis suggests productivity gains in only two periods (1995–2001 and 2006–2010) and loss in productivity during the period of stock market collapse and recovery. The first productivity growth pattern, which occurred during the phase of growth, has to be ascribed to gains in both efficiency and technology. This is in line with the previous studies that focused on the period of post-deregulation for countries outside the EU (e.g., Avkiran, 2000 and Rezvanian, Rao, and Mehdian, 2008). The second pattern of productivity

Table 5.3   Bias-corrected efficiency change (EC), technical change (TC), and productivity change (M) of the entire sample.

Years	Entire sample		
	EC	TC	M
Panel A			
1995–1996	0.9687	1.0340	1.0038
1996–1997	1.0023	0.9916	0.9944
1997–1998	0.9890	1.0074	0.9973
1998–1999	1.0239	0.9677	0.9920
1999–2000	1.0059	0.9812	0.9880
2000–2001	0.9897	0.9889	0.9798
2001–2002	1.0237	1.0077	1.0328
2002–2003	1.0009	1.0173	1.0202
2003–2004	1.0229	0.9400	0.9636
2004–2005	1.0019	1.0335	1.0377
2005–2006	0.9541	1.0756	1.0281
2006–2007	1.0445	0.9161	0.9653
2007–2008	0.9877	0.8999	0.8912
2008–2009	1.0471	0.9630	1.0131
2009–2010	1.0062	0.9586	0.9675
1995–2010	**1.0046**	**0.9855**	**0.9916**
Panel B			
1995–2001	**0.9966**	**0.9951**	**0.9925**
2001–2003	**1.0123**	**1.0125**	**1.0265**
2003–2006	**0.9929**	**1.0164**	**1.0098**
2006–2010	**1.0214**	**0.9344**	**0.9593**
Panel C			
1995–2006	**0.9985**	**1.0041**	**1.0034**
2006–2010	**1.0214**	**0.9344**	**0.9593**

growth, which occurred during the financial crisis, is driven only by improvement in technical change and particularly remarkable compared to the previous one. It is also interesting to note that the stock market crisis, in contrast with the recent one, was characterized by different patterns, namely, deterioration in both efficiency and technology. Lastly, comparing the productivity changes before and during the financial crisis (Panel C), results show that there exists an improvement in productivity change during the crisis due to an important shift in the frontier, that is, due to an important progress in technology, but at the same time there has been an important deterioration in the efficiency change. On the contrary, prior to the financial crisis, the Luxembourg banking sector reached a moderate improvement in efficiency, jointly with a small technical regress. These results are in line with the previous results obtained by Guarda and Rouabah (2009) who focused their analysis on the period 1994–2007.

Table 5.4 Bias-corrected efficiency change (EC), technical change (TC), and productivity change (M) of banks classified by organizational form.

Years	Subsidiary			Branch		
	EC	TC	M	EC	TC	M
Panel A						
1995–1996	0.9674	1.0348	1.0032	0.9796	1.0273	1.0091
1996–1997	1.0028	0.9919	0.9951	0.9976	0.9887	0.9865
1997–1998	0.9873	1.0082	0.9963	1.0030	1.0011	1.0050
1998–1999	1.0254	0.9643	0.9900	1.0102	0.9994	1.0104
1999–2000	1.0066	0.9768	0.9843	0.9991	1.0221	1.0223
2000–2001	0.9909	0.9921	0.9842	0.9760	0.9510	0.9288
2001–2002	1.0247	1.0157	1.0420	1.0088	0.9021	0.9117
2002–2003	0.9912	1.0217	1.0144	1.1396	0.9605	1.1015
2003–2004	1.0258	0.9479	0.9744	0.9989	0.8775	0.8791
2004–2005	0.9952	1.0275	1.0245	1.0619	1.0871	1.1601
2005–2006	0.9549	1.0632	1.0170	0.9471	1.1855	1.1256
2006–2007	1.0527	0.9100	0.9662	0.9555	0.9891	0.9553
2007–2008	1.0014	0.9175	0.9209	0.8492	0.7270	0.6209
2008–2009	1.0474	0.9856	1.0368	1.0434	0.7354	0.7750
2009–2010	1.0072	0.9411	0.9506	0.9927	1.2221	1.2199
1995–2010	**1.0054**	**0.9866**	**0.9933**	**0.9975**	**0.9784**	**0.9807**
Panel B						
1995–2001	0.9967	0.9947	0.9922	0.9942	0.9983	0.9937
2001–2003	1.0080	1.0187	1.0282	1.0742	0.9313	1.0066
2003–2006	0.9920	1.0129	1.0053	1.0026	1.0501	1.0549
2006–2010	1.0272	0.9386	0.9687	0.9602	0.9184	0.8928
Panel C						
1995–2006	0.9975	1.0040	1.0023	1.0111	1.0002	1.0127
2006–2010	1.0272	0.9386	0.9687	0.9602	0.9184	0.8928

5.5.2 Breaking down productivity growth

As mentioned in Section 5.4, we want to investigate the performance of foreign banks along several dimensions. For this purpose, we discuss our results by breaking down banks into categories. First, we are interested in understanding which organization form of foreign banks (subsidiary vs branch banks) experiences the highest productivity gains in financial center (Table 5.4). Here, there appears higher annual rate of productivity growth for the case of branch banks (1.93%), compared to the rate of subsidiary banks (0.67%) (Table 5.4, Panel A). Overall, the year-on-year TFP growth associated with subsidiary and branch banks is quite different: (a) for subsidiary banks, TFP is negative during the six-year period, while for branch banks it is negative during the eight-year period, and (b) the magnitude of change is different with more volatile behavior for branch banks. Nonetheless, for both groups, productivity evolution seems to be driven by technological changes. Moreover, it seems that subsidiary

Table 5.5 Bias-corrected efficiency change (EC), technical change (TC), and productivity change (M) of banks classified by size.

Years	Big banks			Medium banks			Small banks		
	EC	TC	M	EC	TC	M	EC	TC	M
Panel A									
1995–1996	0.9567	1.0336	0.9908	0.9665	1.0253	0.9931	0.9727	1.0412	1.0151
1996–1997	1.0000	1.0018	1.0021	1.0038	0.9896	0.9938	1.0016	0.9912	0.9932
1997–1998	0.9947	0.9868	0.9821	0.9899	1.0083	0.9990	0.9870	1.0121	1.0000
1998–1999	1.0068	0.9566	0.9650	1.0178	0.9467	0.9645	1.0341	0.9885	1.0233
1999–2000	1.0043	0.9753	0.9804	0.9997	0.9904	0.9910	1.0125	0.9745	0.9880
2000–2001	0.9966	0.9455	0.9434	0.9917	0.9881	0.9807	0.9854	1.0086	0.9952
2001–2002	1.0037	0.9984	1.0034	1.0157	1.0633	1.0811	1.0398	0.9670	1.0068
2002–2003	1.0018	0.9823	0.9852	0.9911	1.0431	1.0364	1.0085	1.0095	1.0198
2003–2004	1.0000	0.9657	0.9672	1.0454	0.9286	0.9734	1.0118	0.9398	0.9529
2004–2005	0.9899	1.0235	1.0154	1.0021	1.0226	1.0269	1.0083	1.0498	1.0608
2005–2006	0.9566	1.0524	1.0081	0.9375	1.0963	1.0297	0.9688	1.0705	1.0391
2006–2007	1.0566	0.8957	0.9547	1.0695	0.9292	1.0027	1.0083	0.9155	0.9311
2007–2008	1.0193	0.9168	0.9376	0.9898	0.9038	0.8965	0.9455	0.8706	0.8253
2008–2009	1.0768	0.9598	1.0387	1.0565	0.9246	0.9820	1.0109	1.0397	1.0551
2009–2010	0.9986	0.9674	0.9690	1.0111	0.9809	0.9947	1.0029	0.9160	0.9216
1995–2010	**1.0042**	**0.9774**	**0.9829**	**1.0059**	**0.9894**	**0.9964**	**0.9999**	**0.9863**	**0.9885**
Panel B									
1995–2001	**0.9932**	**0.9833**	**0.9773**	**0.9949**	**0.9914**	**0.9870**	**0.9989**	**1.0027**	**1.0025**
2001–2003	**1.0027**	**0.9903**	**0.9943**	**1.0034**	**1.0532**	**1.0587**	**1.0242**	**0.9882**	**1.0133**
2003–2006	**0.9822**	**1.0139**	**0.9969**	**0.9950**	**1.0158**	**1.0100**	**0.9963**	**1.0200**	**1.0176**
2006–2010	**1.0378**	**0.9349**	**0.9750**	**1.0317**	**0.9346**	**0.9690**	**0.9919**	**0.9354**	**0.9333**
Panel C									
1995–2006	**0.9919**	**0.9929**	**0.9857**	**0.9965**	**1.0093**	**1.0063**	**1.0028**	**1.0048**	**1.0086**
2006–2010	**1.0378**	**0.9349**	**0.9750**	**1.0317**	**0.9346**	**0.9690**	**0.9919**	**0.9354**	**0.9333**

Each cell contains the geometric mean of the individual observation scores.

and branch banks suffer from the highest loss in productivity in different periods (Table 5.4, Panel B): subsidiary banks registered the highest loss during the stock market crisis, while branch banks between 2003 and 2006. Specifically, while subsidiary banks experience loss in technology during the stock market crisis (2001–2003), branch banks experience innovation. Whereas branch banks experience loss both in technology and efficiency between 2003 and 2006, subsidiary banks gain in efficiency. For branches, progress in technology is positive for almost all the periods. During the financial crisis, both subsidiary and branch banks improved their productivity, but with a very different growth rate: for subsidiary banks, TFP increases at an annual growth rate of 3.13%, driven by technical change (an annual growth rate of 6.14%) against a negative efficiency change (−2.72%). Differently, branch banks improve their TFP, and this improvement was driven by a positive change in efficiency (3.98%) as well as in technology (8.16%). Comparing the period prior to the crisis with the period of the

Table 5.6 Bias-corrected efficiency change (EC), technical change (TC), and productivity change (M) of banks classified by home country origin, neighboring countries.

Years	Belgium			France			Germany		
	EC	TC	M	EC	TC	M	EC	TC	M
Panel A									
1995–1996	0.9113	1.0603	0.9668	0.9551	1.0370	0.9920	0.9844	1.0207	1.0079
1996–1997	0.9925	0.9848	0.9775	1.0050	0.9921	0.9972	0.9984	1.0250	1.0240
1997–1998	0.9836	1.0225	1.0064	0.9705	1.0340	1.0043	0.9902	0.9614	0.9531
1998–1999	1.0299	0.9533	0.9822	1.0254	0.9573	0.9825	1.0252	0.9786	1.0048
1999–2000	1.0278	0.9494	0.9763	1.0095	0.9799	0.9895	1.0009	0.9859	0.9883
2000–2001	0.9650	0.9846	0.9507	1.0055	0.9952	1.0010	0.9946	0.9877	0.9835
2001–2002	1.0683	0.9853	1.0536	1.0122	0.9921	1.0047	1.0179	1.0131	1.0326
2002–2003	0.9957	1.0253	1.0218	0.9993	1.0036	1.0036	0.9977	1.0360	1.0363
2003–2004	1.0010	0.9856	0.9876	1.0042	0.9604	0.9655	1.0305	0.9303	0.9608
2004–2005	0.9342	1.0343	0.9674	1.0012	1.0444	1.0470	1.0200	0.9974	1.0198
2005–2006	0.9268	1.0400	0.9645	0.9648	1.1063	1.0684	0.9601	1.0925	1.0512
2006–2007	1.1488	0.8969	1.0345	1.0488	0.8917	0.9422	1.0420	0.9536	1.0067
2007–2008	0.9973	0.9994	0.9973	1.0593	0.9273	0.9837	0.9961	0.8384	0.8387
2008–2009	1.0637	0.9270	0.9874	1.0572	0.9765	1.0355	1.0317	0.8874	0.9217
2009–2010	1.0481	0.9717	1.0194	0.9748	0.9305	0.9086	1.0268	1.0059	1.0367
1995–2010	**1.0063**	**0.9880**	**0.9929**	**1.0062**	**0.9886**	**0.9951**	**1.0078**	**0.9809**	**0.9911**
Panel B									
1995–2001	**0.9850**	**0.9925**	**0.9766**	**0.9952**	**0.9992**	**0.9944**	**0.9990**	**0.9932**	**0.9936**
2001–2003	**1.0320**	**1.0053**	**1.0377**	**1.0058**	**0.9979**	**1.0041**	**1.0078**	**1.0245**	**1.0344**
2003–2006	**0.9540**	**1.0200**	**0.9732**	**0.9901**	**1.0371**	**1.0270**	**1.0035**	**1.0067**	**1.0106**
2006–2010	**1.0645**	**0.9488**	**1.0096**	**1.0350**	**0.9315**	**0.9675**	**1.0242**	**0.9213**	**0.9510**
Panel C									
1995–2006	**0.9851**	**1.0023**	**0.9868**	**0.9957**	**1.0093**	**1.0051**	**1.0018**	**1.0026**	**1.0057**
2006–2010	**1.0645**	**0.9488**	**1.0096**	**1.0350**	**0.9315**	**0.9675**	**1.0242**	**0.9213**	**0.9510**

financial crisis, branches benefited from gains in both efficiency and technology, while subsidiaries benefited from technical change but they became less efficient. Thus, it seems that in terms of productivity (and their components), branches have approached the financial crisis better than the subsidiaries did.

Given the information revealed by the descriptive statistics, banks operating in financial centers could experience different patterns of productivity growth depending also on their size. We therefore explore our results by investigating whether big banks are more productive than medium and small banks in the financial center. The results are reported in Table 5.5. The analysis leads to additional important insights into how the consolidation process in the financial center has spurred banks' productivity. In particular, the results show that during the 15-year period, big banks reach a higher productivity growth, followed by small and medium banks (Panel A). While the productivity growth of big and medium banks is driven by technical progress, the productivity growth of small banks is due to both technical progress and slow

Table 5.7 Bias-corrected efficiency change (EC), technical change (TC), and productivity change (M) of banks classified by home country origin, other countries.

Years	Switzerland			Italy			Other countries		
	EC	TC	M	EC	TC	M	EC	TC	M
Panel A									
1995–1996	0.9741	1.1261	1.0988	0.9205	1.0501	0.9681	0.9954	1.0009	0.9988
1996–1997	1.0054	1.0196	1.0253	0.9985	0.9569	0.9557	1.0058	0.9750	0.9812
1997–1998	0.9906	1.0179	1.0092	1.0057	1.0191	1.0257	0.9856	1.0271	1.0133
1998–1999	1.0181	0.9976	1.0165	1.0282	0.9315	0.9594	1.0211	0.9781	0.9995
1999–2000	1.0072	0.9716	0.9793	1.0020	0.9562	0.9594	1.0019	1.0059	1.0085
2000–2001	1.0161	0.9453	0.9617	0.9834	0.9580	0.9431	0.9820	1.0050	0.9882
2001–2002	1.0158	1.0303	1.0477	1.0249	1.0333	1.0609	1.0235	0.9960	1.0207
2002–2003	0.9552	1.0655	1.0206	1.0126	0.9953	1.0105	1.0138	1.0020	1.0173
2003–2004	1.0560	0.8987	0.9516	0.9908	1.0361	1.0291	1.0274	0.9195	0.9474
2004–2005	1.0155	1.0599	1.0793	1.0152	1.0271	1.0447	0.9880	1.0658	1.0556
2005–2006	0.9488	1.0917	1.0378	0.9716	1.0386	1.0105	0.9466	1.0598	1.0052
2006–2007	1.0559	0.8023	0.8541	1.0629	0.9237	0.9873	1.0185	0.9209	0.9438
2007–2008	0.9604	0.7724	0.7458	0.9856	0.9856	0.9728	0.9541	0.9197	0.8792
2008–2009	1.0009	1.0208	1.0295	1.0201	0.9103	0.9320	1.0480	1.0390	1.0938
2009–2010	0.9986	0.9106	0.9121	1.0271	0.9184	0.9457	1.0037	0.9382	0.9453
1995–2010	**1.0012**	**0.9820**	**0.9846**	**1.0033**	**0.9827**	**0.9870**	**1,0010**	**0.9902**	**0.9932**
Panel B									
1995–2001	**1.0019**	**1.0130**	**1.0151**	**0.9897**	**0.9787**	**0.9686**	**0.9986**	**0.9987**	**0.9983**
2001–2003	**0.9855**	**1.0479**	**1.0341**	**1.0188**	**1.0143**	**1.0357**	**1.0187**	**0.9990**	**1.0190**
2003–2006	**1.0068**	**1.0168**	**1.0229**	**0.9926**	**1.0340**	**1.0281**	**0.9873**	**1.0150**	**1.0027**
2006–2010	**1.0039**	**0.8765**	**0.8854**	**1.0239**	**0.9345**	**0.9594**	**1.0061**	**0.9544**	**0.9655**
Panel C									
1995–2006	**1.0002**	**1.0204**	**1.0207**	**0.9958**	**1.0002**	**0.9970**	**0.9992**	**1.0032**	**1.0032**
2006–2010	**1.0039**	**0.8765**	**0.8854**	**1.0239**	**0.9345**	**0.9594**	**1.0061**	**0.9544**	**0.9655**

improvements in efficiency. Regarding the performance of these groups during the different periods (Panel B), it is interesting to note that big banks reach productivity growth in each period, while medium and small banks attend to a decline in productivity during the collapse of the stock market period and the recovery period. Small banks experience productivity decline also during the consolidation period. Finally, prior to the financial crisis (Panel C), on average, big banks are more productive than medium and small banks. Their productivity grows at an annual growth rate of 1.42% compared to the negative annual growth rate of medium and small banks, respectively, of 0.63% and 0.86%. The productivity growth seems to be driven by improvement in technology and efficiency change. Differently from big banks, medium and small banks experience slight changes in efficiency with an annual growth rate of 0.35% and −0.28%, respectively, and technological regress with an annual rate of 0.93% and 0.48%, respectively. Overall, before the financial crisis, big banks seem to perform better compared to medium and small banks. During the financial crisis, all groups of banks improve

Table 5.8 Bias-corrected efficiency change (EC), technical change (TC), and productivity change (M) of domestic banks.

Years	Luxembourg		
	EC	TC	M
Panel A			
1995–1996	0.9593	1.0204	0.9813
1996–1997	1.0096	0.9601	0.9696
1997–1998	1.0024	1.0652	1.0683
1998–1999	1.0181	0.9358	0.9535
1999–2000	1.0367	0.9609	0.9984
2000–2001	0.9833	1.1039	1.0881
2001–2002	1.0385	0.9734	1.0111
2002–2003	0.9617	0.9953	0.9577
2003–2004	1.0235	1.0196	1.0560
2004–2005	0.9849	0.9419	0.9288
2005–2006	0.8920	1.0873	0.9711
2006–2007	1.0293	0.9775	1.0182
2007–2008	1.0136	1.0005	1.0147
2008–2009	1.3492	0.9438	1.2751
2009–2010	0.8821	0.9693	0.8557
1995–2010	**1.0123**	**0.9970**	**1.0098**
Panel B			
1995–2001	**1.0016**	**1.0077**	**1.0099**
2001–2003	**1.0001**	**0.9843**	**0.9844**
2003–2006	**0.9668**	**1.0163**	**0.9853**
2006–2010	**1.0685**	**0.9728**	**1.0409**
Panel C			
1995–2006	**0.9918**	**1.0058**	**0.9985**
2006–2010	**1.0685**	**0.9728**	**1.0409**

their productivity. However, small banks perform better than the others with an annual growth rate of 6.67% driven by high progress in technology (6.46%) and slight improvement in efficiency (0.81%). Big and medium banks respond with high improvement in technology but, contemporaneously, with deterioration in efficiency of −3.78% and −3.17%, respectively.

Lastly, we investigate the results by analyzing separately the productivity growth of banks according to their nationality. Considering banks that belong to the neighboring countries, namely, Belgium, France, and Germany (Table 5.6, Panel A), it is evident that German banks are more productive than the others with a rate of productivity growth equal to 0.89%. Over time, the productivity of German banks seems to be driven by technological progress, which is also the case with French and Belgian banks. In all cases, efficiency changes are modest, and German banks show a more volatile pattern in technological progress. During the financial crisis (Table 5.6, Panel B), both French and German banks gain improvement in productivity, respectively of 3.25% and 4.90%. These changes are driven by important progress in technology with an annual growth rate of 6.85% and 7.87%, respectively. However, Belgian banks also experience important technical changes (5.12%), but, contemporaneously, the highest deterioration in efficiency (−6.45%) compared to −3.50% and −2.42% of French and German banks, respectively. In terms of productivity path during the different periods, French and German banks show a similar performance: they experience productivity decline during the collapse of the stock market crisis and the phase of recovery, while Belgian banks, although their productivity deteriorates during the collapse of the stock market crisis, improve productivity during the phase of recovery. Among the other two countries (Italy and Switzerland) analyzed separately (Table 5.7), Swiss banks experience the highest productivity change. However, both Swiss and Italian banks drive their productivity growth through technical progress, which is higher for Swiss banks than for Italian banks, 1.80% and 1.73%, respectively. Prior to the financial crisis, Swiss banks experience deterioration in productivity with an annual rate of −2.07%. Deterioration in productivity occurs over the entire period, except for the period of maximum expansion of the sector (1999–2001 and 2003–2004). Deterioration in efficiency is more negligible with an annual growth rate of −0.02%. During the financial crisis, Swiss banks experience increase in productivity, driven by important changes in technology. The TFP pattern of Italian banks is different: after a positive trend in productivity from 1998 to 2000, Italian banks experience a decline in TFP, driven almost by deterioration in technology. However, during the financial crisis, banks improve their technology and worsen their efficiency. Other banks (Table 5.7) show an increase in TFP (0.68%), driven by a positive change in technology. Prior to the financial crisis, this set of banks attends technical regress; however, during the financial crisis, improvements in technology enhance the productivity growth rate to positive values. These results are in line with Casu, Girardone, and Molyneux (2004), according to which European countries follow different and mixed patterns of performance behavior. Lastly, domestic banks (Table 5.8): compared to foreign banks, they seem neither more productive nor more efficient nor more innovative. During the financial crisis, they also innovate but at the lowest annual rate when compared to other bank groups. Moreover, they experience the highest deterioration in efficiency.

Overall, the results show that, on average, over the 15-year period, foreign banks from Switzerland are those with the highest rate of productivity growth, mainly due to the important improvement in technical change during the financial crisis. However, prior to the financial crisis, Belgian banks were the most productive. Interesting enough about the results is the fact that it seems that while during the financial crisis foreign banks in Luxembourg (except for Belgian banks) experience growth in productivity mainly due to an important technological progress (since in most of the cases the efficiency change deteriorates), domestic banks, jointly with the Belgian banks, experience a decline in productivity due to an important decline in efficiency change. Moreover, if we compare the productivity of foreign banks during the financial crisis to the productivity during the collapse of the stock market crisis, it seems that during the latter, foreign banks did not respond well to it. Actually, with independence of the origin of the foreign banks, during the collapse of the stock market crisis, there exists a decline in productivity. However, during that period, domestic banks show the opposite outcomes, an improvement in productivity.

5.6   Conclusions

The Luxembourg financial center presents an interesting case to study the impact on productivity evolution of the post-deregulation and also the recent financial crisis from the point of view of a host country. Moreover, it is a unique case where the number of foreign banks exceeds the number of domestic banks, and it also allows quantifying the impact of different phenomena on productivity change over time and across groups of banks with different organizational forms (branch vs subsidiaries) and different origin. We analyze the problem over the years 1995–2010, an interesting time period characterized by several structural changes in the banking industry. Given that, we analyze the evolution of productivity change over different time periods, focusing on the period before and during the current financial crisis. Lastly, the analysis on performance evaluation of foreign banks is developed by accounting for distinguishing aspects of foreign banks, namely organizational form, size, and nationality. This enables us to control for heterogeneity in the sample used, a peculiarity of a financial center. To measure the productivity change and its components, the nonparametric method based on DEA and on M decomposition is used, where the Simar and Wilson (1999) procedure is conducted to derive bias-corrected estimates.

On average, for the whole sample, without breaking out productivity growth, the results show that the productivity evolution was characterized by slow and negative growth during the period 2001–2006, driven by efficiency gain only during the period 2003–2006. The financial crisis, however, played a positive effect on the performance of the Luxembourg financial center as banks responded with high improvement in technology, reversing the previous negative productivity pattern to a positive one. Technological improvement has been the main driver of this new trend.

In terms of nationality, foreign banks seem to have responded to the financial crisis better than the domestic banks did, gaining their performance advantage from the high technical progress. Prior to the financial crisis, domestic banks seem to perform better than the others, except for Belgian and Italian banks. An interesting result is that while during the financial crisis foreign banks in Luxembourg (except the Belgium banks) show productivity growth mainly due to an important technological progress (since in most of the cases the efficiency change deteriorates), domestic banks, jointly with the Belgian banks, show a decline in productivity due to an important decline in efficiency change. Moreover, banks from Belgium show higher productivity growth prior to the financial crisis. Thus, it seems that the foreign banks in the financial center approach the financial crisis through important technical progress.

In terms of organizational form, the results show that branches seem to be more productive than subsidiaries on average during the financial crisis. Thus, it seems that branches approach better the financial crisis than subsidiaries. However, before the financial crisis, both types of organizational forms show similar patterns in the evolution of productivity change, although branches show more volatility.

Finally, in terms of size, it seems that big banks are those that experienced a positive trend in productivity over time, driven by technical change, while medium and small banks seem to be less innovative. However, during the financial crisis, banks from all types of size categories respond with technical changes, and only small and medium banks respond with improvement in efficiency.

Overall, the results show that irrespective of the organizational form, size, and nationality, foreign banks in the financial center have responded to the financial crisis with an important improvement in technical change, while domestic banks did not. In general, it seems that in normal time (without crisis), both types of organizational form show barely equal paths in productivity, mainly due to similarities in the technological path. Big banks show higher productivity growth than medium and small banks. Moreover, depending on the nationality, Belgium banks are those able to get higher productivity when they are developing their activity in the Luxembourg financial center.

To conclude, this chapter makes an important contribution to the appraisal of foreign bank productivity in financial center offering evidence on the possible effects of organizational form, size, and nationality of the foreign bank on productivity. Moreover, the chapter gives insight into the performance of foreign banks and the source of TFP growth, particularly during the financial crisis. We are therefore able to provide insights with regard to the impact that organizational form, size, and home origin of foreign banks have on productivity as well as to know which type of foreign bank approaches better to the financial crisis. Our exercise might be of interest to make decision about the form of entry of foreign banks in financial centers and the optimal bank size to take advantages in terms of productivity in financial centers.

Acknowledgments

The authors are grateful to Paolo Guarda, Roland Nockels, and Jean-Pierre Schoder of the BCL. Claudia Curi also thanks the BCL for supporting her research, where this work took shape, and her current employer, where it was completed. Ana Lozano-Vivas acknowledges financial support from Ministerio de Ciencias e Innovacion grant reference ECO2011-26996. Any remaining errors are solely our responsibility.

References

Abbott, M., Wu, S., and Wang, W.C. (2011) The productivity and performance of Australian's major bask since deregulation. *Journal of Economics and Finance*, **1**, 1–14.

Alam, I.M.S. (2001) A nonparametric approach for assessing productivity dynamics of large U.S. banks. *Journal of Money, Credit and Banking*, **33** (1), 121–139.

Avkiran, N. (2000) Rising productivity of Australian trading banks under deregulation 1986–1995. *Journal of Economics and Finance*, **24**, 122–140.

Battese, G.E., Heshmati, A., and Hjalmarsson, L. (2000) Efficiency of labour use in the Swedish Banking Industry: a stochastic frontier approach. *Empirical Economics*, **25**, 623–640.

Bauer, P.W., Berger, A.N., and Humphrey, D.B. (1993) Efficiency and productivity growth in US banking, in *The Measurement of Productive Efficiency: Techniques and Applications* (eds H.O. Fried, C.A. Knox Lovell, and S.S. Schmidt), Oxford University Press, New York.

Berg, S.A., Forsund, F.R., and Jansen, E.S. (1992) Malmquist indices of productivity growth during the deregulation of Norwegian banking, 1980–89. *Scandinavian Journal of Economics*, **94**, 211–228.

Berger, A.N. (2007) International comparisons of banking efficiency. *Financial Markets, Institutions & Instruments*, **16** (3), 119–144.

Berger, A.N. and Humphrey, D.B. (1997) Efficiency of financial institutions: international survey and directions for future research. *European Journal of Operational Research*, **98**, 175–212.

Berger, A.N. and Mester, L. (2003) Explaining the dramatic changes in performance of US banks: technological change, deregulation, and dynamic changes in competition. *Journal of Financial Intermediation*, **12**, 57–95.

Berger, A.N., DeYoung, R., Genay, H., and Udell, G.F. (2000) Globalization of financial institutions: evidence from cross-border banking performance. *Brookings-Wharton Papers on Financial Services*, **3**, 23–120.

Berger, A.N., Buch, C.B., DeLong, G., and DeYoung, R. (2004) Exporting financial institutions management via foreign direct investment mergers and acquisitions. *Journal of International Money and Finance*, **23**, 333–366.

Berger, A.N., Miller, N.H., and Petersen, M.A. (2005) Does function follow organizational form? Evidence from the lending practices of large and small banks. *Journal of Financial Economics*, **76**, 237–269.

Burgstaller, J. and Cocca, T.D. (2011) Efficiency in private banking: evidence from Switzerland and Liechtenstein. *Financial Markets and Portfolio Management*, **25**, 75–93.

Canhoto, A. and Dermine, J. (2003) A note on banking efficiency in Portugal, new vs. old banks. *Journal of Banking & Finance*, **27**, 2087–2098.

Casu, B., Girardone, C., and Molyneux, P. (2004) Productivity change in European banking: a comparison of parametric and non-parametric approaches. *Journal of Banking & Finance*, **28**, 2521–2540.

Caves, D., Christensen, L., and Diewert, E. (1982) The economic theory of index numbers and the measurement of input, output, and productivity. *Econometrica*, **50**, 1393–1414.

Cerruti, E., Dell'Ariccia, G., and Peria, M.S.M. (2007) How banks go abroad: branches or subsidiaries? *Journal of Banking & Finance*, **31**, 1669–1692.

Chen, S.H. and Liao, C.C. (2011) Are foreign banks more profitable than domestic banks? Home- and host-country effects of banking market structure, governance and supervision. *Journal of Banking & Finance*, **35**, 819–839.

Clark, A.J. and Siems, T.F. (2002) X-efficiency in banking: looking beyond the balance sheet. *Journal of Money, Credit and Banking*, **34**, 987–1013.

Claessens, S. and Van Horen, N. (2012a) Being a foreigner among domestic banks: asset or liability? *Journal of Banking & Finance*, **36** (5), 1276–1290.

Claessens, S. and Van Horen, N. (2012b) Foreign banks: trends, impact and financial stability. IMF working paper, January.

Curi, C. (2007) *An improved procedure for bootstrapping Malmquist indices*, http://hdl.handle.net/2108/725 (accessed 21 November 2012).

Curi, C., Guarda, P., and Zelenyuk V. (2011) Changes in bank specialization: comparing foreign subsidiaries and branches in Luxembourg. Banque Centrale du Luxembourg working paper no. 67.

Curi, C., Guarda, P., Lozano-Vivas, A., and Zelenyuk V. (2012) Is foreign-bank efficiency in financial centers driven by home or host country characteristics? *Journal of Productivity Analysis*. doi: 10.1007/s11123-012-0294-y

Curi, C., Guarda, P., Lozano-Vivas, A., and Zelenyuk V. (2013) The impact of the crisis on the foreign banks business model and performance. Banque Centrale du Luxembourg working paper, forthcoming.

Dermine, J. (2006) European banking integration: don't put the cart before the horse. *Financial Markets, Institutions & Instruments*, **15** (2), 57–106.

European Central Bank (ECB) (2004) Report on EU Banking Structure, November.

ECB (2006) EU Banking Structure, September.

ECB (2010) EU Banking Structure, September.

Färe, R., Grosskopf, S., Norris, M., and Zhang, Z. (1994) Productivity growth, technical progress, and efficiency change in industrialized countries. *American Economic Review*, **84** (1), 66–83.

Farrell, M.J. (1957) The measurement of productive efficiency. *Journal of the Royal Statistical Society, Series A* **120**, 253–281.

Fiechter, J.F., Otker-Robe, I., and Ilyina, A. (2011) Subsidiaries or branches: does one size fit all? IMF Staff discussion note, March.

Fiorentino, E., De Vincenzo, A., and Heid, F. (2010) The effects of privatization and consolidation on bank productivity: comparative evidence from Italy and Germany. Deutsche Bundesbank, discussion paper series 2: banking and financial studies no. 03/2009.

Fixler, D. (1993) Measuring financial service output and prices in commercial banking. *Applied Economics*, **25** (7), 983–993.

Fixler, D. and Zieschang, K. (1992) User costs, shadow prices, and the real output of banks, in *Output Measurement in the Service Sectors* (ed. Z. Griliches), University of Chicago Press, Chicago.

Gilbert, R.A. and Wilson, P.W. (1998) Effect of deregulation on the productivity of Korean banks. *Journal of Economics and Business*, **50**, 133–155.

Grifell-Tatjé, E. and Lovell, C.A.K. (1996) Deregulation and productivity decline: the case of Spanish savings banks. *European Economic Review*, **40** (6), 1281–1303.

Grifell-Tatjé, E. and Lovell, C.A.K. (1997) The sources of productivity change in Spanish banking. *European Journal of Operational Research*, **98**, 364–380.

Guarda, P. and Rouabah, A. (2006) Measuring banking output and productivity: a user cost approach to Luxembourg data, in *Measuring Productivity in Services: New Dimensions* (eds P.K. Bandyopadhyay and G.S. Gupta), ICFAI University Press, Hyderabad.

Guarda, P. and Rouabah, A. (2007) Banking output & price indicators from quarterly reporting data. Banque Centrale du Luxembourg working paper no. 27.

Guarda, P. and Rouabah, A. (2009) Bank productivity and efficiency in Luxembourg: Malmquist indices from a parametric output distance function, in *Productivity in the Financial Services Sector*, SUERF – The European Money and Finance Forum, Vienna, pp. 151–166.

Hadad, M.D., Hall, M.J.B., and Kenjegalieva, K.A. (2011) Banking efficiency and stock market performance: an analysis of listed Indonesian banks. *Review of Quantitative Finance and Accounting*, **37** (1), 1–20.

Humphrey, D.B. (1992) Flow versus stock indicators of banking output: effects on productivity and scale economy measurement. *Journal of Financial Services Research*, **6**, 115–135.

Isik, I. and Hassan, M.K. (2003) Financial deregulation and total factor productivity change: an empirical study of Turkish commercial banks. *Journal of Banking & Finance*, **27**, 1455–1485.

Kneip, A., Simar, L., and Wilson, P.W. (2008) Asymptotics and consistent bootstraps for DEA estimators in non-parametric frontier models. *Econometric Theory*, **24**, 1663–1697.

Korostelev, A., Simar, L., and Tsybakov, A. (1995) Efficient estimation of monotone boundaries. *Annals of Statistics*, **23** (2), 476–489.

Kumbhakar, S.C. and Lozano-Vivas, A. (2005) Deregulation and productivity: the case of Spanish banks. *Journal of Regulatory Economics*, **27**, 331–351.

Kwan, S.H. (2006) The X-efficiency of commercial banks in Hong Kong. *Journal of Banking & Finance*, **30**, 1127–1147.

Lozano-Vivas, A. (1998) Efficiency and technical change for Spanish banks. *Applied Financial Economics*, **8**, 289–300.

Lozano-Vivas, A. and Pasiouras, F. (2010) The impact of non-traditional activities on the estimation of bank efficiency: international evidence. *Journal of Banking & Finance*, **34**, 1436–1449.

Mendes, V. and Rebelo, J. (1999) Productive efficiency, technological change and productivity in Portuguese banking. *Applied Financial Economics*, **9**, 513–521.

Park, Y.S. and Essayyad, M. (1989) *International Banking and Financial Centers*, Kluwer Academic Publishers, Boston.

Park, B.U., Jeong, S.O., and Simar, L. (2010) Asymptotic distribution of conical-hull estimators of directional edges. *Annals of Statistics*, **38** (3), 1320–1340.

Rezvanian, R., Rao, N., and Mehdian, S.M. (2008) Efficiency change, technological progress and productivity growth of private, public and foreign banks in India: evidence from the post-liberalization era. *Applied Financial Economics*, **18**, 701–713.

Rime, B. and Stiroh, K.J. (2003) The performance of universal banks: evidence from Switzerland. *Journal of Banking & Finance*, **27**, 2121–2150.

Rouabah, A. (2002) Economie d'echelle, economies de diversification et efficacité productive des banques luxembourgeoises: une analyse comparative des frontières stochastiques sur données en panel. Banque Centrale du Luxembourg working paper no. 3.

Sealey, C.W., Jr and Lindley, J.T. (1977) Inputs, outputs, and the theory of production and cost at depository financial institutions. *Journal of Finance*, **32**, 1251–1266.

Silverman, B.W. (1986) *Density Estimation for Statistics and Data Analysis*, Chapman & Hall, London.

Simar, L. and Wilson, P.W. (1999) Estimating and bootstrapping Malmquist indices. *European Journal of Operational Research*, **115**, 459–471.

Sturm, J.E. and Williams, B. (2008) Characteristics determining the efficiency of foreign banks in Australia. *Journal of Banking & Finance*, **32**, 2346–2360.

Sufian, F. and Majid, M.Z.A. (2007) Deregulation, consolidation and banks efficiency in Singapore: evidence from an event study window approach and Tobit analysis. *International Review of Economics*, **54** (2), 261–283.

Tirtiroglu, D., Daniels, K.N., and Tirtiroglu, E. (2005) Deregulation, intensity of competition, industry evolution and the productivity growth of US commercial banks. *Journal of Money, Credit and Banking*, **37**, 339–360.

Tortosa-Ausina, E., Grifell-Tatjé, E., Armero, C., and Conesa, D. (2008) Sensitivity analysis of efficiency and Malmquist productivity indices: an application to Spanish savings banks. *European Journal of Operational Research*, **184**, 1062–1084.

Tschoegl, A.E. (2000) International banking centers, geography, and foreign banks. *Financial Markets, Institutions & Instruments*, **9** (1), 1–32.

Wheelock, D.C. and Wilson, P.W. (1999) Technical progress, inefficiency, and productivity change in US banking, 1984–1993. *Journal of Money, Credit and Banking*, **31** (2), 212–234.

Zofio, J.L. (2007) Malmquist productivity index decompositions: a unifying framework. *Applied Economics*, **39** (18), 2371–2387.

6

The impact of merger and acquisition on efficiency and market power

Franco Fiordelisi[1] and Francesco Saverio Stentella Lopes[2]

[1] Department of Business Studies, University of Rome III, Italy and
Bangor Business School, Bangor University, UK
[2] Department of Business Studies, University of Rome III, Italy and
Finance Department, Tilburg University, The Netherlands

6.1 Introduction

Merger and acquisition (M&A) deals are the two most visible expressions of the functioning of the corporate control market. While M&A refer to two different deals, these are usually analyzed together since both achieve the same goal (i.e., the ownership of an entire company changes hands in a single transaction). Specifically, in a merger deal, two companies agree to combine into a single corporate entity (rather than remain separately owned) by issuing stock of the controlling firm to replace most of the other company's stock. In an acquisition deal, a company purchases a company through another company.[1]

The volume of M&A transactions boosted until the end of 2007, then strongly dropped until the second quarter of 2009 (caused by concerns in the credit markets), and thereafter increased until the end of 2010. According to the Thomson Financial (2010), the worldwide volume of announced M&A in 2010 was USD 2.4 trillion, that is, 22.9% increase from the comparable 2009 levels. The M&A phenomenon concerns all countries worldwide (Table 6.1): in 2010, M&A deals in North America was USD 821 billion (i.e., 34% of M&A deals value worldwide), USD 523 billion in Western Europe (i.e., 21% of M&A deals value

---

[1] For further details, see Fiordelisi (2009).

Efficiency and Productivity Growth: Modelling in the Financial Services Industry, First Edition. Edited by Fotios Pasiouras.
© 2013 John Wiley & Sons, Ltd. Published 2013 by John Wiley & Sons, Ltd.

Table 6.1 Worldwide announced M&As in 2010.

Region	Rank value (in USD billion)	Number of deals	Change in rank value (in %)
America	1 136.3	12 013	23.3
North America	921.2	10 080	13.0
Central America	53.0	335	644.7
South America	143.7	1 383	54.8
Caribbean	18.4	215	184.6
Africa/Middle East	91.0	1 143	84.6
Asia-Pacific	482.0	10 564	49.0
Europe	641.0	14 779	10.3
Eastern Europe	117.7	4 766	125.1
Western Europe	523.3	10 013	−1.0
Worldwide	2 434.2	40 660	22.9

Source: Thomson Financial (2010: 2).

worldwide), and USD 482 billion in the Asian-Pacific area (17% of M&A deals value worldwide). Regarding the type of deals, the M&A cross-border activity value was USD 925.5 billion in 2010 accounting for 39.1% of worldwide activity.

We analyze the M&A phenomenon focusing on the banking industry since the consolidation in that industry was particularly important worldwide. Most deals take place in the energy power (20%), financial (15%), and materials (11%) industries. This data provides evidence that the M&A phenomenon is particularly relevant and it is therefore an important research area. Given the importance of the M&A phenomenon, this is one of the most investigated areas in finance (Fiordelisi, 2009). To have an idea, we simply searched the Google scholar website (www.scholar.google.com, accessed on 1 February 2012) and found 99 600 papers showing the term 'Merger and Acquisition' in the title and 63 200 papers having the word 'M&A' quoted in the title. We also searched the Amazon website (http://www.amazon.com, accessed on 1 February 2012): 18 080 pieces of work quote 'Mergers and Acquisitions'.

One of the main reasons for M&As is to increase bank efficiency or market power. Indeed, extensive literature is available on banking dealing with efficiency by both developing new estimation approaches and measuring inefficiencies in various sectors (for a review, see Berger and Humphrey, 1997, and Hughes and Mester, 2010). Similarly, literature measuring bank market power estimated by the new competition measures, as the Boone Index, the *H*-statistic, and the Lerner index is increasing.

Surprisingly, there is only a handful of studies investigating the impact of M&A deals on bank efficiency (Rhoades, 1998; Calomiris, 1999; Garden and Ralston, 1999; Rezitis, 2008; Fiordelisi, 2009). This chapter aims at linking these three branches of literature by estimating the M&A effect on bank efficiency and market power. We focus on a small sample of M&A deals in Europe and in the United States by estimating how efficiency and market power changed one and two years after the deal announcement.

The rest of this chapter is organized as follows: Section 6.2 has the literature review. Section 6.3 illustrates our empirical approach. Section 6.4 discusses the empirical results and Section 6.5 concludes.

6.2   Literature review

There is a large number of studies dealing with M&As in the financial service industry, but there is little consensus as to the effects of this consolidation on industry performance (DeYoung, Evanoff, and Molyneux, 2009). We review studies assessing the M&A impact on the bank's efficiency and operating performance beyond the short time period (i.e., these papers usually run an event study).

The first papers dealing with this issue go back to the early 1980s (Frieder and Apilado, 1983; Rhoades, 1986; Rose, 1987a, b), while most recent studies are from Huizinga, Nelissen, and Vander Vennet (2001), Cuesta and Orea (2002), Berger and Mester (2003), Wang (2003), Carbo-Valverde and Humphrey (2004), Humphrey and Vale (2004), Koetter (2005), De Guevara and Maudos (2007), Ashton and Pham's (2007), Rezitis (2008), Altunbaş and Marqués-Ibanez (2008), Hagendorff and Vallascas (2011), and Behr and Heid (2011).

Specifically, Huizinga, Nelissen, and Vander Vennet (2001) estimate substantial scale, cost and profit efficiency gains by analyzing 52 M&As in Europe between 1994 and 1998. Cuesta and Orea (2002) show that merged and nonmerged banks have different technical efficiency changes by examining 858 Spanish saving banks (132 of which were involved in M&As) over the period 1985–1998. Berger and Mester (2003) estimate substantial profit efficiency gains by assessing a sample of US banks over the period 1984–1997. According to Carbo-Valverde and Humphrey (2004), M&As produced positive effects both for shareholders and customers for a sample of 20 M&As in Spain between 1986 and 1998. Humphrey and Vale (2004) estimated an average cost reduction of 2.81% after assessing 26 M&As in Norway. Koetter (2005) and Ashton and Pham (2007) observe that M&As improve bank cost efficiency in Germany and the United Kingdom, respectively. Rezitis (2008) shows that the M&A effects on technical efficiency and total factor productivity growth of Greek banks (1993–2004) are negative. Altunmaş and Marqués-Ibanez (2008) show that it is costly to integrate dissimilar financial institutions by assessing a sample of European M&As from 1992 to 2001. Hagendorff and Vallascas (2011) analyze 134 European bidding banks and show that bank mergers are generally risk neutral; for relatively safe banks, mergers also generate a significant increase in default risk. Behr and Heid (2011) run a new analysis of German bank mergers (1995–2000) showing a neutral effect of mergers on profitability and cost efficiency.

As far as we are aware, there are no studies that have empirically measured the impact of M&As on the bank's market power.

6.3   Empirical design

This section illustrates the empirical design by presenting our data (Section 6.3.1), variables (Section 6.3.2), and the econometric approach (Section 6.3.3).

6.3.1   Data

Our sample included mergers selected according to the following criteria: (a) merger announced between 1 January 1998 and 2006; (b) acquirers are banks; (c) target firms are banks, insurance companies, or other financial companies; (d) both targets and acquirers are located in Austria, France, Germany, Greece, Italy, and the United States; (e) both the acquiring and the target banks have been publicly quoted on a stock exchange for an entire year prior to the announcement date and at least 20 days after the announcement day; (f) the value of the transaction is greater than or equal to €100 million; (g) the merger was effectively completed, generating a change of control of the target bank (i.e., the acquiring bank has complete corporate control, holding more than 50% of the target company's equity).

Merger data was obtained from the Thompson One bankers' database. In order to estimate the cost, scale, and revenue efficiency, we focus on both listed and nonlisted banks with financial information obtained from Bankscope database. Overall, our sample comprises 1720 observations from unconsolidated commercial banks' balance sheets. One hundred and seventeen of those banks were implicated in large mergers involving target(s?) from the United States and five European countries.

Table 6.2 (Panel A) shows the dimensions of banks in our sample. Return on equity (ROE) is quite dispersed in the sample when we group our observation on countries. However, looking at the overall sample average by year (Table 6.2, Panel B), we can observe that the banks' dimensions, in terms of total asset and total loan, have increased year by year.

Table 6.2   Descriptive statistics.

Country	Total asset (billion)	Total loan (billion)	Return on equity (%)	Number observation
Panel A: Descriptive statistics grouped by country				
Austria	4	2	8.00	141
France	62	18	3.00	351
Germany	16.5	8	9.00	287
Greece	12	7	2.00	173
Italy	27	12	8.00	217
United States	10	6	7.00	551
Total	22	9	6.00	1720

Year	Total asset (billion)	Total loan (billion)	Return on equity (%)	Number observation
Panel B: Full sample descriptive statistics grouped by year				
1998	10	5	8.00	145
1999	12	5	7.00	144
2000	16	7	8.00	177
2001	19	8	6.00	142
2002	18	8	4.00	137
2003	19	8	7.00	135
2004	22	9	6.00	150
2005	27	11	8.00	169
2006	32	12	9.00	156
2007	29	12	9.00	140
2008	37	13	0.00	151
2009	53	18	−8.00	74
Total	25	10	5.00	1720

In particular, focusing on the increased dimension in terms of total asset, we can observe decreases in the average dimension on two occasions only, the first time between 2001 and 2002 and the second between 2006 and 2007.

6.3.2 Variables

First, we estimate bank cost efficiency (CE) by the ratio of operating cost and operating income.

Second, we estimate bank market power using the Lerner index of Monopoly Power (LER). In this case, we use one single output in the cost function (Shaffer, 1993; Berg and Kim, 1994; Angelini and Cetorelli, 2003; Fernandez de Guevara, Maudos, and Perez, 2005; Casu and Girardone, 2009). The Lerner index measures the extent to which market power allows firms to fix a price above marginal cost (MC):

$$\text{LERNER} = \frac{p_{it} - \text{MC}_{it}}{p_{it}}, \tag{6.1}$$

where p is the price of output Q and is calculated as total revenue (interest plus noninterest income) divided by total assets. Marginal costs are obtained by estimating the cost function (TC). We use the stochastic frontier approach, originally proposed by Aigner, Lovell, and Schmidt (1977), assuming half-normal distribution for the inefficiency and pooling the data at the country level. The final specification is as follows:

$$
\begin{aligned}
\ln \text{TC}_{it} = {} & \beta_0 + \beta_1 \ln Q_1 + \beta_2 \ln Z_2 + \tau_1 T \\
& + \frac{1}{2}\left[\beta_3 \ln Q_1^2 + \beta_4 \ln Z_1^2 + \sum_{j=1}^{3}\sum_{i=1}^{3}\beta_{ij} \ln P_j \ln P_i + \tau_2 T^2 \right] \\
& + \beta_5 \ln Z_1 \ln Q_1 \sum_{j=1}^{3}\beta_{1j} \ln Q_1 \ln P_j \\
& + \sum_{j=1}^{3}\beta_{2j} \ln Z_1 \ln P_j + \theta_1 T \ln Q_1 + \theta_2 T \ln Z_1 + \sum_{j=1}^{3}\psi_j T \ln P_j + \varepsilon_{it},
\end{aligned} \tag{6.2}
$$

where TC is the sum of personnel expenses, other administrative expenses, other operating expenses, and price of funds; α, β, δ, γ, ρ, t, θ, ψ are coefficients to be estimated; and ε_{it} is a two-component error term $\varepsilon_{it} = u_{it} + v_{it}$, where v_{it} is a two-sided error term.[2]

We posit that banks' inputs are the price of labor calculated as personnel expenses over total assets (P_1); price of funds, measured as interest expenses over total deposits plus money market funding and other funding (P_2); and one additional input, the price of physical capital, measured as other administrative expenses plus other operating expenses over total assets (P_3). On the other hand, and despite the multi-output nature of the investment banking business, we define total assets (Q_1) as one single output. By doing so, we assume that the flow of services produced by an investment bank is proportional to its total assets. Finally, to

---

[2] The v_{it} are assumed to be independently and identically normal distributed with mean 0 and variance σ_v^2 and independent of u_{it}, where the latter is a one-sided error term capturing the effects of inefficiency and assumed to be half-normally distributed with mean 0 and variance σ_u^2. We apply the common restrictions of standard symmetry and homogeneity in prices to the translog functional form.

account for capitalization, we introduce the ratio of equity on total asset (Z_1), and we include a time trend to account for technological shifts (T).

Marginal costs are derived from the following equation:

$$\text{MC}_{it} = \frac{\text{TC}_{it}}{Q_{it}}\left(\beta_1 + \beta_3 \ln Q_{1t} + \beta_5 \ln Z_{1t} + \beta_{1j} \ln P_j + \theta_1 T\right). \tag{6.3}$$

We also control for various factors at the bank level (for both acquired and target banks), such as income diversification (measured by the nonoperating income over the operating income), liquidity reserves (measured by the ratio of liquid assets over total assets), credit risk (measured by the loan loss provision over the total loans), and asset size (measured by the total assets). A detailed summary of the variables used for the empirical investigation is provided in Table 6.3.

Table 6.3 Variables definition.

Variables[a]	Symbol	Definition and calculation method
Lerner index	MP	This represents the extent to which market power allows the bank to fix a price (P) above its marginal cost (MC)
Output price	P	Following recent studies (Berger et al., 2009; Turk-Ariss, 2010) and assuming that the banks produce a heterogeneous flow of services that is proportional to their dimension, we use the banks' total asset as a proxy of their overall activity (Cetorelli, 2003), and we estimate the average price as total revenue (interest and noninterest income) on total asset
Marginal costs	MC	The marginal costs of the product are obtained by estimating a single output translog cost function and using firm-fixed effect to handle the average heterogeneity among banks and a technology shift trend to capture the average change in production technology in our sample period
Bank asset size	Size	This is measured by the natural logarithm of total assets
Income diversification	ID	This is measured by the nonoperating income over the operating income
Cost inefficiency	INEFF	This is measured by the ratio between operating costs and operating income
Cross border	CB	This is a dummy variable to capture if the M&A is a cross-border (CB = 1) or domestic (CB = 0) deal
Bank liquidity reserve	LIQ	This is measured by the ratio of liquid assets over the total assets
Bank credit risk	CR	This is measured by the loan loss provision over the total loans
Bank interest rate exposure	IRR	This is measured by the ration between liquid asset and demand deposits
Bank equity	EQ	This is the ratio between the book value of total equity and the total asset

[a] All variables are calculated for both the acquirer and the target banks.
Source: Bankscope.

6.3.3 The econometric approach

We estimate the effect of M&A on bank market power by running the following linear multiple regression models:

$$\Delta MP_{(Tj;t-1,t)} = \alpha_0 + \sum \beta_i X_{(i,t-1)} + \sum \gamma_j Z_{(j,t-1)} + ID_{(Tj;t-1)} + CI_{(Tj;t-1)} + CB + \varepsilon_{t-1,t}, \quad (6.4)$$

$$\Delta MP_{(Tj;t-2,t)} = \alpha_0 + \sum \beta_i X_{(i,t-2)} + \sum \gamma_j Z_{(j,t-2)} + ID_{(Tj;t-2)} + CI_{(Tj;t-2)} + CB + \varepsilon_{t-1,t}, \quad (6.5)$$

$$\Delta MP_{(Aj;t-1,t)} = \alpha_0 + \sum \beta_i X_{(i,t-1)} + \sum \gamma_j Z_{(j,t-1)} + ID_{(Aj;t-1)} + CI_{(Aj;t-1)} + CB + \varepsilon_{t-1,t}, \quad (6.6)$$

$$\Delta MP_{(Aj;t-2,t)} = \alpha_0 + \sum \beta_i X_{(i,t-2)} + \sum \gamma_j Z_{(j,t-2)} + ID_{(Aj;t-2)} + CI_{(Aj;t-2)} + CB + \varepsilon_{t-1,t}, \quad (6.7)$$

where $\Delta MP_{(Aj;t-1,t)}$ is the Lerner index change for the acquirer bank j between the time period t and $t-1$; $\Delta MP_{(Aj;t-2,t)}$ is the Lerner index change for the acquirer bank j between the time period t and $t-2$; $\Delta MP_{(Tj;t-1,t)}$ is the Lerner index change for the target bank j between the time period t and $t-1$; $\Delta MP_{(Tj;t-2,t)}$ is the Lerner index change for the target bank j between the time period t and $t-2$; Xi ($i=1, ..., 5$) is a set of features of the target bank (that is also considered for the acquirer bank), Zi ($i=1, ..., 5$) is a set of features of the acquirer bank (that is also considered for the target bank); ID is the income diversification (measured by the nonoperating income over the operating income); CI is the cost income ration (measured by the ratio of operating income over operating costs); CB is a dummy variable to capture if the M&A is a cross-border (CB=1) or domestic (CB=0) deal. Specifically, we include as bank variables the following three items: liquidity reserves (measured by the ratio of liquid assets over total assets), credit risk (measured by the loan loss provision over the total loans), and asset size. A detailed summary of the variables used for the empirical investigation is provided in Table 6.3.

6.4 Results

Focusing on a sample of large M&A deals, we analyze the relationship between market power changes (after one and two years) and various bank characteristics for both target and bidder banks.

First, we analyze the M&A effects for the target bank. As shown in Panel A of Table 6.4, we find that target banks' market power variation between two consecutive years is negatively related to their asset size (at the 1% confidence level), their liquidity reserves (at the 1% confidence level), and their credit risk (at the 5% confidence level): this shows that target banks achieve larger market power changes in one year if they are smaller, with lower credit risk and smaller liquid assets. The negative link between bank asset size and market power seems to support the existence of increasing returns to scale: as a bank increases its size by merging with another bank, the latter increases its market power. The negative link between bank credit risk and market power shows that the bank's ability in screening and managing loans is positively related to the bank's market power. The negative link between liquidity reserves and market power may appear surprising since one may expect that safer banks will also have a larger market power. There are various reasons that explain this result:

Table 6.4 The merger effects on bank market power: the target.

	Panel A (one year) $y=\Delta MP_{(Tj;t-1,t)}$		Panel B (two years) $y=\Delta MP_{(Tj;t-2,t)}$	
	Coefficient	Standard error	Coefficient	Standard error
$Size_T$	−0.6297*	0.3291	−0.6422	0.3793
LIQ_T	−0.0260*	0.0148	−0.0247	0.0250
CR_T	−2.7845**	1.2507	−4.0360**	1.9511
$Size_A$	−0.0010	0.0028	−0.0030	0.0045
LIQ_A	−0.0033	0.0356	0.0024	0.0528
CR_A	−0.0211	0.0247	−0.0015	0.0361
CB	0.0074	0.0135	−0.0154	0.0179
ID_T	−0.0132	0.0118	−0.0008	0.0308
$INEFF_T$	−0.0340	0.0419	−0.0463	0.0834
$INEFF_T$	0.0842	0.2849	−0.1107	0.4188
Intercept	0.0756	0.0604	0.1487	0.0944
Number of observations		56		56
Adjusted R^2		0.3256		0.3494

The subscript T denotes that the variable refers to the target bank; A denotes that the variable refers to the acquirer bank.
* Statistically significant at the 1% level.
** Statistically significant at the 5% level.

first, liquidity reserves have a high opportunity cost, and this affects the bank's market power negatively; second, our empirical analysis is based on a period of banking stability (1998–2006) – the recent financial crisis reminded bankers of the fact that liquidity management is critical in banking, but this point was underestimated (and taken for granted) in times of banking prosperity. Surprisingly, we find that the acquirer bank features do not display a statistically significant (at the 10% confidence level or less) link with the target banks' market power changes.

When we extend our analysis over a two-year time period (Panel B of Table 6.4), we find that only the target bank credit risk is statistically significantly related to the target banks' market power changes: this shows that the bank's ability in screening and managing loans is the only factor that is (statistically significant at 10% or less) related to its market power changes.

Focusing on the acquirer bank, we find that the banks' market power variation between two consecutive years is negatively related to their asset size (at the 5% confidence level) and cross-border deals (at the 5% confidence level). We also observe a positive link between banks' market power variation and income diversification (at the 10% confidence level). Our findings suggest that acquirer banks achieve larger market power changes in one year if they are smaller, involved in domestic deals, and more diversified. The negative link between the bank asset size and market power seems to support the existence of increasing returns to scale: as a bank increases its size by merging with another bank, this bank increases its market power. The negative link between cross-border merger deals and market power is consistent with the previous studies showing that these deals do not create value for acquirer banks.

Table 6.5   The merger effects on bank market power: the acquirer.

	Panel A $y=\Delta MP_{(Tj;t-1,t)}$		Panel B $y=\Delta MP_{(Tj;t-2,t)}$	
	Coefficient	Standard error	Coefficient	Standard error
$Size_T$	−0.9529	2.1112	−0.5267	2.0001
LIQ_T	−0.1312	0.1062	−0.1391	0.1049
CR_T	9.5915	11.3126	6.1812	10.4669
$Size_A$	−0.0391*	0.0179	−0.0201*	0.0104
LIQ_A	−0.2376	0.2201	−0.2717	0.2159
CR_T	−0.0909	0.1358	−0.1331	0.1383
CB	−0.1963**	0.0830	−0.1448*	0.0762
ID_A	0.1867*	0.1070	0.1642*	0.0870
$INEFF_A$	−0.0423	0.4206	0.0119	0.4310
$INEFF_A$	1.8546	1.9692	2.3783	2.0114
Intercept	0.6220	0.4466	0.2844	0.3946
Number of observations		56		56
Adjusted R^2		0.2659		0.2179

* Statistically significant at the 1% level.
** Statistically significant at the 5% level.

The positive link between income diversification and market power is not surprising since more diversified banks achieve higher benefit than less diversified banks by merging with other banks. Surprisingly, we find that target bank features do not display a statistically significant (at the 10% confidence level or less) link with the acquirer banks' market power changes.

When we extend our analysis over a two-year time period (Panel B of Table 6.5), we find very consistent results with the ones discussed earlier. As such, we find that the estimated relationships between market power changes are more time-persistent for acquirer banks than for target banks.

6.5   Conclusions

We analyzed the relationship between market power changes (after one and two years) and various bank characteristics for both target and bidder banks by using a sample of large M&A deals in Europe and the United States, prior to the financial crisis. Overall, our sample comprises 1720 observations from unconsolidated commercial banks' balance sheets. One hundred and seventeen of those banks were implicated in large mergers involving targets from the United States and five European countries.

Despite the economic literature dealing with M&As in banking being vast, there are no studies that have empirically measured the impact of M&As on bank's market power.

We show that target banks achieve larger market power changes in one year if these banks are smaller, with lower credit risk and smaller liquid assets. These results suggest the existence of increasing returns to scale and that the bank's ability in screening and managing loans is positively related to the bank market power. When we extend our analysis over a two-year period, the target bank credit risk is statistically significantly related to the target banks' market power changes. Focusing on the acquirer banks, we show that they achieve larger market power changes in one year if these banks are smaller, involved in domestic deals, and more diversified. Surprisingly, we find that the target bank features do not display a statistically significant link with the acquirer banks' market power changes and the reverse.

References

Aigner, D.J., Lovell, C.A.K., and Schmidt, P. (1977) Formulation and estimation of stochastic frontier production function models. *Journal of Econometrics*, **6**, 21–37.

Altunbaş, Y. and Marqués-Ibanez, D. (2008) Mergers and acquisitions and bank performance in Europe: the role of strategic similarities. *Journal of Economics and Business*, **60**, 204–222.

Angelini, P. and Cetorelli, N. (2003) The effects of regulatory reform on competition in the banking industry. *Journal of Money, Credit and Banking*, **35**, 663–684.

Ashton, J.K. and Pham, K. (2007) Efficiency and price effects of horizontal bank mergers. CCP working paper no. 07-9. Available at SSRN: http://ssrn.com/abstract=997995 (accessed on 12 December 2012).

Behr, A. and Heid, F. (2011) The success of bank mergers revisited: an assessment based on a matching strategy. *Journal of Empirical Finance*, **18**, 117–135.

Berg, S.A. and Kim, M. (1994) Oligopolistic interdependence and the structure of production in banking: an empirical evaluation. *Journal of Money, Credit and Banking*, **26**, 309–322.

Berger, A.N. and Humphrey, D.B. (1997) Efficiency of financial institutions: international survey and directions for future research. *European Journal of Operational Research*, **98**, 175–212.

Berger, A.N. and Mester, L.J. (2003) Explaining the dramatic changes in performance of US banks: technological change, deregulation, and dynamic changes in competition. *Journal of Financial Intermediation*, **12**, 57–95.

Berger, A.N., Klapper F.L. and Turk-Ariss, R. (2009) Bank competition and financial stability. *Journal of Financial Services Research*, **35**, 99–118.

Calomiris, C.W. (1999) Gauging the efficiency of bank consolidation during a merger wave. *Journal of Banking and Finance*, **23**, 615–621.

Carbo-Valverde, S. and Humphrey, D.B. (2004) Predicted and actual costs from individual bank mergers. *Journal of Economics and Business*, **56**, 137–157.

Casu, B. and Girardone, C. (2009) Testing the relationship between competition and efficiency in banking: a panel data analysis. *Economics Letters*, **105**, 134–137.

Cetorelli, N. (2003) Life-cycle dynamics in industrial sectors: the role of banking market structure. *The Federal Reserve Bank of St. Louis Review*, **85**, 135–148.

Cuesta, R.A. and Orea, L. (2002) Time varying efficiency and stochastic distance functions: the effect of mergers on Spanish savings banks. *Journal of Banking and Finance*, **26**, 2231–2247.

De Guevara, J.F. and Maudos, J. (2007) Explanatory factors of market power in the banking system. *The Manchester School*, **75**, 275–296.

DeYoung, R., Evanoff, D., and Molyneux, P. (2009) Mergers and acquisitions of financial institutions: a review of the post-2000 literature. *Journal of Financial Service Research*, **36**, 87–110.

Fernandez de Guevara, J., Maudos, J., and Perez, F. (2005) Market power in European banking sector. *Journal of Financial Services Research*, **27**, 109–137.

Fiordelisi, F. (2009) *M&A in Banking,* Palgrave Macmillan, Studies in Banking and Financial Institution, UK.

Frieder, L.A. and Apilado, V.P. (1983) Bank holding company expansion: a refocus on its financial rationale. *Journal of Financial Research*, **6**, 67–81.

Garden, K.A. and Ralston, D.E. (1999) The x-efficiency and allocative efficiency effects of credit union mergers. *Journal of International Financial Markets, Institutions and Money*, **9**, 285–301.

Hagendorff, J. and Vallascas, F. (2011) CEO pay incentives and risk-taking: evidence from bank acquisitions. *Journal of Corporate Finance*, **17** (4), 1078–1095.

Hughes, J.P. and Mester, L.J. (2010) Efficiency in banking: theory, practice, and evidence, in *The Oxford Handbook of Banking* (eds A.N. Berger, P. Molyneux, and J. Wilson), Oxford University Press, Oxford.

Huizinga, H.P., Nelissen, J.H.M., and Vander Vennet, R. (2001) Efficiency effects of bank mergers and acquisitions in Europe. Tinbergen Institute, discussion paper no. 2001-088/3.

Humphrey, D.B. and Vale, B. (2004) Scale economies, bank mergers, and electronic payments: a spline function approach. *Journal of Banking and Finance*, **28**, 1671–1696.

Koetter, M. (2005) Evaluating the German bank merger wave. Deutsche Bundesbank, discussion paper, series 2: banking and financial studies no. 12.

Rezitis, A.N. (2008) Efficiency and productivity effects of bank mergers: evidence from the Greek banking industry. *Economic Modelling*, **25**, 236–254.

Rhoades, S.A. (1986) *The Operating Performance of Acquired Firms in Banking Before and After Acquisition*. Staff Studies 149. Washington: Board of Governors of the Federal Reserve System, 1986. A version with alternative statistical tests is, *The Operating Performance of Acquired Firms in Banking* (eds R. Wills, J.A. Caswell, and J.D. Culbertson), Issues after a Century of Federal.

Rhoades, S.A. (1998) The efficiency effect of bank mergers: an overview of case studies of nine mergers. *Journal of Banking and Finance*, **22**, 273–291.

Rose, P.S. (1987a) Improving regulatory policy for mergers: an assessment of bank merger motivations and performance effects. *Issues in Bank Regulation*, **10** (3, Winter), 32–39.

Rose, P.S. (1987b) The impact of mergers in banking: evidence from a nationwide sample of federally chartered banks. *Journal of Economics and Business*, **39**, 289–312.

Shaffer, S. (1993) A test of competition in Canadian banking. *Journal of Money, Credit and Banking*, **25**, 49–61.

Thomson Financial (2010) Mergers & Acquisitions Review, Fourth Quarter 2010. http://banker.thomsonib.com (accessed on 20 December 2012).

Turk-Ariss, R. (2010) On the implications of market power in banking: evidence from developing countries. *Journal of Banking and Finance*, **34**, 765–775.

Wang, J.C. (2003) Merger-related cost savings in the production of bank services. Federal Reserve Bank of Boston, working paper no. 03-8.

7

Backtesting superfund portfolio strategies based on frontier-based mutual fund ratings

Olivier Brandouy,[1] Kristiaan Kerstens,[2] and
Ignace Van de Woestyne[3]

[1] *IAE – Sorbonne Graduate Business School, Université Paris 1, France*
[2] *CNRS-LEM (UMR 8179), IESEG School of Management, France*
[3] *Hogeschool-Universiteit Brussel, Belgium*

7.1 Introduction

Professional mutual fund managers and individual investors face an increasingly large set of investment opportunities. Among these opportunities, the subset of mutual funds (MF) offers an amazing heterogeneity. For example, Morningstar, one of the best established rating agencies in this business (one can also cite Lipper, Standard & Poor's, Fitch Ratings, EuroPerformance, etc.), offers access to more than 160 000 MF all over the world through its database 'Morningstar Direct' (about 77 000 in Europe and about 26 000 in the United States). Consequently, the ratings offered by these rating agencies, and particularly their reliability, are of major interest for investors considering selecting some of these MF shares for their portfolios.

For example, Blake and Morey (2000) examine the Morningstar rating as a predictor of US domestic equity MF performance: these authors find little evidence that Morningstar's top-rated funds outperform the second- and third-rated funds. More recently, Kräussl and Sandelowsky (2007) offer a much more extensive study: not only US domestic equity MF, but also international equities, taxable bonds, and municipal bonds as well, with not a 5-year but a 10-year horizon. Their sample allows assessing Morningstar's revised rating since July 2002 when ratings got based on 64 categories rather than four broad asset classes till then.

Efficiency and Productivity Growth: Modelling in the Financial Services Industry, First Edition. Edited by Fotios Pasiouras.
© 2013 John Wiley & Sons, Ltd. Published 2013 by John Wiley & Sons, Ltd.

The upshot of their study is that Morningstar's rating till July 2002 can predict severe underperformance, but cannot discriminate between 3- to 5-star-rated MF. But, the results of the revised rating system are even worse: there is no significant performance difference in out of sample periods between 1- and 5-star MF. The merits of the 2011 Morningstar announcement of a new so-called analyst rating that intends to look forward rather than backward and that will coexist with the current ratings obviously remains to be evaluated.

Since about two decades, there have been successive attempts to transpose the successful frontier estimation methodologies from production theory to the analysis of financial problems. Sengupta (1989) is probably the first to introduce an explicit efficiency measure into a mean-variance (MV) portfolio model. In a MF rating context, the seminal article proposing some efficiency measure is Murthi, Choi, and Desai (1997). But, it is likely the article by Morey and Morey (1999) proposing both a mean-return expansion and a risk-contraction function that has in fact triggered a series of new developments in the use of efficiency measures in portfolio theory and in MF rating in particular (e.g., Briec, Kerstens, and Lesourd, 2004; Lamb and Tee, 2012; Lozano and Guttiérez, 2008, among others). Anyway, these developments have led to a burgeoning literature of about 40 or so related articles, published in a variety of journals and covering a collection of MF types (ethical MF, hedge funds, pension funds, etc.). Somewhat in parallel, starting with the seminal article of Alam and Sickles (1998) one can mention an emerging literature providing frontier-based asset selection and integration in portfolio models (e.g., Abad, Thore, and Laffarga, 2004 or Nguyen and Swanson, 2009).

Starting from the foundations of Modern Portfolio Theory, an enormous literature on portfolio performance evaluation has been developed using total-risk foundations (e.g., the standard deviation or variance of returns). Among the wide variety of financial performance indexes, one classic is the Sharpe ratio (also known as the reward-to-volatility). A recent survey summarizing these and more recent developments of financial and portfolio performances gauges is found in the book by Bacon (2008).

While a MV utility-maximizing agent should aim at a portfolio with the highest reward-to-risk ratio (a tangency portfolio or maximum Sharpe ratio portfolio), very few operational procedures in finance are in fact available that guarantee such a position on the MV frontier. For instance, market-cap-weighted indices are known to provide inefficient risk–return trade-offs. This has recently led to a stream of articles aimed at making hitherto inefficient benchmarking indices efficient (see, e.g., Martellini, 2008 or Clark, Jokung, and Kassimatis, 2011). Independently of the above frontier-based literature applied to financial topics, in the finance literature some authors have proposed to define a relative performance measure using the portfolio frontier as benchmark. For instance, the work of Cantaluppi and Hug (2000) proposes an efficiency ratio in relation to the MV-efficient frontier and contests the arbitrary and nonfrontier nature of most current proposals that define performance with respect to some other, supposedly relevant, portfolio or index. These authors suggest looking for the maximum performance that could have been achieved by a given portfolio relative to a relevant portfolio frontier. Also Broihanne, Merli, and Roger (2008) stress the importance of measuring performance relative to a portfolio frontier and underline that such an approach avoids the choice of a risk-free rate and a market portfolio as some kind of absolute benchmark. While this review may be incomplete, it is an understatement to conclude that the use of relative performance efficiency measures evaluated relative to some frontier is marginal at best in the current finance literature.

The purpose of this chapter is not to offer a complete overview of these rather recent frontier applications in a MF rating context. For instance, it completely ignores the question of the determinants of MF performance (examples include sector- and country-specific factors, such as different time paths of technological innovations, fluctuations in the macro-economic environment, etc.). Also the question about the precise attribution of the role of the MF manager, traditionally focusing on stock picking and market timing capabilities, is sidestepped. Instead, the goal is to offer the first detailed backtesting analysis of these frontier-based MF ratings compared to the Morningstar rating (representative for the fund rating agencies) and some traditional financial performance measures. While this is not the first backtesting analysis of frontier MF ratings (see, e.g., Matallín, Soler, and Tortosa-Ausina, 2011), it is – to the best of our knowledge – the most extensive backtesting analysis provided in the literature focusing on the relative merits of different backward-looking performance rating tools in predicting future MF performance.

7.2   Frontier-based mutual funds rating models

7.2.1   A taxonomy

Given the rather widespread criticisms of traditional financial performance measures, several authors have recently been introducing nonparametric frontier methods to assess MF performance.[1] Following Tsolas (2011), it is useful to distinguish between four different modeling approaches:

(i)   Models directly transposed from production theory;

(ii)   Models combining traditional performance measures (e.g., Sharpe ratio) with additional dimensions;

(iii) Models directly transposed from portfolio theory;

(iv)   Hedonic price models (a new proposal launched in Kerstens, Mounir, and Van de Woestyne, 2011b).

We discuss each of these modeling approaches in turn to develop our own selection of models put at a test in the Empirical section.

    (i) Frontier models can handle multiple dimensions and yield a single efficiency measure with respect to a frontier composed of similar units. Directly transposing these basic models from production theory to financial applications, there seems to be a wide variation of specifications around in the literature. A rather up-to-date and fairly comprehensive review of this literature is offered in Glawischnig and Sommersguter-Reichmann (2010). A mix of traditional and frontier-based performance studies as well as an international perspective is provided in the Gregoriou (2007) book.

    Without any ambition of completeness or representativeness, Table 7.1 reviews a limited selection of articles. Several observations can be made. First, apart from a variety of costs,

---

[1] Stochastic frontier models have been more rarely employed for MF rating purposes: an example is Annaert, van den Broeck, and Vander Vennet (2003). However, notice that little of what is developed in this chapter is really conditioned by the use of some specific frontier estimation method.

Table 7.1 Frontier models of MF performance: a selection.

Article	Inputs	Outputs	Model
Basso and Funari (2003)	Subscription and redemption cost Beta Standard deviation returns Standard semi-deviation returns	Return (or excess return) Ethical score	VRS
Choi and Murthi (2001)	Standard deviation returns Expense ratio Loads Turnover ratio	Gross return	VRS
Galagedera and Silvapulle (2002)	1, 2, 3, and 5 year standard deviation returns Sales charges Expense ratio Minimum investment	1, 2, 3, and 5 year gross return	VRS
Glawischnig and Sommersguter-Reichmann (2010)	Mean lower return Lower mean semi-variance Lower mean semi-skewness	Mean upper return Upper semi-variance Upper semi-skewness	VRS
McMullen and Strong (1998)	Standard deviation returns Sales charges Expense ratio Minimum investment	1, 3, and 5 year annual returns	VRS
Murthi, Choi, and Desai (1997)	Standard deviation returns Expense ratio Loads Turnover	Gross excess return	CRS
Wilkens and Zhu (2001)	Standard deviation returns % Negative monthly returns per year	Average return Skewness Minimum return	VRS

there seems to be a need for models accounting for higher-order moments (especially for hedge funds, and the like). Most articles we are aware of include at most up to the third moment (e.g., Glawischnig and Sommersguter-Reichmann, 2010, or Wilkens and Zhu, 2001). Second, some articles use standard moments (e.g., Galagedera and Silvapulle, 2002), while other articles use lower or upper partial moments or some combination of these partial moments (e.g., Glawischnig and Sommersguter-Reichmann, 2010), or even combine standard and partial moments (e.g., Basso and Funari, 2003). Third, there are quite a few attempts to offer assessments over several time horizons simultaneously. For instance, McMullen and Strong (1998) include one-, three-, and five-year annualized returns, but do not apply these same time horizons to the other factors in their model. A more consistent model in this respect is probably the article by Morey and Morey (1999) (see also (iii)) who evaluate both returns and standard deviations over three time horizons simultaneously.

This literature suffers in our view from a variety of problems. First, by defining some frontier on a sample of MF, one in fact seems to target at defining a series of superfunds to assess in turn each of the underlying MF. But, such an approach ignores diversification and interaction effects among moments. It could at best be considered a type of linear approximation of a possibly nonlinear portfolio model. An open question of how good such approximation turns out to be then remains.

Second, these nonparametric frontier-based articles assessing MF raise three crucial specification issues (following Kerstens, Mounir, and Van de Woestyne, 2011b). First, the very notion of returns to scale may not necessarily be directly transposed to the finance context. Second, the inclusion of higher-order moments is sometimes rather arbitrarily stopped at some order and rarely is based on a systematic moment approximation strategy. And third, related to the first problem, the hypothesis of convexity may or may not provide a suitable linear approximation. But, if it offers just an approximation, then an open question is the meaning of the resulting projection points. Depending on the closeness of the approximation, such projection points can or cannot in fact be obtained by an investor. If so, then how can these projection points be of any help to the investor looking for advice from MF ratings?

(ii) Some frontier models combine a traditional financial performance measure with some additional variables. There is some variation of specifications: for instance, Basso and Funari (2001, 2003), Chang (2004), and Sengupta and Zohar (2001) add the Beta coefficient from the Capital Asset Pricing Model (CAPM), while Haslem and Scheraga (2003) include a Sharpe index. In fact, such models can be interpreted as a special case of the first category.

Being just a special case of the first category, these models in our opinion share all of their defects. In addition, there is the risk of one major interpretational problem: what does the efficiency measure in such a frontier, being a combination of a traditional financial performance measure and some additional variables, mean? For instance, Haslem and Scheraga (2003) define an input-oriented frontier model with the Sharpe index as the single output combined with a series of inputs (% cash, % stocks, total assets, expense ratio, price/earnings, and price/book ratios). By contrast, the articles of Basso and Funari (2001, 2003) pay a lot of attention to the interpretation of the resulting efficiency measure.

Specifically for the use of Beta in some of these articles, it is widely recognized that CAPM does a poor job at explaining gross returns (even in a MV world). This has led to alternative multifactor models, some of which are theoretically grounded (e.g., Arbitrage Pricing Theory) and others that are just empirically validated (so-called Fama-French models).[2] Furthermore, CAPM presupposes normality and relates a MF to a market portfolio

---

[2] See, e.g., Peterson Drake and Fabozzi (2010) for more background information.

benchmark approximated by a factually suboptimal market index. Beta as such measures undiversifiable risk with respect to this supposedly universal market benchmark. Thus, it has little meaning in a nonnormal world that many of these frontier models aim to focus on. Obviously, depending on the number of moments to be integrated in the frontier model, one could eventually consider looking at 3CAPM and 4CAPM (e.g., Jurczenko and Maillet, 2006). The latter possibilities remain to be explored, subject to the caveats already mentioned.

(iii) Starting with Sengupta (1989) and especially Morey and Morey (1999), portfolio models with either return-expansion or risk-contraction efficiency measures have been developed that can handle diversification and interaction effects among eventually multiple moments and/or multiple time horizons. For instance, Zhao, Wang, and Lai (2011) develop a special MV model that includes Beta and that aims at increasing both return and return above the market benchmark, while Lamb and Tee (2012) develop an iterative approximation procedure to deal with the traditional failure of frontier models to cope with diversification.

These efficiency measures have been generalized by Briec, Kerstens, and Lesourd (2004) who propose transposing the shortage function as an efficiency measure into the MV model and who also develop a dual framework to assess the satisfaction of MV preferences. By analogy to other fields, a decomposition of portfolio performance into allocative and portfolio (technical) efficiency is proposed. The key advantage is that this shortage function is perfectly compatible with a general mixed risk aversion preference structure (i.e., a preference for augmenting odd moments and reducing even moments). Therefore, the approach can be extended to higher-dimensional portfolio spaces: for example, MV-skewness (MVS) space (see Briec, Kerstens, and Jokung, 2007) or even higher-order models (see Briec and Kerstens, 2010). Another advantage is that a slight variation on the shortage function (see Kerstens and Van de Woestyne, 2011) can handle negative data that occur naturally in a financial context, while maintaining a proportional interpretation − a convenience for practitioners.

The article by Morey and Morey (1999) on multihorizon return-expansion or risk-contraction MV frontier models has been extended into a more general time-discounted model based on the shortage function in Briec and Kerstens (2009). Brandouy et al. (2010a) propose a Luenberger portfolio productivity indicator to estimate changes in the relative positions of portfolios with respect to the traditional MV frontier, and the eventual shifts of the frontier over time. Analyzing the local changes relative to these MV (or higher-moment) portfolio frontiers, this productivity indicator separates performance changes due to portfolio strategies and performance changes due to the market evolution. Applications of this new portfolio modeling approach include reconstructions of MVS portfolio sets (see Kerstens, Mounir, and Van de Woestyne (2011a) for the general-moment case and Brandouy, Kerstens, and Van de Woestyne (2010b) for the lower partial-moment case), comparisons with alternative portfolio models (e.g., Lozano and Guttiérez, 2008), and applications to hedge fund industry (see Jurczenko and Yanou, 2010).

While the promise of this new multimoment approach to portfolio analysis remains to be thoroughly assessed, the application of such portfolio approach to rate MF could face at least the following two problems. First, there is the choice of the MF universe that is unclear. Should one be comparing MF among themselves in different classes? Or, should one be comparing the MF relative to some supposedly common underlying asset universe for a given class of MF (as if the investor could pick his/her own MF from the underlying assets)? Second, for even small classes of MF, the computational burden is very high when adding higher moments and/or multiple time horizons. For some of the larger MF classes of considerable size, the computational burden could be prohibitive, even with today's computing power. Probably, this makes these models currently unsuitable to implement in practice.

(iv) A new proposal launched in Kerstens, Mounir, and Van de Woestyne (2011b) is to found MF rating on a hedonic pricing model. Based on the characteristics' approach to consumer theory, utility is a function of the characteristics of products and services, not a function of the goods vector itself. The study of market equilibria for heterogeneous goods differing along multiple characteristics derives an implicit price for the vector of observed characteristics aggregating these into a single measure of market value. While the estimation of price–qualities functions and frontiers has become quite common in consumer and marketing applications, the use of this characteristics' approach in the finance literature remains rather rare.[3] In finance, authors like Blake (1990), Dodds (1986), Heffernan (1992), among others, have argued in favor of interpreting several financial products in terms of a variety of characteristics.

In essence, the proposal of Kerstens, Mounir, and Van de Woestyne (2011b) is to consider MF as fee-based (e.g., various loads) financial products characterized by some distributional characteristics as summarized by a combination of moments. Thus, the traditional price–qualities frontier of consumer goods and services simply becomes a fee–moments frontier for MF: the investor pays a series of loads and fees to benefit of some future, uncertain return when he/she resells his/her shares depending on the investment horizon.

This viewpoint disregards any repercussions at the portfolio level. In fact, given investor preferences for higher-order moments and the current lack of any summary performance measure in this multimoment portfolio theory, one could argue that any private or professional investor can only judge the suitability of any financial product, such as a MF, by explicitly testing whether it could improve his/her particular portfolio. The MF rating based on the hedonic pricing model then only has the modest aim to offer a list of potentially suitable candidates of MF to consider for inclusion in any given portfolio.

As argued at length in Kerstens, Mounir, and Van de Woestyne (2011b), there are important issues to tackle when constructing hedonic rating models. Firstly, there is the issue of specifying the nature of returns to scale to be used in the model. From the existing literature, we observe that variable returns to scale (VRS) and constant returns to scale (CRS) are quite common assumptions. Kerstens, Mounir, and Van de Woestyne (2011b) discuss this issue of choosing the proper returns to scale in detail and conclude that VRS is more meaningful from a theoretical perspective. Furthermore, their empirical tests also favor VRS to CRS. Secondly, one can opt for either convex or nonconvex models. Kerstens, Mounir, and Van de Woestyne (2011b) raise the issue of indivisibilities, that is, the impossibility of combining some dimensions of several MF because this ignores diversification effects (see also the discussion in (i)). This leads to a preference for a basic nonconvex model avoiding the interpretation problems raised above that projection points should be

---

[3] Friedman (1983) offers a detailed survey and critique on the use of the characteristics' approach in various industries (see his ch. 4).

feasible. On occasion, also frontier studies from the first category apply a nonconvex model (see, e.g., Matallín, Soler, and Tortosa-Ausina, 2011).

While such a basic nonconvex model in a production context sometimes leads to a majority of efficient units, yielding a lack of discrimination, the potentially huge amount of observations around in most financial settings creates the hope to mitigate this problem. Anyway, Kerstens, Mounir, and Van de Woestyne (2011b) preliminary empirical tests favor nonconvex compared to convex models with a variety of higher moments.

In summary of these four different modeling approaches, the upshot of this discussion is that especially the first and the fourth approaches show some promise for practical MF rating purposes. Therefore, we consider both the traditional convex and nonconvex VRS models when developing our empirical research strategy.

7.2.2   MF frontier rating models retained

Following this conclusion, we now introduce the convex and nonconvex VRS-based shortage function (sometimes also known under the name directional distance function). Assume the set of n MF under evaluation is indexed by $j=1,\ldots,n$. Each MF is characterized by m input-like values x_{ij}, $i=1,\ldots,m$ and s output-like values y_{rj}, $r=1,\ldots,s$. Assume that MF $o\in\{1,\ldots,n\}$ needs to be evaluated. Then, its inefficiency determined by the convex VRS-based shortage function is obtained by solving a mathematical programming problem:

$$\max \lambda \text{ s.t. } \sum_{j=1}^{n} z_j y_{rj} \geq y_{ro} + \lambda |y_{ro}|, \qquad r=1,\ldots,s,$$

$$\sum_{j=1}^{n} z_j x_{ij} \leq x_{io} - \lambda |x_{io}|, \qquad i=1,\ldots,m, \tag{7.1}$$

$$\lambda \geq 0, \forall j=1,\ldots,n: z_j \geq 0, \sum_{j=1}^{n} z_j = 1.$$

The inefficiency determined by the nonconvex VRS-based shortage function can be computed with the help of the following mathematical program:

$$\max \lambda \text{ s.t. } \sum_{j=1}^{n} z_j y_{rj} \geq y_{ro} + \lambda |y_{ro}|, \qquad r=1,\ldots,s,$$

$$\sum_{j=1}^{n} z_j x_{ij} \leq x_{io} - \lambda |x_{io}|, \qquad i=1,\ldots,m, \tag{7.2}$$

$$\lambda \geq 0, \forall j=1,\ldots,n: z_j \in \{0,1\}, \sum_{j=1}^{n} z_j = 1.$$

Notice that both models specify the direction vector following the proposal made by Kerstens and Van de Woestyne (2011). This specification allows to have negative data values, which is indeed the case when examining periods involving a financial crisis (e.g., negative returns). Furthermore, both models project the MF o under evaluation in the direction $g=(-|x_{1o}|,\ldots,-|x_{mo}|,|y_{1o}|,\ldots,|y_{so}|)$. Consequently, all output-like values, y_{ro}, $(r=1,\ldots,s)$,

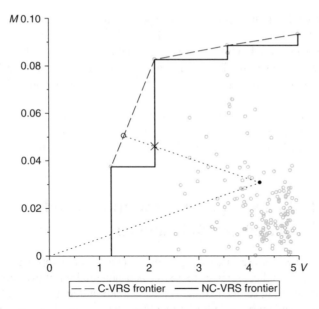

Figure 7.1 Visualization of convex and nonconvex frontiers determined by 814 MF in the main database: Expected return and variance are considered as output and input, respectively.

are increased and the input-like values, $x_{io}, (i=1,...,m)$, are decreased simultaneously in proportion to their initial values. Furthermore, λ measures the amount of inefficiency (i.e., larger values indicate less efficient MF), while an efficient MF has a shortage function value of 0 ($\lambda = 0$). Finally, note that model (7.1) results in a linear programming (LP) problem, while model (7.2) leads to a mixed integer programming (MIP) problem.

As an example, Figure 7.1 visualizes the convex (dashed line) and nonconvex frontiers (solid line) determined by the 814 MF (gray circles) from the main database of this research (see Section 7.3.3). For these funds, the expected return and variance computed over the first three-year period are considered as output and input, respectively. This time window of three years used in these computations is identical to the one employed by Morningstar in its three-year rating. This particular restriction to one input and one output is considered here only for visualization purposes. In the procedure described later (see Section 7.3.1), multiple inputs and/or outputs are considered. In order to focus on the interesting region of the frontiers, Figure 7.1 only shows the area with positive returns and a variance limited to 5. Consequently, not all initial MF are visible here.

To show the projecting capabilities of the shortage function, MF number 502 ('LBBW Dividenden Strategie Euroland R') is located as a black solid circle. For this fund, the expected return is equal to 0.0307 while the variance equals 4.2238. The optimization program determined by model (7.1) returns the inefficiency value of $\lambda_C=0.6457$. From this value, the location of the optimal frontier position (visible as the black circled point) can be derived easily following the right-hand sides of the inequality constraints in model (7.1). This resulting optimal point has an expected return of 0.0505 and a variance of 1.4964. Note that the return is increased while the variance is decreased compared to its initial position. Analogously, when considering the nonconvex model (7.2), one obtains an inefficiency value of $\lambda_{NC}=0.4973$ leading to the optimal frontier point with an expected return of 0.0459 and a variance of 2.1235. This point is shown in Figure 7.1 as a diagonal cross. Also here, the expected return is increased, while the variance is reduced. Note that, due to the shape of the nonconvex frontier in relation with that of the convex frontier, the 'distance' from the initial position to the optimal frontier point is larger in the convex case compared with the nonconvex case: this explains the relation $\lambda_C > \lambda_{NC}$.

Intensive comparative testing of these two selected approaches, a selection of more traditional financial performance measures, as well as rating agencies ratings are needed to assess which MF rating methodology turns out to be better. It is exactly to this exercise that we now turn.

7.3   Backtesting setup, data description, and frontier-based portfolio models

In this section, we evaluate several strategies using a backtesting approach. Backtesting consists in simulating financial strategies to assess their potential merit over historical data sets. On the one hand, in doing so one maintains the assumption that their factual implementation during past periods and market conditions would not have had an impact on historical prices. This conception is very neoclassical: it suggests a price taker framework describing how prices emerge in thick financial markets. For example, such a framework is widely used in risk management (e.g., for evaluating Value at Risk models: see Campbell, 2005), or in portfolio management for comparing several investment strategies (for instance, DeMiguel, Garlappi, and Uppal (2009) or Tu and Zhou (2011) compare a naïve diversification strategy against several more or less sophisticated optimization strategies). On the other hand, this neoclassical approach appears to be inconsistent for those claiming that prices may well be influenced by factual implementation of the backtested investment strategies (see, for a general example of a microscopic simulation, Levy, Levy, and Solomon, 1995). In line with this conception, an ecological competition approach is more suitable (see, e.g., Sprott, 2004 or Brandouy, Mathieu, and Veryzhenko, 2012).

In the following, we present a large number of strategies that are backtested in this chapter. Note that we have opted for a simple, direct backtesting approach illustrating the outcomes of different models over a given time window rather than, for example, exploring massive simulations of a single strategy over different time windows. Apart from Matallín, Soler, and Tortosa-Ausina (2011) to which we turn in some detail below, we are unaware of any other frontier study of MF involving backtesting.

7.3.1   Backtesting setup

We now consider two major backtesting strategies. The first strategy is called 'EWP without optimization' (where EWP stands for equally weighted portfolio) and composes for each period considered the EWP of the m best performing MF. For obtaining these best performing funds, a ranking is made according to some performance measure. This can be a particular hedonic rating model, or one of the traditional performance indicators (such as Morningstar's stars or, for example, the Sharpe, Sortino, Treynor, Kappa, or Omega ratios).

The second strategy is called 'EWP with optimization' and starts off by identifying an EWP just like in the first strategy. Rather than using these portfolios directly as a guideline for investment, these EWPs are now considered as starting points for an additional portfolio optimization process based on the position-dependent shortage function. This shortage function then projects the EWP onto an optimal portfolio frontier derived from a universe that is limited to only those MF present in the EWP. Three portfolio models are considered: the traditional MV model, a multimoment MVS model, and a MVS-kurtosis (MVSK) model. Since for each period the universe is restricted to only those MF considered in the EWP, the shortage function ensures that the projected optimal portfolio is also composed of only these same underlying MF. Consequently, this additional optimization process only changes the weights of the MF present in the EWP, not the set of MF considered.

An overview of all detailed backtesting scenarios considered further on in Section 7.3 is found in the first column of Table 7.3. The notation consists of two parts, that is, the parts before and after 'to'. The first part of the notation indicates what model is used for ranking the MF for selecting the m best ones. This can be done using a convex (indicated with 'C') hedonic rating model or a nonconvex (indicated with 'NC') model. For example, 'MVS+L-C' refers to the convex model with expected returns, variance, skewness, and the loads/fees selected. Apart from a hedonic rating system, more traditional indicators are considered as well. In particular, we include the Morningstar rating and the traditional performance measures: Sharpe, Sortino, Treynor, Kappa, and Omega ratios.[4] For example, 'Stars' refers to a ranking based on the three-year Morningstar ratings. The second part of the notation refers to one of the three portfolio models additionally used in the case of EWP with optimization (i.e., MV, MVS, or MVSK). Since the first strategy, that is, EWP without optimization, involves no portfolio model in the second stage, this is indicated with 'NoOpt' in the second part. Proceeding in this way, in total 48 backtesting scenarios can be identified for further investigation.

Note that in the backtesting scenarios considered in Section 7.3, a selection of the 10, 20, 30, 40, or 50 best open-end funds is considered. In the case of ties (e.g., in the Morningstar ratings or particularly when using nonconvex hedonic models), MF are selected randomly among the tied observations.

The performance of all these backtesting scenarios is tested first and foremost by evaluating and ranking the realized terminal value starting with a capital of unity, with and without transaction costs. In addition, some representative traditional performance measures in finance, such as the Sharpe and Omega ratios, are computed and interpreted as performance gauges. The Sharpe ratio is traditionally conceived as suitable for the MV world, while the Omega ratio is supposedly capable of assessing a nonnormal world.

Notice that Matallín, Soler, and Tortosa-Ausina (2011) essentially follow our 'EWP without optimization' strategy. These authors focus on the impact of some robustness parameters in the estimation of robust versions of the nonconvex frontier model (i.e., Free Disposal Hull (FDH), and its order-m and order-α robust versions) on some basic backtesting strategies. In fact, three backtesting strategies based on past efficiency are considered: (a) buying past top quantile and selling past bottom quantile; (b) buying past top quantile; and (c) selling bottom quantile. Thus, their study mainly differs from ours in that we also employ traditional convex frontier models and that we also apply an 'EWP with optimization' strategy.[5]

---

[4] For definitions and detailed presentations of these ratios: see, for example, Bacon (2008).

[5] There do exist some backtesting studies using frontier models for asset selection in terms of a financial strength indicator and then assessing the resulting portfolios over historic time (see, e.g., Edirisinghe and Zhang, 2007, 2010).

7.3.2   Frontier-based portfolio models

We now briefly describe the position-dependent shortage function in the portfolio context. This shortage function is first introduced in Briec, Kerstens, and Lesourd (2004) with respect to a MV universe. In Briec, Kerstens, and Jokung (2007), this model is then adapted to the MVS world. For a discussion of methods capable of visualizing the corresponding MVS-frontier, we refer to Kerstens, Mounir, and Van de Woestyne (2011a). Briec and Kerstens (2009) show that this adaptation from MV to MVS can be generalized to an arbitrary multi-moment universe. To save space, we restrict ourselves to only mentioning the position-dependent shortage function with respect to the MVSK universe.

We introduce some basic notations. A portfolio consisting of n MF available in the financial universe can be considered as a vector of weights $x=(x_1,...,x_n)$ indicating the individual propor-tions of each MF. Obviously, $\sum_{i=1}^{n} x_i = 1$. If short selling is excluded, then the condition $x_i \geq 0$ for all $i \in \{1,...,n\}$ must be satisfied. All MF in the financial universe are characterized by their raw returns registered over a given time window. From this information, the expected return vector, covariance matrix, and the skewness and kurtosis tensors can be derived. Additionally, either directly from the initial raw returns or indirectly from the statistical vectors, matrices, and ten-sors computed, the expected return, variance, skewness and kurtosis of an individual portfolio x can be computed. We denote these by Ret(x), Var(x), Skew(x), and Kurt(x), respectively.

Consider a portfolio x_o under evaluation. Then, the position-dependent shortage function identifies an inefficiency value λ for x_o obtained from solving the following nonlinear optimi-zation model:

$$\max \lambda \text{ s.t. } \text{Ret}(x) \geq \text{Ret}(x_o) + \lambda|\text{Ret}(x_o)|,$$
$$\text{Var}(x) \leq \text{Var}(x_o) - \lambda\text{Var}(x_o),$$
$$\text{Skew}(x) \geq \text{Skew}(x_o) + \lambda|\text{Skew}(x_o)|,$$
$$\text{Kurt}(x) \leq \text{Kurt}(x_o) - \lambda\text{Kurt}(x_o), \tag{7.3}$$
$$\lambda \geq 0, \forall i = 1,...,n : 0 \leq x_i \leq 1, \sum_{i=1}^{n} x_i = 1.$$

Note that if one needs the MV-based position-dependent shortage function, then the inequal-ity constraints involving skewness and kurtosis must be dropped from model (7.3). Analogously, dropping only the kurtosis inequality constraint in (7.3) leads to the MVS-based position-dependent shortage function.

The position-dependent shortage function results in an inefficiency value λ. Similar with the shortage functions introduced in models (7.1) and (7.2), the portfolio x_o is more efficient if its inefficiency value is closer to 0. Moreover, the optimal portfolio x corresponding with the optimal value of λ for x_o is located on the corresponding MVSK-frontier. Clearly, this optimal portfolio has a higher return and skewness than x_o, while its variance and kurtosis are lower than that of x_o. Put differently: even moments are increased, while odd moments are decreased.

7.3.3   Data description

For the empirical analysis in Section 7.4, we use several data sets extracted from the 'Morningstar Direct' database. First, we extract a set of 814 open-end MF for which we have collected weekly returns from 9 October 2005 to 2 October 2011. This yields a total of 156 weeks. These open-end MF are rather homogeneous in that they all belong to the large caps European universe (Eurozone and the United Kingdom).

Second, running a MF business induces a variety of costs that are linked either to regular operations (brokerage, consultancy, marketing and distribution expenses, etc.), or to specific investors' transactions (mainly purchases and redemptions). The former costs are gathered under the 'Annual Fund Operating Expenses' denominator, while the latter costs are usually gathered under the heading 'Shareholder Fees'. The first cost category is indirectly covered by the fund shareholders with the proceeds of asset transactions decided by the fund manager and cash inflows coming from new investors. The second cost category is directly charged to the investor for each transaction of a fund's share, or on a regular (e.g., yearly) basis. In the MF prospectus, these elements of information are available. Since we need part of this information in the backtesting setup, we have downloaded for the 814 funds the 'Front Load' (entry fee), 'Redemption Load' (exit fee), and the 'Annual Report Net Expense Ratio'. The 'Front Load' and 'Redemption Load' are examples of shareholder fees and are fixed throughout the whole period under evaluation. The 'Annual Report Net Expense Ratio' (which reflects the actual fees charged during a particular fiscal year) proxies the total annual fund operating expenses. Clearly, this information is only available for each particular year. Therefore, it is extracted for the years 2006 till 2010.

Third, since some of the backtesting models involve Morningstar ratings, we extract for the period October 2008 till September 2011 the three-year Morningstar rating, which is available on a monthly basis. Detailing the Morningstar rating system, it is based on the risk-adjusted rate of return of a MF relative to its 'Morningstar Category'. This rating system for each MF is based on the total return in excess of a 90 days T-Bill, adjusted for front-end loads, deferred loads, and redemptions fees, and decreased by a risk penalty that particularly takes into account downward deviations. Thus, the risk considered by Morningstar consists in the lower semi-standard deviation for the excess return (to be explicit: $\sigma[\min((r_i - r_f),0)])$.[6] The best funds obtain 5 stars, the worst ones, a single star. In fact, the distribution of these stars in each category is a priori fixed and nonuniform. In other words, 10% of the best performing funds receive 5 stars, 22.5% 4 stars, 35% 3 stars, 22.5% 2 stars, and 10% 1 star.

This Morningstar rating is calculated on subsets of MF belonging to the same Morningstar category.[7] Furthermore, the composition of these Morningstar categories evolves over time following changes in the asset basket held by the MF. Notice that the time window of three years used in our MF frontier rating computations is identical to the one employed by Morningstar in determining its three-year Morningstar rating.

Fourth, since some of the backtesting models make use of traditional financial performance measures, the values for the Sharpe, Sortino, Treynor, Kappa, and Omega ratios are also extracted. These data are again available on a monthly basis. Therefore, these are extracted for the same period from October 2008 till September 2011 as the one used for the Morningstar ratings data.

Fifth, backtesting also requires prices for the individual MF. Again, these prices are extracted from 'Morningstar Direct' for the selected MF. These data are available on a daily basis from 1 January 2007 till 1 December 2011. Evidently, all price data are expressed in euros and converted when needed at the ongoing exchange rate. However, this data set turns out to be incomplete. Whenever possible, this missing data is completed (e.g., by using the price of the previous transaction day). For 103 MF in the initial sample, completing

---

[6] One can also refer to Sharpe (1998) for a more detailed discussion of this measure.

[7] These Morningstar categories are similar in nature to the Standard & Poor's 'GICS'.

missing data is impossible. Therefore, these MF are removed from the data set of prices. This reduces the number of MF to a total of 711. When prices are not needed, then computations are done on the data set containing the original 814 MF. Otherwise, the restriction to the 711 MF with full price information applies. Note that due to the monthly frequency with which the backtesting scenarios are executed, in fact we only need monthly prices from this daily prices database. Thus, this implies that monthly prices are used from 5 October 2008 till 4 September 2011: this is a period of 36 months. However, note that all statistics necessary for the hedonic rating models are computed using the available weekly data (see the first point).

In conclusion, somewhat in contrast to Blake and Morey (2000) and Kräussl and Sandelowsky (2007), notice that our backtesting setup is particular in three respects. First, the sample is more homogeneous in that it focuses on a limited set of Morningstar categories. Second, while the other studies focus on testing the internal validity of the Morningstar ratings (across stars), we backtest the external validity of solely the three-year Morningstar ratings compared to various alternatives in selecting top performers. Last but not least, we on purpose create the harshest possible testing environment by computing our ratings over a normal-to-bull market period (2005–2008), while we backtest all strategies over one of the worst financial crisis periods ever (2008–2011).

7.4   Empirical analysis

7.4.1   Descriptive statistics

We analyze the characteristics of the returns distribution for the sample consisting of 814 MF over the whole period from 9 October 2005 to 2 October 2011. Figure 7.2a to d reports some basic descriptive statistics: in particular, the distributions of the expected return, standard deviation, skewness, and excess kurtosis are displayed. The most prominent features are situated at the third and fourth moments: there is a noticeable negative skewness and a substantial excess kurtosis. Both phenomena can be linked to the financial crisis resulting in a negative trend and a high volatility regime during the period under investigation.

Probably for the same reason, the potential diversification effects in this universe are relatively weak due to rather high levels of correlation. Figure 7.3 reports a gray tone mapping of the covariance matrix computed for all 814 MF over the same period. Notice that off-diagonal correlations vary widely between 0.097 (darkest gray tone) and 1 (lightest gray tone). Typically, several MF belonging to the same investment firm have very similar (or even exactly the same) composition and hence correlate very strongly. For example, two fund shares with a close-to-unity correlation coefficient are 'Vanguard European Stock Idx Inst EUR' and 'Vanguard European Stock Idx Inst USD', the first one being listed in euros while the second in US dollars. These substantial correlation levels may well mitigate the scope for diversification. The extent to which these may compromise the exploitation of higher-order moments and co-moments remains to be explored.

7.4.2   Analysis of both hedonic rating models

Since hedonic frontier rating models are considered in the backtesting procedures, we first compare the efficiency distributions obtained for some of these frontier models over all 814 selected MF. For these comparisons, we use nonparametric Li (1996) tests. This test statistic

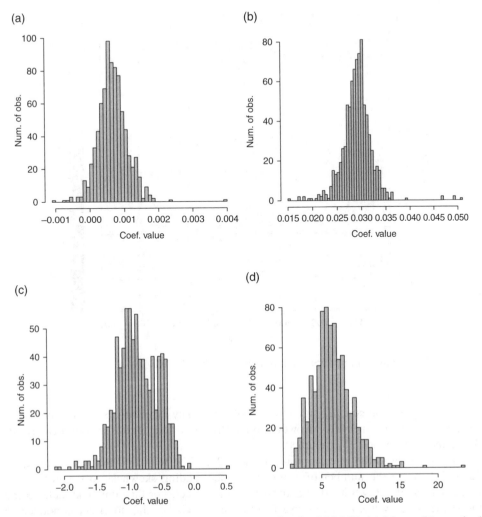

Figure 7.2 Distribution of the first four moments for the 814 MF. (a) Mean, (b) standard deviation, (c) skewness, and (d) excess kurtosis.

basically compares two entire distributions and it is valid for both dependent and independent variables. Notice that independency is not a valid assumption in frontier models where efficiency estimates depend, among others, on the relative size of the sample, the dimensionality of the space under analysis, etc.

Table 7.2 reports the result of in total seven comparisons. The first two (columns 2 and 3) look for the effect of adding an additional moment to the nonconvex model. Similarly, the next two comparisons (columns 4 and 5) demonstrate the effect of adding an additional moment to the convex model. Finally, the last three comparisons (columns 6–8) evaluate the effect of switching between the convex and nonconvex models for an identical number of moments.

Notice that a comparison is made for all 36 months present in the time window under evaluation: each period is represented in a row. In terms of notation, the comparison between

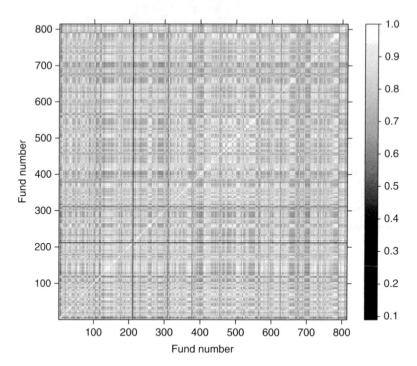

Figure 7.3   Visualization of correlation matrix for the 814 MF.

two models yields an equality sign ('=') if both efficiency distributions can be considered equal at a 5% significance level. However, if the efficiency distributions are significantly different, then the symbol '*' is returned.

We now turn to the interpretation of these 252 (7 × 36) Li (1996) test statistics in Table 7.2. As to the number of moments needed, adding skewness to the nonconvex model has an effect, while adding kurtosis has no effect. For the convex model, adding the skewness only has an effect in some of the periods. Again, no effect is noticeable from adding the kurtosis to this model. Finally, the comparison of the effect of switching between convex and nonconvex models in the last three columns demonstrates that these models are in a large majority of periods significantly different across the three multimoment models. This corroborates the results in Kerstens, Mounir, and Van de Woestyne (2011b).

A second step in the analysis consists in verifying whether the hedonic ratings are correlated with the other performance measures available. To answer this question, we run a nonparametric rank correlation analysis using Kendall's τ statistics. Therefore, consider the rating pairs of rating matrices each of dimension (814×36) (i.e., funds per time windows) containing the efficiency values for all 814 MF and all periods present in the monthly backtesting calendar (i.e., 36 periods). The first rating, hereafter denominated R_1, systematically consists of a hedonic rating. The second rating, denominated R_2 is either the Morningstar rating, or the Sharpe ratio matrix. The correlation estimates are computed over all time periods as follows: $\text{Cor}_{i \in [1,36]}(R_1(.,i), R_2(.,i))$. Literally, this means that we use the set of 814 MF time window after time window (36 times) to assess how our hedonic rankings are in line

Table 7.2 Comparison of different hedonic rating models by means of Li (1996) tests.

Period	MV+L-NC / MVS+L-NC	MVS+L-NC / MVSK+L-NC	MV+L-C / MVS+L-C	MVS+L-C / MVSK+L-C	MV+L-C / MV+L-NC	MVS+L-C / MVS+L-NC	MVSK+L-C / MVSK+L-NC
1	=	=	=	=	*	*	*
2	=	=	=	=	*	*	*
3	*	=	=	=	*	*	*
4	*	=	=	=	*	*	*
5	*	=	=	=	=	*	*
6	*	=	=	=	=	*	*
7	*	=	*	=	*	*	*
8	*	=	*	=	*	*	*
9	*	=	*	=	*	*	*
10	*	=	*	=	*	*	*
11	*	=	=	=	*	*	*
12	*	=	=	=	*	*	*
13	*	=	=	=	*	*	*
14	*	=	=	=	*	*	*
15	*	=	=	=	*	*	*
16	*	=	=	=	*	*	*
17	*	=	=	=	*	*	*
18	*	=	=	=	*	*	*
19	*	=	=	=	*	*	*
20	=	=	=	=	*	*	*
21	*	=	=	=	*	*	*
22	*	=	=	=	*	*	*
23	*	=	=	=	*	*	*
24	*	=	=	=	*	*	*

(continued overleaf)

Table 7.2 (*continued*)

Period	MV+L-NC / MVS+L-NC	MVS+L-NC / MVSK+L-NC	MV+L-C / MVS+L-C	MVS+L-C / MVSK+L-C	MV+L-C / MV+L-NC	MVS+L-C / MVS+L-NC	MVSK+L-C / MVSK+L-NC
25	*	=	=	=	*	*	*
26	*	=	=	=	*	*	*
27	*	=	=	=	*	*	*
28	*	=	=	=	*	*	*
29	*	=	=	=	*	*	*
30	*	=	=	=	*	*	*
31	*	=	=	=	*	*	*
32	*	=	=	=	*	*	*
33	*	=	=	=	*	*	*
34	*	=	*	=	*	*	*
35	*	=	*	=	*	*	*
36	*	=	*	=	*	*	*

*, Means both efficiency distributions can be considered not equal at 5% significance level.
=, Means both efficiency distributions can be considered equal at 5% significance level.

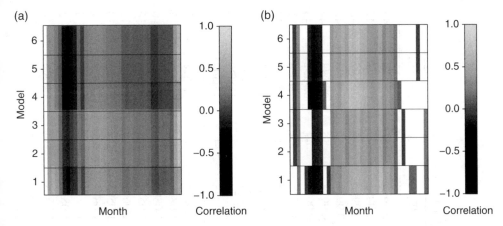

Figure 7.4 Rank correlations by model over time (Models 1: MV+L-NC, 2: MVS+L-NC, 3: MVSK+L-NC, 4: MV+L-C, 5: MVS+L-C, 6: MVSK+L-C). (a) Hedonic versus Morningstar's and (b) Hedonic versus Sharpe.

with other traditional performance ratios. The resulting Kendall's τ statistics are reported in Figure 7.4a and 7.4b. In these figures, a white patch denotes that the correlation is statistically insignificant. The horizontal axis reports the time windows, while the vertical axis indicates with which model the ratings are compared (1: MV+L-NC, 2: MVS+L-NC, 3: MVSK+L-NC, 4: MV+L-C, 5: MVS+L-C, 6: MVSK+L-C).

Starting with Figure 7.4a, most of the time the hedonic ratings delivered by the convex models (models 4–6 in the figure) obtain a negative correlation coefficient when compared to the Morningstar ratings. This result is to some extent expected, since the lower the hedonic ratings the better the MF efficiency levels. Thus, if our approach delivers ratings similar to the Morningstar stars (which increases in performance level), then one should observe a negative correlation coefficient. This is actually the case for most of the periods considered in the backtesting time window, except month 1 (October 2008) and months 35 and 36 (September and October 2011). However, this rather systematic negative correlation is not evident when analyzing nonconvex models. Turning now to Figure 7.4b, that is, the correlation between the hedonic ratings and the Sharpe ratios, no clear relationship can be discerned at all.

Thus, considering the potential impacts of these correlations on the backtesting strategy being tested, one could imagine that significant differences might be observed when using nonconvex models in conjunction with the Morningstar ratings, or when comparing with the Sharpe ratios.

7.4.3   Backtesting results for 48 different strategies

First, we start out by commenting on the first two Tables 7.3 and 7.4 containing terminal values for the different backtesting portfolio strategies in terms of the unity initial wealth without and with transaction costs respectively. Thus, terminal wealth is expressed as a percentage of unity initial wealth. Thereafter, we present some tables evaluating the same strategies in terms of a series of traditional financial performance measures. All of these tables share the same basic structure. The different sample sizes are in the column heading. The different names of the backtesting portfolio strategies following the coding defined in

Table 7.3 Terminal wealth and rank (R) when investing unity for 48 backtesting scenarios without transaction costs.

	MF(10)	R	MF(20)	R	MF(30)	R	MF(40)	R	MF(50)	R
MV+L-C to NoOpt	0.9751	27	0.9999	28	0.9828	47	0.9679	48	0.9600	48
MVS+L-C to NoOpt	1.0050	23	0.9931	30	1.0106	34	1.0035	43	0.9871	47
MVSK+L-C to NoOpt	0.9886	24	1.0135	23	0.9938	43	0.9996	44	0.9892	46
MV+L-NC to NoOpt	1.0151	22	0.9981	29	0.9889	44	0.9978	45	0.9993	43
MVS+L-NC to NoOpt	0.9659	31	0.9821	35	0.9889	45	0.9927	46	0.9983	44
MVSK+L-NC to NoOpt	0.9741	28	0.9751	37	0.9730	48	0.9922	47	0.9932	45
Stars to NoOpt	1.0803	5	1.1344	1	1.0935	1	1.1225	1	1.1094	1
Sharpe to NoOpt	1.0757	6	1.0507	11	1.0518	23	1.0614	14	1.0437	27
Sortino to NoOpt	1.0371	15	1.0630	7	1.0582	20	1.0504	21	1.0391	34
Omega to NoOpt	1.0206	18	1.0155	22	1.0206	27	1.0267	35	1.0238	42
Treynor to NoOpt	1.0583	9	1.0215	20	1.0327	25	1.0279	34	1.0439	26
Kappa to NoOpt	1.0180	21	1.0338	19	1.0265	26	1.0345	32	1.0407	32
MV+L-C to MV	1.0844	3	1.0752	4	1.0680	14	1.0678	12	1.0628	12
MVS+L-C to MV	1.0881	2	1.1038	3	1.0810	6	1.0846	5	1.0723	7
MVSK+L-C to MV	1.1101	1	1.1085	2	1.0815	5	1.0894	2	1.0741	6
MV+L-NC to MV	1.0639	8	1.0505	13	1.0447	24	1.0524	18	1.0568	16
MVS+L-NC to MV	1.0383	14	1.0207	21	1.0798	7	1.0695	10	1.0628	14
MVSK+L-NC to MV	1.0389	13	1.0029	26	1.0899	2	1.0535	17	1.0655	10
MV+L-C to MVS	1.0844	4	1.0752	6	1.0680	14	1.0678	13	1.0628	12
MVS+L-C to MVS	0.9738	30	1.0443	16	1.0636	16	1.0881	3	1.0840	4
MVSK+L-C to MVS	1.0341	16	1.0348	18	1.0605	18	1.0831	7	1.0846	2
MV+L-NC to MVS	1.0510	10	1.0505	13	1.0573	22	1.0524	18	1.0568	16
MVS+L-NC to MVS	0.9659	32	0.9822	34	1.0749	8	1.0696	8	1.0558	18
MVSK+L-NC to MVS	0.9803	26	0.9643	38	1.0892	4	1.0561	16	1.0656	8

MV+L-C to MVSK	1.0754	7	1.0752	4	1.0680	14	1.0678	11	1.0628	12
MVS+L-C to MVSK	0.9738	30	1.0443	16	1.0636	17	1.0881	4	1.0840	5
MVSK+L-C to MVSK	1.0341	17	1.0348	18	1.0605	19	1.0831	6	1.0846	2
MV+L-NC to MVSK	1.0510	10	1.0505	13	1.0573	21	1.0524	20	1.0568	16
MVS+L-NC to MVSK	0.9659	32	0.9822	34	1.0749	8	1.0696	8	1.0558	18
MVSK+L-NC to MVSK	0.9803	26	0.9643	38	1.0892	4	1.0561	15	1.0656	8
Stars to MV	1.0397	12	1.0595	10	1.0700	12	1.0363	31	1.0400	33
Stars to MVS	1.0181	20	1.0598	8	1.0712	11	1.0380	27	1.0425	28
Stars to MVSK	1.0181	20	1.0598	8	1.0712	10	1.0380	28	1.0425	28
Sharpe to MV	0.9576	34	0.9561	44	1.0071	35	1.0257	37	1.0280	41
Sharpe to MVS	0.9569	36	0.9590	43	1.0037	38	1.0202	41	1.0349	38
Sharpe to MVSK	0.9569	36	0.9590	42	1.0037	38	1.0162	42	1.0349	37
Sortino to MV	0.9340	43	0.9500	48	1.0054	36	1.0266	36	1.0307	40
Sortino to MVS	0.9481	38	0.9500	46	1.0023	40	1.0217	39	1.0375	36
Sortino to MVSK	0.9481	38	0.9500	46	1.0023	40	1.0217	40	1.0375	35
Omega to MV	0.9460	39	0.9558	45	0.9834	46	1.0248	38	1.0339	39
Omega to MVS	0.9409	40	0.9614	40	0.9978	41	1.0366	30	1.0409	31
Omega to MVSK	0.9409	40	0.9614	41	0.9978	42	1.0366	29	1.0409	30
Treynor to MV	0.9319	44	1.0017	27	1.0194	30	1.0396	24	1.0448	24
Treynor to MVS	0.9305	48	1.0118	24	1.0195	28	1.0477	23	1.0529	22
Treynor to MVSK	0.9305	48	1.0118	24	1.0195	28	1.0477	22	1.0529	22
Kappa to MV	0.9363	42	0.9755	36	1.0194	31	1.0292	33	1.0444	25
Kappa to MVS	0.9315	46	0.9852	32	1.0118	33	1.0394	25	1.0547	21
Kappa to MVSK	0.9315	46	0.9852	32	1.0118	32	1.0394	26	1.0547	20

Table 7.4 Terminal wealth and rank (R) when investing unity for 48 backtesting scenarios with transaction costs.

	MF(10)	R	MF(20)	R	MF(30)	R	MF(40)	R	MF(50)	R
MV+L-C to NoOpt	0.8447	15	0.9102	7	0.9067	9	0.9022	6	0.8946	7
MVS+L-C to NoOpt	0.8840	8	0.9233	6	0.9523	1	0.9492	3	0.9402	3
MVSK+L-C to NoOpt	0.8308	18	0.9416	4	0.9421	2	0.9572	2	0.9437	2
MV+L-NC to NoOpt	0.9294	5	0.9246	5	0.9223	4	0.9168	4	0.9166	4
MVS+L-NC to NoOpt	0.8146	19	0.8810	12	0.9010	11	0.8948	11	0.9078	5
MVSK+L-NC to NoOpt	0.8436	16	0.8824	11	0.8965	12	0.9025	12	0.9042	6
Stars to NoOpt	0.9531	2	1.0070	1	0.9371	3	0.9575	3	0.9562	1
Sharpe to NoOpt	0.9352	3	0.8254	15	0.8409	20	0.8745	16	0.8597	11
Sortino to NoOpt	0.8853	7	0.8521	13	0.8527	19	0.8629	17	0.8527	17
Omega to NoOpt	0.8473	14	0.8029	19	0.8164	25	0.8345	24	0.8459	18
Treynor to NoOpt	0.8962	6	0.7926	20	0.8332	21	0.8309	25	0.8551	16
Kappa to NoOpt	0.8337	17	0.8257	14	0.8227	24	0.8557	18	0.8585	12
MV+L-C to MV	0.8563	12	0.8079	16	0.8152	27	0.8109	29	0.8023	28
MVS+L-C to MV	0.9557	1	0.9538	3	0.9022	10	0.8789	15	0.8559	15
MVSK+L-C to MV	0.9297	4	0.9563	2	0.9110	6	0.8850	14	0.8645	10
MV+L-NC to MV	0.6961	31	0.8843	9	0.8077	31	0.7693	32	0.7879	32
MVS+L-NC to MV	0.8738	9	0.6788	25	0.8854	13	0.8909	12	0.8793	9
MVSK+L-NC to MV	0.8649	10	0.5999	33	0.8822	16	0.8896	13	0.8842	8
MV+L-C to MVS	0.8627	11	0.8079	18	0.8152	27	0.8109	30	0.8023	30
MVS+L-C to MVS	0.7339	24	0.7798	22	0.8105	29	0.8264	26	0.8383	24
MVSK+L-C to MVS	0.6540	34	0.7655	24	0.7868	32	0.8397	22	0.8393	20
MV+L-NC to MVS	0.7168	29	0.8843	9	0.8233	23	0.7693	32	0.7879	32
MVS+L-NC to MVS	0.7937	22	0.6706	26	0.8671	18	0.8926	10	0.8393	22
MVSK+L-NC to MVS	0.8079	20	0.6277	28	0.8835	14	0.8952	7	0.8562	14
MV+L-C to MVSK	0.8496	13	0.8079	16	0.8152	27	0.8109	28	0.8023	28

MVS+L-C to MVSK	0.7339	24	0.7798	21	0.8105	30	0.8264	27	0.8383	24
MVSK+L-C to MVSK	0.6540	35	0.7655	23	0.7868	33	0.8397	23	0.8393	20
MV+L-NC to MVSK	0.7168	28	0.8843	9	0.8233	22	0.7693	33	0.7879	32
MVS+L-NC to MVSK	0.7937	22	0.6706	27	0.8671	18	0.8926	10	0.8393	22
MVSK+L-NC to MVSK	0.8079	20	0.6277	28	0.8835	14	0.8952	8	0.8562	14
Stars to MV	0.7093	30	0.6222	30	0.9145	5	0.8526	19	0.8418	19
Stars to MVS	0.7197	26	0.6126	32	0.9072	8	0.8487	20	0.8350	26
Stars to MVSK	0.7197	26	0.6126	31	0.9072	7	0.8487	21	0.8350	26
Sharpe to MV	0.6551	33	0.5806	42	0.6055	40	0.6642	44	0.6724	48
Sharpe to MVS	0.6435	38	0.5848	37	0.5893	42	0.6611	47	0.6809	45
Sharpe to MVSK	0.6435	38	0.5848	38	0.5893	42	0.6566	48	0.6809	46
Sortino to MV	0.6095	45	0.5873	36	0.5728	43	0.6645	43	0.6822	44
Sortino to MVS	0.6109	44	0.5840	40	0.5687	44	0.6628	45	0.6906	40
Sortino to MVSK	0.6109	44	0.5840	40	0.5687	44	0.6628	46	0.6906	41
Omega to MV	0.6646	32	0.4865	48	0.5182	48	0.6646	42	0.6744	47
Omega to MVS	0.6491	36	0.4960	46	0.5438	46	0.6782	39	0.6825	42
Omega to MVSK	0.6491	36	0.4960	47	0.5438	46	0.6782	40	0.6825	43
Treynor to MV	0.5937	46	0.5842	39	0.6086	39	0.6992	36	0.7122	38
Treynor to MVS	0.5827	48	0.5939	34	0.6136	38	0.7110	34	0.7210	34
Treynor to MVSK	0.5827	48	0.5939	34	0.6136	38	0.7110	35	0.7210	34
Kappa to MV	0.6336	40	0.5702	45	0.6244	36	0.6765	41	0.7063	39
Kappa to MVS	0.6216	42	0.5787	44	0.6282	35	0.6869	38	0.7177	36
Kappa to MVSK	0.6216	42	0.5787	44	0.6282	34	0.6869	38	0.7177	37

Section 7.3.1 are in the first column. On the right-hand side of each terminal value as percentage of initial wealth one finds a rank in descending order over all possible backtesting strategies within the same column.

Starting with the 'EWP without optimization' strategies, the following tentative conclusions can be drawn. First, the frontier MF ratings are performing rather poorly without transaction costs (rank 22 at best and rank 48 at worst), with a tendency of performance to worsen as the size of the number of included MF increases. Second, the traditional financial performance measures slightly outperform these frontier MF ratings in terms of rankings. In terms of terminal wealth, the results are even more marked: traditional financial performance measures all gain money, while some of the frontier MF ratings lose money overall. Third, the Morningstar ratings systematically outperform all other strategies, except for the smallest sample size (column MF(10)).

In the case of the net terminal value with transaction costs, this picture changes rather drastically. First, the frontier MF ratings are performing rather well in terms of ranks (rank 1 at best and rank 19 at worst). There is now some tendency for performance to improve with the size of the number of included MF. Except for the smallest size classes (MF(10) and MF(20)), the average ranks are very decent to rather exceptional. In particular, the best ranks are obtained by some multimoment (in particular, MVS and MVSK) convex frontier MF ratings: for MF(30), these models occupy ranks 1 and 2. The nonconvex model manages to obtain somewhat lower ranks for the larger size classes (rank 4 and 5 at best). Second, in terms of terminal wealth, all strategies lose money in the end, but traditional financial performance measures now tend to do worse than frontier MF ratings. Third, the Morningstar ratings still tend to outperform all other strategies, except for two sample sizes (i.e., MF(10) and MF(30)). Furthermore, for one sample size (MF(20)), Morningstar is the only strategy generating a positive net terminal gain at all.

One probable explanation for the contrasting results when comparing frontier MF ratings vs traditional financial performance measures is that the former yield relatively more stable portfolios. This seems to result in an edge in terms of economizing transaction costs. Perhaps, due to their multidimensional nature compared to the two-dimensional ratio nature of traditional financial measures, these frontier MF ratings seem intrinsically more stable. This probably requires further exploration.

Switching now to 'EWP with optimization' strategies, one can observe the following. Without transaction costs, the frontier MF ratings are clearly outperforming by large the traditional financial performance measures. There is always a whole series of MF rating strategies that also beat any of the Morningstar-based strategies. Morningstar is now even beaten in a traditional MV world. This basic result holds fundamentally true over all subsamples.

When including transaction costs to look at net terminal wealth, the same basic conclusions can be drawn. However, there are some particular differences worth highlighting. First, the best ranks are obtained by some multimoment (in particular, MVS and MVSK) convex frontier MF ratings for the two smallest size classes (ranks 1–4 at best) and by the same nonconvex models for the two largest size classes (ranks 7–9 at best). Second, one Morningstar rating strategy beats these frontier MF ratings for the class MF(30) solely in a MV context.

Notice that when comparing optimization strategies without and with transaction costs, there is a noticeable difference in the type of portfolio optimization models that seem to perform well. While without transaction costs, some strategies dominate using either MV, MVS, or MVSK portfolio optimization, with transaction costs only MV-based optimization strategies perform well, except for the MF(40) size class where a nonconvex model combined with MVS and MVSK portfolio optimization yields the best ranks (ranks 7–8 at best). A clear reason for this remarkable difference in results is currently wanting. We return to this issue in Section 7.4.4.

Next, we turn to the evaluation of the same basic portfolio strategies in terms of some traditional financial performance measures in Table 7.5 and Table 7.6 reporting the Sharpe and the Omega ratios, respectively. While the Sharpe ratio is a standard MV-oriented performance measure, the Omega ratio by definition aims at capturing substantial parts of the whole return distribution. It can be interpreted as a ratio of gains over losses relative to some loss threshold (see Bacon, 2008 for details). Therefore, it should ideally be able to function in a nonnormal portfolio world. Notice that while the portfolio strategies have been phrased in terms of Sharpe, Sortino, Omega, Treynor, and Kappa ratios, we limit the evaluation of all 48 backtesting scenarios due to lack of space to just a selection of two representative ratios. Fundamentally, the other ratios do not deliver any significant differences in basic conclusions.

Again starting with the 'EWP without optimization' strategies, according to the Sharpe ratio one can make the following observations. First, the Morningstar ratings overall beat the frontier MF ratings, though not in all size classes. In particular, in two size classes, convex frontier MF rating models do better. Second, the traditional financial performance measures do worse than the frontier MF ratings, except perhaps in the first size class. Notice that even some of the nonconvex frontier MF rating models perform decently well, often just behind their convex counterparts. Remark that the strategy based on the Sharpe ratio is beaten by both the Morningstar rating and the frontier MF rating models, except in the smallest size class.

Now considering the 'EWP with optimization' strategies, one immediately can reach the following conclusions. First, these strategies in general do poorer than the nonoptimized strategies, except for the smallest size classes. Second, in general, the frontier MF ratings do a better job than both the traditional financial performance measure and the Morningstar ratings.

Turning now finally to the Omega ratio, starting with the 'EWP without optimization' strategies one can conclude the following. First, the ranks are identical to the ones resulting from the Sharpe ratio, except for the MF(40) size class. Hence, the same basic conclusions can be transposed. One remark to make is that the strategy based on the Omega ratio is beaten by both the Morningstar rating and the frontier MF rating models, except again in the smallest size class for some of the frontier MF rating models. Changing focus to the 'EWP with optimization' strategies, one can quickly reach consensus that the findings very closely follow the ones obtained for the Sharpe ratio previously.

Hence, both traditional financial performance measures yield close to the same conclusions. But, since the strategies constructed on both of these traditional financial performance measures do worse than most of the other strategies, one may wonder to what extent these traditional financial performance measures are really suitable to gauge the performance of such wildly varying portfolio strategies. Perhaps, this question merits thorough further reflection.

7.4.4   Backtesting results for MF rating models: Some plausible explanations

This section aims to shed light on two plausible mechanisms driving some of the empirical results reported in Section 7.4.3. First, one may wonder why frontier MF ratings seem to do rather well compared to well-established financial performance measures that have a long

Table 7.5 Performance backtesting using the Sharpe ratios and rank (R) for 48 backtesting scenarios.

	MF(10)	R	MF(20)	R	MF(30)	R	MF(40)	R	MF(50)	R
MV+L-C to NoOpt	−0.0846	12	−0.0476	7	−0.0495	5	−0.0510	6	−0.0552	7
MVS+L-C to NoOpt	−0.0565	8	−0.0391	6	−0.0244	1	−0.0263	3	−0.0311	3
MVSK+L-C to NoOpt	−0.0795	9	−0.0296	2	−0.0302	2	−0.0222	1	−0.0295	2
MV+L-NC to NoOpt	−0.0359	4	−0.0380	5	−0.0390	4	−0.0426	4	−0.0426	4
MVS+L-NC to NoOpt	−0.0986	18	−0.0627	9	−0.0511	6	−0.0540	7	−0.0476	5
MVSK+L-NC to NoOpt	−0.0809	10	−0.0622	8	−0.0537	7	−0.0506	5	−0.0494	6
Stars to NoOpt	−0.0269	1	0.0039	1	−0.0358	3	−0.0244	2	−0.0250	1
Sharpe to NoOpt	−0.0322	3	−0.0979	14	−0.0859	20	−0.0657	8	−0.0726	8
Sortino to NoOpt	−0.0564	7	−0.0819	13	−0.0796	17	−0.0718	11	−0.0766	11
Omega to NoOpt	−0.0827	11	−0.1102	16	−0.1009	23	−0.0894	19	−0.0831	14
Treynor to NoOpt	−0.0562	6	−0.1192	17	−0.0898	21	−0.0900	20	−0.0758	10
Kappa to NoOpt	−0.0891	14	−0.0993	15	−0.0984	22	−0.0770	16	−0.0750	9
MV+L-C to MV	−0.0945	17	−0.1356	19	−0.1251	28	−0.1273	29	−0.1327	29
MVS+L-C to MV	−0.0292	2	−0.0322	4	−0.0704	12	−0.0861	18	−0.1016	19
MVSK+L-C to MV	−0.0430	5	−0.0305	3	−0.0643	11	−0.0820	17	−0.0957	15
MV+L-NC to MV	−0.2069	43	−0.0796	10	−0.1314	31	−0.1561	32	−0.1432	32
MVS+L-NC to MV	−0.0859	13	−0.2039	27	−0.0736	13	−0.0743	14	−0.0824	13
MVSK+L-NC to MV	−0.0914	16	−0.2608	41	−0.0766	16	−0.0750	15	−0.0791	12
MV+L-C to MVS	−0.0902	15	−0.1356	20	−0.1251	30	−0.1273	30	−0.1327	30
MVS+L-C to MVS	−0.1462	25	−0.1407	22	−0.1215	26	−0.1204	26	−0.1103	22
MVSK+L-C to MVS	−0.1907	32	−0.1469	24	−0.1430	32	−0.1106	24	−0.1105	24
MV+L-NC to MVS	−0.1886	28	−0.0796	12	−0.1193	25	−0.1561	32	−0.1432	32
MVS+L-NC to MVS	−0.1286	23	−0.2034	25	−0.0848	19	−0.0728	13	−0.1120	26
MVSK+L-NC to MVS	−0.1198	20	−0.2374	36	−0.0750	15	−0.0702	9	−0.1002	18

MV+L-C to MVSK	-0.0997	19	-0.1356	18	-0.1251	30	-0.1273	28	-0.1327	28
MVS+L-C to MVSK	-0.1462	24	-0.1407	21	-0.1215	27	-0.1204	27	-0.1103	22
MVSK+L-C to MVSK	-0.1907	33	-0.1469	23	-0.1430	33	-0.1106	25	-0.1105	25
MV+L-NC to MVSK	-0.1886	29	-0.0796	12	-0.1193	24	-0.1561	33	-0.1432	31
MVS+L-NC to MVSK	-0.1286	22	-0.2034	26	-0.0848	18	-0.0728	12	-0.1120	27
MVSK+L-NC to MVSK	-0.1198	21	-0.2374	36	-0.0750	14	-0.0702	10	-0.1002	18
Stars to MV	-0.1943	38	-0.2708	43	-0.0549	8	-0.0938	21	-0.0999	16
Stars to MVS	-0.1906	30	-0.2796	47	-0.0584	10	-0.0951	22	-0.1026	20
Stars to MVSK	-0.1906	30	-0.2796	46	-0.0584	9	-0.0951	23	-0.1026	21
Sharpe to MV	-0.1836	27	-0.2246	30	-0.2278	34	-0.2074	43	-0.2044	47
Sharpe to MVS	-0.1911	34	-0.2225	28	-0.2403	35	-0.2103	47	-0.1980	42
Sharpe to MVSK	-0.1911	34	-0.2225	29	-0.2403	36	-0.2139	48	-0.1980	43
Sortino to MV	-0.2048	40	-0.2326	31	-0.2524	43	-0.2081	44	-0.2020	46
Sortino to MVS	-0.2058	41	-0.2367	34	-0.2556	44	-0.2100	45	-0.1956	40
Sortino to MVSK	-0.2058	42	-0.2367	35	-0.2556	44	-0.2100	46	-0.1956	41
Omega to MV	-0.1824	26	-0.2837	48	-0.3030	48	-0.2056	42	-0.2056	48
Omega to MVS	-0.1928	36	-0.2786	44	-0.2828	46	-0.1957	39	-0.1984	44
Omega to MVSK	-0.1928	37	-0.2786	45	-0.2828	47	-0.1957	40	-0.1984	45
Treynor to MV	-0.2182	46	-0.2406	38	-0.2475	42	-0.1840	36	-0.1784	36
Treynor to MVS	-0.2263	48	-0.2362	32	-0.2443	39	-0.1756	34	-0.1713	35
Treynor to MVSK	-0.2263	47	-0.2362	33	-0.2443	40	-0.1756	35	-0.1713	34
Kappa to MV	-0.1986	39	-0.2629	42	-0.2464	41	-0.2015	41	-0.1882	39
Kappa to MVS	-0.2083	45	-0.2586	39	-0.2440	38	-0.1939	38	-0.1790	37
Kappa to MVSK	-0.2083	44	-0.2586	40	-0.2440	37	-0.1939	38	-0.1790	38

Table 7.6 Performance backtesting using the Omega ratios and rank (R) for 48 backtesting scenarios.

	MF(10)	R	MF(20)	R	MF(30)	R	MF(40)	R	MF(50)	R
MV+L-C to NoOpt	0.7990	12	0.8834	7	0.8797	5	0.8761	5	0.8650	7
MVS+L-C to NoOpt	0.8588	8	0.9029	6	0.9387	1	0.9346	3	0.9229	3
MVSK+L-C to NoOpt	0.8064	10	0.9261	2	0.9248	2	0.9441	1	0.9268	2
MV+L-NC to NoOpt	0.9096	4	0.9044	5	0.9016	4	0.8942	4	0.8947	4
MVS+L-NC to NoOpt	0.7753	19	0.8493	9	0.8745	6	0.8689	6	0.8834	5
MVSK+L-NC to NoOpt	0.8104	9	0.8504	8	0.8669	7	0.8755	7	0.8782	6
Stars to NoOpt	0.9303	1	1.0103	1	0.9108	3	0.9385	3	0.9367	1
Sharpe to NoOpt	0.9193	3	0.7729	14	0.7988	18	0.8437	8	0.8294	8
Sortino to NoOpt	0.8606	7	0.8043	13	0.8118	17	0.8310	13	0.8209	11
Omega to NoOpt	0.8041	11	0.7557	16	0.7663	23	0.7909	20	0.8045	14
Treynor to NoOpt	0.8658	6	0.7292	17	0.7924	21	0.7921	19	0.8216	10
Kappa to NoOpt	0.7955	14	0.7710	15	0.7727	22	0.8189	16	0.8233	9
MV+L-C to MV	0.7891	16	0.7074	18	0.7264	30	0.7214	29	0.7129	29
MVS+L-C to MV	0.9287	2	0.9221	4	0.8372	12	0.8032	18	0.7701	18
MVSK+L-C to MV	0.8926	5	0.9258	3	0.8517	11	0.8123	17	0.7830	15
MV+L-NC to MV	0.5629	45	0.8177	10	0.7171	31	0.6718	32	0.6946	32
MVS+L-NC to MV	0.7929	15	0.5574	25	0.8229	13	0.8284	14	0.8105	13
MVSK+L-NC to MV	0.7807	17	0.4764	42	0.8157	16	0.8249	15	0.8170	12
MV+L-C to MVS	0.7972	13	0.7074	20	0.7264	28	0.7214	30	0.7129	30
MVS+L-C to MVS	0.6836	24	0.6994	22	0.7276	26	0.7351	26	0.7491	26
MVSK+L-C to MVS	0.6168	28	0.6757	24	0.6944	32	0.7550	24	0.7504	22
MV+L-NC to MVS	0.5806	37	0.8177	12	0.7396	25	0.6718	32	0.6946	32
MVS+L-NC to MVS	0.6877	23	0.5518	26	0.7962	20	0.8313	12	0.7498	24
MVSK+L-NC to MVS	0.7094	20	0.5038	38	0.8173	15	0.8351	9	0.7740	16

MV+L-C to MVSK	0.7787	18	0.7074	19	0.7264	28	0.7214	28	0.7129	28
MVS+L-C to MVSK	0.6836	25	0.6994	21	0.7276	27	0.7351	27	0.7491	26
MVSK+L-C to MVSK	0.6168	29	0.6757	23	0.6944	33	0.7550	25	0.7504	23
MV+L-NC to MVSK	0.5806	38	0.8177	12	0.7396	24	0.6718	33	0.6946	31
MVS+L-NC to MVSK	0.6877	22	0.5518	27	0.7962	19	0.8313	11	0.7498	25
MVSK+L-NC to MVSK	0.7094	21	0.5038	38	0.8173	14	0.8351	10	0.7740	16
Stars to MV	0.5751	42	0.4313	46	0.8643	8	0.7789	21	0.7650	19
Stars to MVS	0.5845	36	0.4263	48	0.8557	10	0.7753	22	0.7585	20
Stars to MVSK	0.5845	36	0.4263	47	0.8557	9	0.7753	23	0.7585	21
Sharpe to MV	0.6188	27	0.5336	30	0.5212	34	0.5692	43	0.5635	47
Sharpe to MVS	0.6094	30	0.5360	28	0.5021	38	0.5625	47	0.5721	42
Sharpe to MVSK	0.6094	30	0.5360	29	0.5021	39	0.5558	48	0.5721	43
Sortino to MV	0.5799	39	0.5205	31	0.4871	43	0.5685	44	0.5659	46
Sortino to MVS	0.5792	40	0.5128	35	0.4834	44	0.5636	45	0.5745	40
Sortino to MVSK	0.5792	40	0.5128	36	0.4834	44	0.5636	46	0.5745	41
Omega to MV	0.6220	26	0.4351	45	0.4126	48	0.5693	42	0.5623	48
Omega to MVS	0.6070	32	0.4425	43	0.4264	46	0.5820	39	0.5715	44
Omega to MVSK	0.6070	33	0.4425	44	0.4264	47	0.5820	40	0.5715	45
Treynor to MV	0.5540	46	0.5141	34	0.5015	42	0.6004	36	0.6134	36
Treynor to MVS	0.5452	47	0.5198	32	0.5020	40	0.6126	34	0.6236	35
Treynor to MVSK	0.5452	48	0.5198	32	0.5020	40	0.6126	35	0.6236	34
Kappa to MV	0.5871	34	0.4835	41	0.5073	35	0.5768	41	0.5939	39
Kappa to MVS	0.5747	44	0.4878	39	0.5066	37	0.5855	38	0.6075	37
Kappa to MVSK	0.5747	43	0.4878	40	0.5066	36	0.5855	38	0.6075	38

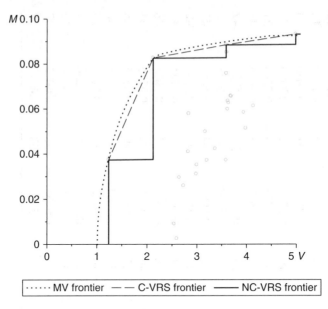

Figure 7.5 Visualization of MV-frontier based on the 40 best MF and the convex and nonconvex frontiers determined earlier in Figure 7.1.

tradition in finance. The second question is why some strategies dominate using either MV, MVS, or MVSK portfolio optimization without transaction costs, while only MV-based optimization strategies seem to perform well when including transaction costs.

The answer to the first question is probably situated in the approximative nature of these frontier estimators compared to some of the underlying portfolio models these are related to in the second stage. For illustrative purpose, Figure 7.5 shows part of the MV portfolio frontier obtained from selecting the 40 best performing MF from the main database. For this example, the MF have been ranked according to the convex shortage function. For comparison, also the convex and nonconvex frontiers obtained earlier using models (7.1) and (7.2) are included in the image.

Clearly, for this example, the MV portfolio frontier and the convex frontier of MF correspond quite well. By contrast, the fit with the nonconvex frontier model is a bit poorer with regard to this convex MV portfolio frontier.

In general, however, the goodness-of-fit of convex and nonconvex frontier MF rating models is an empirical issue. While it is hard or close to impossible to illustrate, one may speculate that a similar argument can be developed to explain why in some cases a multimoment nonconvex rating model can outperform a convex model. While MV portfolio frontiers are convex, it is well-known that MVS and MVSK portfolio frontiers can be highly nonconvex (see, e.g., Kerstens, Mounir, and Van de Woestyne (2011a) for illustrations of the MVS case). In such nonconvex portfolio worlds, the nonconvex rating models may well prove to have better approximation capacity than their currently more widely applied, traditional convex rating models. Obviously, it would be highly desirable to find ways to shed more light on this conjecture.

The answer to the second question is not straightforward. We currently have at best some elements of an answer. First, recall that the period under analysis is characterized by strong

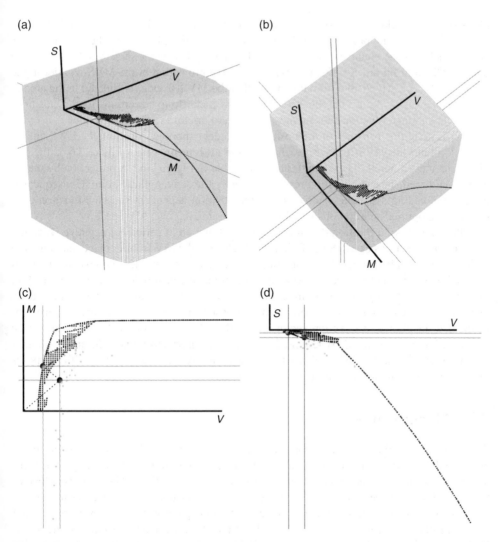

Figure 7.6 Visualization of MVS-frontier based on the 40 best MF. (a) Overview 1, (b) overview 2, (c) MV projection of the MVS model, and (d) VS projection of the MVS model.

negative skewness for most MF (recall Figure 7.2). Second, in periods with strong negative skewness, the way in which the position-dependent shortage function projects relative to multimoment portfolio models seems to lead to optimal frontier portfolios close to the ones situated on a basic MV portfolio frontier. We can illustrate this plausible explanation with the help of Figure 7.6.

Figure 7.6 presents different views of the MVS-frontier corresponding to the same 40 best performing MF used in Figure 7.5. Figure 7.6a and b shows two general views of the MVS-frontier. The black-colored part of the frontier refers to the strongly efficient

MVS-frontier, while the gray-colored part represents the weakly efficient MVS-frontier.[8] Figure 7.6c shows the projection of the MVS-frontier in the MV-plane. The MV-frontier is clearly visible as the boundary of the black-colored area. This is the same MV-frontier as the one visualized in Figure 7.5. Figure 7.6d presents the projection of the MVS-frontier in the VS-plane. It is clearly visible from this figure that the MVS-frontier is situated in the area of negative skewness. As a consequence, projecting a portfolio according to the position-dependent shortage function ends up most likely onto or close to the MV-frontier.

To illustrate this behavior of the position-dependent shortage function, the EWP and its projection onto the MVS-frontier are visualized in all four images of Figure 7.6 as the two big circle-shaped points. These points are best visible in Figure 7.6c. A projection by means of the MV-based shortage function results in exactly the same optimal point. Consequently, in this case adding a skewness constraint to the position-dependent shortage function does not alter the optimal solution obtained by merely optimizing over MV-space.

Thus, the use of a position-dependent shortage function in periods of negative skewness can lead to projections of MVS portfolio models close to the MV-frontier plane, making both models close to indistinguishable in practice. Consequently, the weak or nonexisting gains of using a higher-moment instead of a MV portfolio model are insufficient in combination with the seemingly slightly higher transaction costs to guarantee any strategic advantage. Obviously, this reasoning could benefit from further scrutiny. While it has already been established that the choice of direction vector affects the relative ranking of MV portfolios (see Kerstens, Mounir, and Van de Woestyne, 2012), the current backtesting results are yet another reason to start further exploring alternative choices of direction vector for specific purposes (*in casu*, coping with a period of negative skewness).

7.5 Conclusions

We basically test the traditional convex vs nonconvex frontier MF rating models in an extensive backtesting framework against, on the one hand, traditional financial performance measures and, on the other hand, the three-year Morningstar ratings. The basic conclusions from this first intensive and extensive backtesting analysis can be summarized as follows.

First, without portfolio optimization strategies and ignoring transaction costs, frontier MF ratings are performing poorly compared to traditional financial performance measures. With transaction costs, the previous conclusion reverses. However, in both cases the Morningstar ratings tend to outperform all other strategies, except for some of the smaller sample sizes. In conjunction with portfolio optimization strategies without and with transaction costs, the frontier MF ratings clearly outperform traditional financial performance measures and even do better than the Morningstar-based strategies (except for one particular sample size). Overall, naïve diversification strategies seem to be working better than strategies involving portfolio optimization. However, it is clear that especially MVS and MVSK multimoment portfolio models may currently suffer from the choice of period of analysis that is deeply marked by the financial crisis. Thus, for the time being, our results fall perfectly in line with those of DeMiguel, Garlappi, and Uppal (2009) and others.

Second, convex frontier MF ratings seem to work better than nonconvex ones, though the capability of the latter to better approximate nonconvex higher-moment portfolio models

---

[8] We refer the reader to Kerstens, Mounir, and Van de Woestyne (2011a) for more information on these notions.

remains to be further investigated in view of the difficulty of multimoment portfolio models to operate properly during the current sample period.

Thus, somewhat in contrast to Blake and Morey (2000) and Kräussl and Sandelowsky (2007), our results do offer some benefit of the doubt to the three-year Morningstar ratings to pick MF shares that prove to perform decently in the nearby future, relatively speaking. But, some new alternative gauges founded on frontier estimation prove to function at least equally well. In particular, there seems to be some promise in combining multimoment frontier-based MF ratings in conjunction with corresponding multimoment portfolio optimization models using the same basic methodology. Obviously, there is plenty of room for further refinements and intensive backtesting to substantiate this preliminary conclusion.

In view of some of the difficulties encountered and conjectures made, there is a series of methodological refinements one can think about. First, while we have focused on standard moments in the entire analysis, the Morningstar rating is at least partially influenced by some elements of a lower partial-moments approach. Thus, the whole analysis may eventually benefit from being redone using partial rather than standard moments.

Second, while the frontier MF ratings as well as the traditional financial performance measures are applied to the whole sample, the Morningstar ratings in fact apply only to the same Morningstar category. There happen to be several Morningstar categories in our sample. Furthermore, these Morningstar categories evolve over time following changes in the asset composition of the MF. To have a more refined comparison, one could envision computing frontier MF ratings and traditional financial performance measures on the same Morningstar categories as they evolve over time. Obviously, apart from complicating the analysis, this may also compromise the nonparametric frontier MF rating estimations depending on the size of these Morningstar categories, because of the curse of dimensionality characteristic for nonparametric estimators.

Third, while an equally weighted portfolio is a standard approach to backtest in a MV world, it remains an open question whether this starting point is useful in a nonnormal MVS or MVSK portfolio approach. This question is somewhat related to the issue of the choice of projection direction developed in Section 7.4.4.

In terms of further refinements of the backtesting setup, one can think of the following issues. First, it could be important to select a sample period that includes both clear boom and bust periods. Second, some of the doubts raised with respect to the traditional financial performance measures in relation to the multimoment portfolio models certainly require further investigation. We have hardly been able to scratch the surface in this chapter. Third, the current results seem to vary somewhat in nonmonotonous ways across sample sizes. This is currently very hard to explain and probably merits some deeper investigation.

To some extent, one could also imagine that this experience could inspire another related literature using frontier models for asset selection purposes rather than MF rating (e.g., Abad, Thore, and Laffarga, 2004; Alam and Sickles, 1998; and Nguyen and Swanson, 2009). It clearly remains an open question whether the current approach attempting to define a superfund starting from some subsets of well-rated candidate MF offers the same scope as composing a simple portfolio based on some well-rated candidate assets. A much more fundamental methodological issue is to what extent the use of two-stage approaches, with some frontier model in the first stage and another kind of portfolio selection model in the second stage, can be justified in view of some criticisms of such approaches in the frontier literature (Simar and Wilson, 2007).

Acknowledgments

We thank the editor of this volume for valuable comments. Ignace Van de Woestyne acknowledges financial support by the National Bank of Belgium.

References

Abad, C., Thore, S.A., and Laffarga, J. (2004) Fundamental analysis of stocks by two-stage DEA. *Managerial and Decision Economics*, **25** (5), 231–241.

Alam, I. and Sickles, R. (1998) The relationship between stock market returns and technical efficiency innovations: evidence from the US airline industry. *Journal of Productivity Analysis*, **9** (1), 35–51.

Annaert, J., van den Broeck, J., and Vander Vennet, R. (2003) Determinants of mutual fund underperformance: a Bayesian stochastic frontier approach. *European Journal of Operational Research*, **151** (3), 617–632.

Bacon, C. (2008) *Practical Portfolio Performance Measurement and Attribution*, 2nd edn, John Wiley & Sons, Ltd, Chichester.

Basso, A. and Funari, S. (2001) A data envelopment analysis approach to measure the mutual fund performance. *European Journal of Operational Research*, **135** (3), 477–492.

Basso, A. and Funari, S. (2003) Measuring the performance of ethical mutual funds: a DEA approach. *Journal of the Operational Research Society*, **54** (5), 521–531.

Blake, C. and Morey, M. (2000) Morningstar ratings and mutual fund performance. *Journal of Financial and Quantitative Analysis*, **35** (3), 451–483.

Blake, D. (1990) Portfolio behaviour and asset pricing in a characteristics framework. *Scottish Journal of Political Economy*, **37** (4), 343–359.

Brandouy, O., Briec, W., Kerstens, K., and Van de Woestyne, I. (2010a) Portfolio performance gauging in discrete time using a Luenberger productivity indicator. *Journal of Banking & Finance*, **34** (8), 1899–1910.

Brandouy, O., Kerstens, K., and Van de Woestyne, I. (2010b) Exploring bi-criteria versus multi-dimensional lower partial moment portfolio models. *International Journal of Technology, Modelling and Management*, **1** (1), 25–39.

Brandouy, O., Mathieu, P., and Veryzhenko, I. (2012) Optimal portfolio diversification? A multi-agent ecological competition analysis, in *Highlights in Practical Applications of Agents and Multi-Agent Systems. 10th International Conference on Practical Applications of Agents and Multi-Agent Systems*, vol. **156** of Advances in Intelligent and Soft Computing (eds J. Bajo Pérez, J. Corchado, E. Adam, A. Ortega, M. Moreno, E. Navarro, B. Hirsch, H. Lopes Cardoso, V. Julián, M. Sánchez, and P. Mathieu), Springer, Berlin, pp. 323–332.

Briec, W. and Kerstens, K. (2009) Multi-horizon Markowitz portfolio performance appraisals: a general approach. *Omega*, **37** (1), 50–62.

Briec, W. and Kerstens, K. (2010) Portfolio selection in multidimensional general and partial moment space. *Journal of Economic Dynamics and Control*, **34** (4), 636–656.

Briec, W., Kerstens, K., and Jokung, K. (2007) Mean-variance-skewness portfolio performance gauging: a general shortage function and dual approach. *Management Science*, **53** (1), 135–149.

Briec, W., Kerstens, K., and Lesourd, J. (2004) Single-period Markowitz portfolio selection, performance gauging, and duality: a variation on the Luenberger shortage function. *Journal of Optimization Theory and Applications*, **120** (1), 1–27.

Broihanne, M.H., Merli, M., and Roger, P. (2008) On the robustness of mutual funds ranking with an index of relative efficiency. *Bankers, Markets & Investors*, **94**, 32–43.

Campbell, S. (2005) A review of backtesting and backtesting procedures. Finance and Economics discussion series no. 2005-21. Federal Reserve Board, Washington.

Cantaluppi, L. and Hug, R. (2000) Efficiency ratio: a new methodology for performance measurement. *Journal of Investing*, **9** (2), 1–7.

Chang, K.P. (2004) Evaluating mutual fund performance: an application of minimum convex input requirement set approach. *Computers and Operations Research*, **31** (6), 929–940.

Choi, Y. and Murthi, B. (2001) Relative performance evaluation of mutual funds: a nonparametric approach. *Journal of Business Finance and Accounting*, **28** (7–8), 853–876.

Clark, E., Jokung, O., and Kassimatis, K. (2011) Making inefficient market indices efficient. *European Journal of Operational Research*, **209** (1), 83–89.

DeMiguel, V., Garlappi, L., and Uppal, R. (2009) Optimal versus naive diversification: how inefficient is the 1/n portfolio strategy? *Review of Financial Studies*, **22**, 1915–1953.

Dodds, C. (1986) Portfolio modelling and the characteristics approach. *Managerial Finance*, **12** (3), 16–18.

Edirisinghe, N. and Zhang, X. (2007) Generalized DEA model of fundamental analysis and its application to portfolio optimization. *Journal of Banking & Finance*, **31** (11), 3311–3335.

Edirisinghe, N. and Zhang, X. (2010) Input/output selection in DEA under expert information, with application to financial markets. *European Journal of Operational Research*, **207** (3), 1669–1678.

Friedman, J. (1983) *Oligopoly Theory*, Cambridge University Press, Cambridge.

Galagedera, D. and Silvapulle, P. (2002) Australian mutual fund performance appraisal using data envelopment analysis. *Managerial Finance*, **28** (9), 60–73.

Glawischnig, M. and Sommersguter-Reichmann, M. (2010) Assessing the performance of alternative investments using non-parametric efficiency measurement approaches: is it convincing? *Journal of Banking & Finance*, **34** (2), 295–303.

Gregoriou, G. (ed.) (2007) *Performance of Mutual Funds: An International Perspective*, Palgrave, New York.

Haslem, J. and Scheraga, C. (2003) Data envelopment analysis of Morningstar's large-cap mutual funds. *Journal of Investing*, **12** (4), 41–48.

Heffernan, S. (1992) A computation of interest equivalences for non-price features of bank products. *Journal of Money, Credit and Banking*, **24** (2), 162–172.

Jurczenko, E. and Maillet, B. (2006) The four-moment capital asset pricing model: between asset pricing and asset allocation, in *Multi-Moment Asset Allocation and Pricing Models* (eds E. Jurczenko and B. Maillet), John Wiley & Sons, Ltd, Chichester, pp. 113–163.

Jurczenko, E. and Yanou, G. (2010) Fund of hedge funds portfolio selection: a robust non-parametric multi-moment approach, in *The Recent Trend of Hedge Fund Strategies* (ed. Y. Watanabe), Nova Science, New York, pp. 21–56.

Kerstens, K. and Van de Woestyne, I. (2011) Negative data in DEA: a simple proportional distance function approach. *Journal of the Operational Research Society*, **62** (7), 1413–1419.

Kerstens, K., Mounir, A., and Van de Woestyne, I. (2011a) Geometric representation of the mean-variance-skewness portfolio frontier based upon the shortage function. *European Journal of Operational Research*, **210** (1), 81–94.

Kerstens, K., Mounir, A., and Van de Woestyne, I. (2011b) Non-parametric frontier estimates of mutual fund performance using C- and L-moments: some specification tests. *Journal of Banking & Finance*, **35** (5), 1190–1201.

Kerstens, K., Mounir, A., and Van de Woestyne, I. (2012) Benchmarking mean-variance portfolios using a shortage function: the choice of direction vector affects rankings! *Journal of the Operational Research Society*, **63** (9), 1199–1212.

Kräussl, R. and Sandelowsky, R.M.R. (2007) The predictive performance of Morningstar's mutual fund ratings (August 17, 2007). http://dx.doi.org/10.2139/ssrn.963489; http://papers.ssrn.com/sol3/papers.cfm?abstract_id=963489. Accessed on 20 December 2012.

Lamb, J. and Tee, K.H. (2012) Data envelopment analysis models of investment funds. *European Journal of Operational Research*, **216** (3), 687–696.

Levy, M., Levy, H., and Solomon, S. (1995) Microscopic simulation of the stock market: the effect of microscopic diversity. *Journal de Physique I (France)*, **5**, 1087–1107.

Li, Q. (1996) Nonparametric testing of closeness between two unknown distribution functions. *Econometric Reviews*, **15** (3), 261–274.

Lozano, S. and Guttiérez, E. (2008) TSD-consistent performance assessment of mutual funds. *Journal of the Operational Research Society*, **59** (10), 1352–1362.

Martellini, L. (2008) Towards the design of better equity benchmarks: rehabilitating the tangency portfolio from modern portfolio theory. *Journal of Portfolio Management*, **34** (4), 34–41.

Matallín, J., Soler, A., and Tortosa-Ausina, E. (2011) On the informativeness of persistence for evaluating mutual funds performance using partial frontiers. Valencia, Instituto Valenciano de Investigaciones Económicas, Technical report. *IVIE WP-EC 2011-08*. http://www.ivie.es/downloads/docs/wpasec/wpasec-2011-08.pdf. Accessed on 20 December 2012.

McMullen, P. and Strong, R. (1998) Selection of mutual fund using data envelopment analysis. *Journal of Business and Economic Studies*, **4** (1), 1–14.

Morey, M. and Morey, R. (1999) Mutual fund performance appraisals: a multi-horizon perspective with endogenous benchmarking. *Omega*, **27** (2), 241–258.

Murthi, B., Choi, Y., and Desai, P. (1997) Efficiency of mutual funds and portfolio performance measurement: a non-parametric approach. *European Journal of Operational Research*, **98** (2), 408–418.

Nguyen, G. and Swanson, P. (2009) Firm characteristics, relative efficiency, and equity returns. *Journal of Financial and Quantitative Analysis*, **44**, 213–236.

Peterson Drake, P. and Fabozzi, F. (2010) *The Basics of Finance: An Introduction to Financial Markets, Business Finance, and Portfolio Management*, John Wiley & Sons, Inc, Hoboken.

Sengupta, J. (1989) Nonparametric tests of efficiency of portfolio investment. *Journal of Economics*, **50** (1), 1–15.

Sengupta, J. and Zohar, T. (2001) Nonparametric analysis of portfolio efficiency. *Applied Economics Letters*, **8** (4), 249–252.

Sharpe, W.F. (1998) Morningstar's risk-adjusted ratings. *Financial Analysts Journal*, **54** (4), 21–33.

Simar, L. and Wilson, P. (2007) Estimation and inference in two-stage, semi-parametric models of production processes. *Journal of Econometrics*, **136** (1), 31–64.

Sprott, J. (2004) Competition with evolution in ecology and finance. *Physics Letters A*, **325**, 329–333.

Tsolas, I. (2011) Natural resources exchange traded funds: performance appraisal using DEA modeling. *Journal of CENTRUM Cathedra*, **4** (2), 250–259.

Tu, J. and Zhou, G. (2011) Markowitz meets Talmud: a combination of sophisticated and naive diversification strategies. *Journal of Financial Economics*, **99**, 204–215.

Wilkens, K. and Zhu, J. (2001) Portfolio evaluation and benchmark selection: a mathematical programming approach. *Journal of Alternative Investments*, **4** (1), 9–19.

Zhao, X., Wang, S., and Lai, K. (2011) Mutual funds performance evaluation based on endogenous benchmarks. *Expert Systems with Applications*, **38** (4), 3663–3670.

8

Bank efficiency measurement and network DEA: A discussion of key issues and illustration of recent developments in the field

Necmi K. Avkiran
The University of Queensland, UQ Business School, Australia

8.1 Introduction

We begin this chapter by briefly taking a philosophical look at the question, 'Why should we study efficiency?' As world population grows in an environment of global warming, even sceptics would agree that scarce resources will be more intensely contested. Therefore, operating efficiently will continue to become increasingly important to organizations of all kinds if we are to maintain the level of prosperity we take for granted in developed countries. As expectations about quality of life continue to rise among the populations of developing countries, the need for efficient operations will gain universal prominence beyond the profit motivation.

In the dynamic, innovative, global environment of today, organizations are invariably complex. Nevertheless, they need to be versatile enough to deliver their promised outcomes to various stakeholders, in an environment of increasing uncertainty. Furthermore, in the presence of cyclical economic conditions and uncertain government budgets, identifying inefficiencies becomes more critical for long-term survival. The need for identifying inefficiencies holds equally true for non-profit, as well as for profit-making organizations.

Inefficient operations can happen due to information on the most effective processes not being perfect, always easily available or free. As a result, some organizations may not be as quick in responding to changes in market or environmental conditions. In addition to such

Efficiency and Productivity Growth: Modelling in the Financial Services Industry, First Edition. Edited by Fotios Pasiouras.
© 2013 John Wiley & Sons, Ltd. Published 2013 by John Wiley & Sons, Ltd.

imperfections, market uncertainty also influences the operational processes. For example, regulation and other exogenous constraints may introduce further inefficiencies into operations (see Delmas and Tokat, 2005). The inevitable financial and social costs associated with inefficiencies make a compelling case for identifying and quantifying inefficient processes as part of multidimensional benchmarking.

Another argument in support of identifying inefficiencies is its contribution to organizational learning. An organization that is not constantly acquiring knowledge is likely to lose its competitive advantage. Bartlett and Ghoshal (1998: 35) talk about a shift from strategic planning to organizational learning in terms of '…how to develop the organizational capability to sense and respond rapidly and flexibly to change'. Those who measure efficiency posit, explicitly or implicitly, that responding to change can be more effective if management can track the location of operational inefficiencies within an organization's structure. Thus, apart from the short-term motive of annual profits, efficiency analysis of organizational divisions can be seen as an integral part of organizational learning and a source of competitive advantage in the longer term. Data envelopment analysis (DEA), and more specifically network DEA, can be used in working towards this competitive advantage by providing insight into the specific sources of process inefficiency embedded in various levels of an organization's structure.

The rest of the chapter unfolds as follows. In Section 8.2, we briefly discuss the most recent global financial crisis (GFC) and the importance of DEA in bank performance analysis, including a look at variables used in DEA to capture banking risk. This is followed by Section 8.3, where we expand our discussion to how to deal with potential improvements while keeping managerial concerns in mind. Section 8.4 addresses the principal advantages and disadvantages of DEA, while Section 8.5 compares DEA with stochastic frontier analysis (SFA). Section 8.6, the most extensive section in this chapter, starts with an introduction to network DEA that includes a numerical example using retail bank branches and simulated data; Section 8.6 ends with a discussion of jackknifing versus bootstrapping. Section 8.7 is dedicated to a discussion of some key issues in relation to moving forward with DEA. Section 8.8 concludes the chapter.

8.2 Global financial crisis and the importance of DEA in bank performance analysis

The recent GFC of 2007–2008 originated in the US sub-prime residential mortgage market. At a time of low official interest rates, financial institutions and investors sought investment opportunities that offered higher yields. However, in an environment of increasing house prices and investment-grade ratings reliant on the backing of third parties, the market did not fully assess or review the potential default risks. To make matters worse, the first downgrading actions taken by Moody's and Standard & Poor's were as late as in November 2006 and February 2007, respectively (Crouhy, Jarrow, and Turnbull, 2008), casting doubt on the reliability of credit ratings. The severity of this crisis has been particularly evident in the banking sector due to investment banking activities and large off-balance sheet transactions, and contagion.

Given the well-documented impact of the GFC on the global banking sector and its burden on tax payers, we expect efficiency analysis to continue being an important part of identifying profitable and sustainable banking operations, as well as regulating the sector. That is, in what is essentially a mature and highly regulated sector with limited product innovation to set a bank apart from its competition, bank executives cannot take profits for granted. DEA has been the preferred frontier technique in bank efficiency measurement since the original paper by Sherman and Gold (1985) on bank branch operating efficiency. Other popular non-parametric frontier techniques include the free disposal hull, whereas in the parametric camp, we find the stochastic frontier approach or analysis, the thick frontier approach and the distribution-free approach. A quick survey of articles (excluding conference proceedings, book chapters and reviews) in the Web of Science database reveals that DEA has been applied in bank performance analyses in at least 282 articles. DEA partly owes its popularity to the shortcomings of other performance measurement techniques, namely, the requirement of a representative sample to estimate model parameters, limited scope for decision making with financial ratios and normality assumptions that may not hold. DEA's user-friendly nature as a non-parametric method that relies on actual observed data is one of its most endearing qualities.

DEA's application is certainly not limited to measurement of efficiency. In recent years, it has also been applied to predicting performance or lack thereof. For example, a recent study by Avkiran and Cai (2012) illustrates how DEA can be used as a forward-looking alternative method to flag bank holding companies likely to become distressed in the near future. Their results based on pre-GFC data generally support DEA's discriminatory and predictive power, suggesting that DEA can identify distressed banks up to two years in advance. Robustness tests reveal that DEA has a stable efficient frontier and that the technique's discriminatory and predictive powers prevail even after data perturbations. Another study conducted by Avkiran and Morita (2010) iteratively identifies various combinations of financial ratios and then applies DEA to predict Japanese bank stock performance. Thus, Avkiran and Morita inform readers about which financial ratios to monitor in forecasting stock performance over a number of years. Other predictive studies in the finance sector include Pille and Paradi (2002) who develop DEA models along with the modified Z-score model and equity-to-asset ratio to detect financial failure in credit unions.

With financial crises in mind, we pause here to further identify some of the variables used in applications of DEA to capture various banking-related risks. For example, it can be argued that measuring efficiency of banks in simply generating profits is not an adequate model of bank production because there may be inefficiencies embedded in the intermediation process as reflected by risk management and exposure. Some authors have responded to this criticism by including *loan loss provisions* as an input to capture credit risk (e.g. Drake, Hall and Simper, 2006; Pasiouras, 2008; Gaganis et al., 2009); others use *non-performing loans* (e.g. Avkiran and Goto, 2011; Chiu, Chen and Bai, 2011). On the other hand, market risk and operating risk are more complex and difficult to measure (Chiu and Chen, 2009, incorporate all three categories of banking risk in their study of Taiwanese bank efficiency). Other examples of efficiency adjustment include use of capital adequacy and risk capital requirement (Chiu et al., 2008) and a weighted average borrower risk rating (Paradi and Schaffnit, 2004).

8.3 The wider contribution of DEA to bank efficiency analysis and potential improvements

Bank efficiency analysis would be of most interest to management who monitor performance and regulators who are overseeing financial stability while they attempt to detect distress. On the other hand, investors and market analysts are likely to focus on ranking institutions for possible inclusion in investment portfolios. Paradi and Zhu (2013) provide an extensive survey of bank branch efficiency research using DEA, and appendix to their paper summarizes 80 such studies under the sub-headings of number of decision-making units (DMUs); inputs/outputs; DEA formulation; and objectives. Objectives of studies in this field include ranking branches on efficiency estimates, monitoring performance, reallocating resources, embedding value judgements, measuring cost efficiency under price uncertainty, measuring the impact of external factors, determining cross-country differences in efficient frontiers and so on. Instead of reproducing a similar survey here, we encourage the reader to refer to the literature surveys in Paradi and Zhu (2013), as well as Paradi, Yang and Zhu (2011) and Fethi and Pasiouras (2010).

A sub-category of bank efficiency literature that does not receive as much attention is the so-called *rational inefficiency* that can be argued as being necessary in maintaining smooth operations. That is, some resources may be held in place in quantities more than that can be justified for regular operations. Key questions from a managerial perspective are: how useful would it be to target potential improvements suggested by DEA? Might the presence of certain levels of inefficiency be justified on additional grounds of organizational performance? Bogetoft and Hougaard (2003) suggest that the rational view of inefficiency can explain the excess usage of resources as essential to producing unaccounted outputs such as a loyal pool of highly qualified or satisfied employees.

More generally, how can one adjust potential improvements in order to make DEA's recommendations more acceptable to bank management? Avkiran and Parker (2010) suggest that these questions can be tackled by developing satisficing models that allow for improvement objectives to be determined under uncertainty – which would be less than improvements suggested by full efficiency (see Cooper, Huang and Li, 1996). Such *less-than-ideal* improvement objectives can be further shaped by, for example, assigning probabilities to expected outcomes, where probabilities can be determined by managerial concerns.

There is certainly a compelling rationale for studying bank efficiency from the branch perspective because branches and their managers comprise the key organizational units that are involved with everyday operational decisions and contact with customers. The major obstacle in the way of a keen researcher is gaining access to data at the branch level which are not reported in various commercial databases such as BankScope or Compustat. The rest of the chapter deals with technical issues in relation to DEA and an exposition of network DEA in the context of branch banking.

8.4   Principal advantages and disadvantages of DEA

Key strengths of DEA as a relative efficiency technique include the property that no preconceived functional structure is imposed on the data in determining the efficient units. That is, DEA does not assume a particular production technology common to all DMUs. The importance of this feature is that a unit's efficiency can be assessed based on other observed performance by benchmarking similar organizations that are better at executing various processes. As an efficient frontier method, DEA identifies the inefficiency in a particular DMU by comparing it to efficient DMUs, rather than trying to associate a DMU's performance with statistical averages that may not be applicable to that DMU. We also note that input and output weights are endogenously derived, thus avoiding subjective weights or weights collected external to the sample, although weight restrictions can be imposed when justified.

Another one of DEA's strengths is its ability to handle related multiple inputs and multiple outputs in producing a parsimonious scalar estimate where multidimensional interactions are simultaneously captured. As Gelade and Gilbert (2003) point out, individual ratios (measures) looking at different aspects of an organization's effectiveness cannot depict a full picture because ratios are unlikely to be independent. Furthermore, ratio analysis assumes constant returns-to-scale – an assumption that is not always appropriate. The authors state that, '...the way in which performance measures are combined in DEA rests on strong theoretical foundations of production economics, allowing overall efficiency to be computed in circumstances in which additive scales cannot be legitimately constructed' (Gelade and Gilbert, 2003: 497). On the topic of financial ratios and bank branches, Yang and Liu (2012) similarly highlight shortcomings by pointing out that subjectively aggregating ratios are likely to yield misleading indicators of overall performance.

Despite its advantages, DEA has its limitations. These include the assumption that data is free of measurement error, thus making DEA more sensitive to the presence of measurement error than parametric techniques. That is, DEA assumes random variations cancel out one another. There have been many attempts to circumvent this potential drawback of DEA (see Avkiran and Rowlands, 2008 and the literature review therein).

It is thus common practice to scrutinize data for possible outliers to minimize the impact of measurement error. For example, outlier testing can take the form of DEA super-efficiency estimates. Observing a cut-off for super-efficiency estimates, say, between 2 and 3 helps identify DMUs that may be outliers. Super-efficiency estimates are re-calculated after the first round of deletions to make sure the remaining sample does not have any other potential outliers. Another approach is to visually examine for the presence of univariate outliers using stem–leaf diagrams and box plots. Then, the parametric Grubbs' test and the non-parametric Walsh's test (without the assumption of normal distribution) can be used to examine whether there are further outliers in the data. Paradi, Yang and Zhu (2011) sound a sensible warning that very low efficiency estimates in bank branch studies should also be treated with suspicion because such organizational units are tightly controlled and unlikely to deviate from the rest by a large margin. That is, outliers can be found at either end of the score range, and summing up in the words of Paradi, Yang and Zhu (2011: 327), 'Removing outliers is justified only if they can be identified as having erroneous or missing data, or if, even when the data are correct, these DMUs really are in a different business than the others'.

A second concern is the potentially sensitive nature of DEA efficiency estimates to the composition of the sample, suggesting that addition of a new DMU or a variable may change those benchmark units on the efficient frontier and thus the efficiency estimates for the inefficient units. Once again, this limitation, while real, can be overstated and many have tested the sensitivity of DEA estimates to various perturbations. Examples of papers that have attempted to systematically show the key implications for efficiency estimates of various data perturbations include Smith (1997), Ruggiero (1998), Chapparo, Salinas-Jimenez and Smith (1999), Seiford and Zhu (1999), Galagedera and Silvapulle (2003) and Avkiran (2007). Sensitivity analyses often investigate the stability or robustness of results to changes in the sample size, number of variables in the analysis, importance of inputs and correlation between inputs. Findings from such studies indicate that DEA estimates of efficiency are closest to the so-called *true* levels of efficiency estimated using a Cobb–Douglas type function (see Cobb and Douglas, 1928) when production process is simple in terms of the variables used and sample sizes are large in relation to the number of variables.

Currently, the only known paper on sensitivity of network data envelopment analysis (NDEA) is by Avkiran and McCrystal (2012), who test robustness of network slacks-based measure (NSBM) and network range–adjusted measure (NRAM) estimates. Their key findings indicate that (a) as in traditional DEA, greater sample size brings greater discrimination,

(b) removing a relevant input improves discrimination, (c) introducing an extraneous input leads to a moderate loss of discrimination, (d) simultaneously adjusting data in opposite directions for inefficient versus efficient branches shows generally stable estimates, (e) swapping divisional weights produces a substantial drop in discrimination, (f) stacking perturbations has the greatest impact on efficiency estimates with substantial loss of discrimination, and (g) layering suggests that the core inefficient cohort is resilient against omission of benchmark branches. The authors also report some insight gained by comparing NSBM with NRAM along the lines of (a) identical benchmark groups across both formulations, (b) a narrower range of efficiency estimates and a more stable mean across different sample sizes under NRAM, (c) distribution of NRAM efficiency estimates is negatively skewed whereas NSBM estimates are mostly positively skewed, and (d) there is no evidence of inefficient unit size bias among NRAM estimates, whereas larger inefficient units appear more inefficient under NSBM.

We reiterate that DEA captures the interactions among multiple inputs and multiple outputs in operational processes and identifies shortcomings of an inefficient unit against its best performing peers in the sample. Thus, some researchers refer to DEA as a relative multicriteria decision-making method. The reader can also refer to Coelli et al. (2005), Avkiran (2006), Cooper, Seiford and Tone (2007), Cooper, Seiford and Zhu (2011) for more in-depth treatments of DEA.

8.5   DEA versus stochastic frontier analysis

SFA is considered the parametric equivalent of DEA, and it can be used when DEA's properties are seen as a drawback with a particular data set (see Chapter 10 by Michael Koetter and Aljar Meesters in this book). SFA is a type of regression where the researcher can separate the impact of exogenous factors on the dependent variable (e.g. slacks measured by DEA) from that of measurement error and managerial inefficiency captured in the composed error term. That is, parameter estimates from SFA regressions predict slacks attributable to measurement error and exogenous factors, thus facilitating a purged measure of managerial efficiency. The main justification for SFA (rather than Tobit regression) is an asymmetric error term that identifies the one-sided error component (i.e. managerial inefficiency) and the symmetric error term component (i.e. statistical noise).

Combining DEA and SFA in a multiple-stage approach highlights a possible synergistic partnership. For instance, SFA efficiency measures use the estimated average parameter values in the regression equation. Thus, these efficiencies are not very sensitive to large data changes at the unit level, which is an advantage over DEA in the presence of measurement errors. On the other hand, SFA may be inappropriate if the structural form imposed on the analysis does not represent the behaviour of the organization under study, whereas the nonparametric nature of DEA makes the technique less susceptible to specification errors regarding the production technology. Luo and Donthu (2005) report that management prefer DEA and regard it as a more reliable frontier method. Yet, this observation by no means represents a consensus; for example, Fries and Taci (2005) believe SFA is more appropriate with data from transition economies where measurement errors are more frequent.

Clearly, DEA and SFA both have some assumptions that represent the core weaknesses of these techniques. For example, DEA assumes no measurement error, whereas SFA assumes a particular structure. Other literature suggests that neither approach is superior

(Tortosa-Ausina, 2002). Thus, it is somewhat compelling to take advantage of the strengths of both methods rather than exclusively subscribe to one or the other technique. According to Pastor (2002: 896), '…unlike parametric frontier models, the incorporation of environmental variables in DEA models is a field still being researched…'. We return to the discussion on measurement error when we report a numerical example of NDEA in branch banking and discuss jackknifing versus bootstrapping.

8.6   Drilling deeper with network DEA in search of inefficiencies

Summary measures of organizational efficiency, while opening the door to ranking and identifying potential improvements as per DEA theory, may disappoint those interested in developing a more in-depth understanding of underlying sources of operational inefficiency. As discussed previously regarding financial ratio analysis, additive scales may be misleading if the implicit assumption of independent, individual underlying measures does not hold. In examining multidivisional organizations, ability to account for each division's importance and contribution to organizational performance brings added dexterity to managerial decision making in allocating resources and targeting desired outcomes. Thus, the purpose of this section is to introduce the readers to NDEA in the context of branch banking. We can regard NDEA as a peer benchmarking method useful in comparing performance of organizations of similar structure and identifying underlying divisional inefficiencies that may detract from overall performance.

Traditional DEA (TDEA) does not provide sufficient detail for management to identify the specific sources of inefficiency in organizational divisions because of its tendency to treat operational processes as a *black box* (a term coined by Färe and Grosskopf, 1996). That is, the overall inefficiencies identified for a DMU cannot be directly traced to the inefficiencies that may be found in various divisions that comprise the organizational unit of interest. It also does not allow the option of bringing expert or managerial opinion to bear on what should be the weights assigned to divisional contributions. Fortunately, NDEA, which is an extension of TDEA, provides the investigator with a method to pinpoint whether the overall inefficiency observed in a bank branch (i.e. a DMU) resides, say, with its tellers or relationship bankers (i.e. sub-DMUs). Theoretically, we can treat each organizational function as a sub-DMU or a division. For example, intermediate outputs from tellers could become inputs to the sub-DMU of relationship bankers (Färe and Grosskopf, 2000, dubbed this type of interaction an *intermediate product*). An extension of NDEA, namely, dynamic NDEA or intertemporal NDEA, is discussed in Chapter 9 by Fukuyama and Weber in this book.

8.6.1   Definition of 'Network' in banking applications of NDEA

The first illustration of NDEA in banking can be found in Avkiran (2009). Avkiran follows the principles espoused by Färe and Grosskopf (2000), as well as the formulations put forward in Tone and Tsutsui's (2009) adaptation of the slacks-based measure (SBM) in NDEA (also see the original paper on SBM by Tone, 2001). Figure 8.1 shows the links among three main profit centres (conceptualized around product groups) by identifying exogenous inputs, intermediate products and final outputs in a five-node production network.

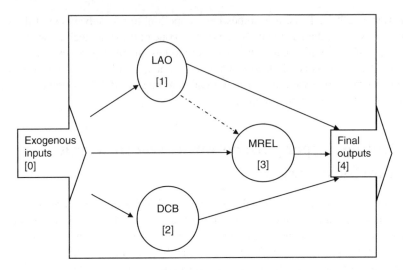

Figure 8.1 Opening the black box of commercial banking. The link between sub-DMUs [1] and [3] indicates an intermediate output that becomes an intermediate input. LAO = loans, advances and overdrafts; MREL = mortgaged real estate loans; DCB = discounted commercial bills. Adapted from Figure 2 in Avkiran (2009: 933).

That is, exogenous inputs (node 0) supply the profit centres (nodes 1–3) whose final outputs are collected in a sink (node 4).

In an effort to lay the groundwork for the illustration of NDEA in branch banking, we expand on Avkiran (2009). Avkiran identifies profit centres (sub-DMUs) and their corresponding expenses and revenues for 15 domestic commercial banks in the United Arab Emirates. Instead of focusing on physical or functional divisions, the author analyses the profit efficiency of a bank from a managerial accounting perspective using actual aggregate data but simulated profit centre data. The two inputs are interest and non-interest expenses, and interest and non-interest income comprise the two outputs in the parsimonious bank performance model.

Referring to Figure 8.1 and opening the black box to expose some of the activities of the network of sub-DMUs, we may encounter instances of cross-selling where business people applying for an overdraft facility are encouraged to take out a mortgage as well. Such referrals represent intermediate outputs from the profit centre [1] that become intermediate inputs to the profit centre [3]. A bank's internal activity-based costing system would recognize such referrals as non-interest revenue for the profit centre [1], and non-interest expense for the profit centre [3] for setting up and maintaining the accounts. In Avkiran (2009), intermediate outputs and intermediate inputs are treated as discretionary variables representing non-core profit centre activities. We now review three more studies of interest.

Kao and Hwang's (2010) technical paper uses an approach to network structure where a unified mathematical relationship is specified between the system efficiency and component process (i.e. sub-DMU) efficiencies. In the so-called *relational* model, the inefficiency of the system is the sum of the values of the process slacks. The authors examine the role of IT budget on bank performance where a first stage collects deposits and the second stage generates profits from loans and securities (i.e. the intermediation approach to bank behaviour where deposits become the intermediate products). IT budget is shared as external inputs between the two stages where the proportions are allowed to vary in order to shed the best efficiency light on the bank investigated.

Fukuyama and Weber (2010) also use a two-stage system but to investigate the network efficiency of Japanese banks. Labour, physical capital and equity are inputs to the first stage of fund or deposit generation (intermediate outputs), which then become inputs to the second stage of generating loans and securities investments. The authors' main contribution is the treatment of non-performing loans as an undesirable output in the second stage. Network estimates show larger inefficiencies compared to inefficiencies revealed by traditional DEA where the bank is treated as a black box, that is, no intermediate products where all the production takes place in one stage.

Finally, Yang and Liu (2012) present a recent study of Taiwanese bank branches – once again – using a two-stage production model that brings in the intermediation function in stage 2. The authors' network framework envisages two stages, namely, the *productivity* stage where various costs are inputs, and deposits are outputs, and the *profitability* stage where deposits become inputs and various types of income are outputs. Results indicate that most branches perform better in stage 1, but those that perform better in stage 2 have higher overall estimates.

8.6.2 Conceptualizing bank branch production

We want to further illustrate network DEA in a business setting where a homogeneous organizational network structure can be easily identified. We fulfil this objective by conceptualizing an intuitive and basic framework describing the main production processes in a retail bank branch. Traditional DEA would treat a branch as a black box, where a set of exogenous inputs (resources) enter a branch, only to emerge as a set of final outputs. Thus, traditional DEA does not explicitly identify any of the key sub-processes engaged by divisions found within a branch.

We would like to re-emphasize that the main motivation here is to illustrate NDEA in a practical branch banking environment, rather than advance existing theories on bank behaviour. In conceptualizing a network structure, let us now imagine what is likely to happen when a customer walks into a branch – based on what has been observed in the author's local branch. Initially, the customer is welcomed by one of the tellers. Once at the window, the customer may initiate one of many simple transactions such as depositing or withdrawing money, opening a term deposit account, purchasing some foreign currency, etc. (see Figure 8.2). Higher-level transactions such as an inquiry about which of the many types of credit cards would be suitable for the customer, taking out a home loan, starting a tax-effective retirement savings account, etc., would normally be referred to one of the relationship bankers who work in a separate room. This key production link between two functional divisions (i.e. tellers and relationship bankers) is known in network DEA literature as an *intermediate product* in acknowledgement of a transaction that starts as an output from one division and becomes an input into another division. The successful sale of products and services that require a more in-depth interview with a customer depends on the resources allocated to the division of relationship bankers, including interpersonal skills acquired in training that is brought to the encounter. Of course, one cannot expect all the referrals from the division of tellers to be turned into sales in the division of relationship bankers.

As part of the exercise of conceptualizing realistic branch production processes, we need to make some basic assumptions about the relationships between the variables before

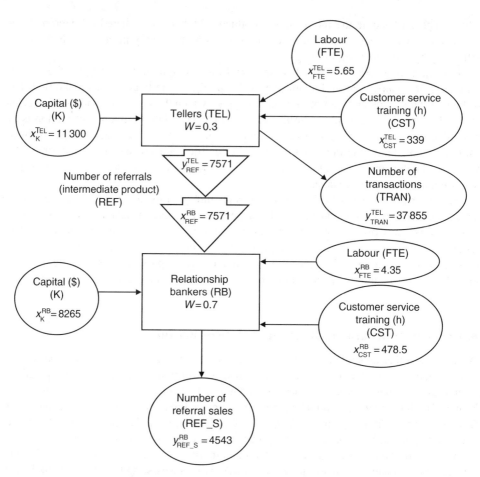

Figure 8.2 A weighted, two-division network structure of a basic bank branch production. The diagram focuses on the main service providers. Values corresponding to the data generation example are shown next to variables. The divisional weights depicted in this figure assume that relationship bankers play a greater role in revenue generation. FTE: Full-time equivalent – a unit that overcomes the difficulty of measuring labour when different mixes of full-time and part-time appointments coexist. Adapted from Figure 1 in Avkiran and McCrystal (2012).

generating data. For example, according to Smith (1997: 236), 'In most practical applications of DEA, inputs are highly correlated because they are all related to the scale of operations of the DMU'. Assuming a performance measurement period of one year, the arguments developed next detail the relationships between the modelled variables and divisions originally outlined in Figure 8.2:

- Various assumed data ranges, while arbitrary for convenience, follow reasonable expectations for bank branch operations.

- Average flow of customer traffic is assumed to manifest itself in labour input measured in full-time equivalent or FTE, and thus labour becomes the key variable driving the rest of the variables. Total labour for each branch is randomly determined in the range of 4 FTE (small branch) to 16 FTE (large branch).

- We assume the division of tellers uses 20–50% more labour than the relationship bankers. For example, if we focus on a 10 FTE branch and assume tellers use 30% more labour, then division-level labour emerges as 5.65 FTE (tellers) and 4.35 FTE (relationship bankers).

- Values for labour and capital within divisions are positively correlated. That is, a branch with higher customer traffic will employ more labour and, thus, higher capital. The relationship between labour and capital is assumed to fall in the range of $1800–2200 per FTE and this rate is allowed to randomly vary to reflect branch differences.

- Similarly, a positive correlation is assumed between labour and customer service training in a given branch, measured in hours. By assuming each unit of labour (FTE) in the division of tellers attracts 50–100 h of training, we account for differences in staff experience and, thus, the need for training. The rate for the division of relationship bankers is set higher at 80–150 h per unit of labour in view of the more complicated products or services this division provides regardless of experience. The rates are allowed to vary randomly across branches and values for customer service training are determined by multiplying rates and labour inputs.

- Since labour is a rough measure of the flow of customers through the division of tellers, we expect the outputs of number of transactions and the number of referrals to be positively correlated with labour. Each unit of labour is assumed to handle, on average, 30 transactions per day or 6600 transactions per year assuming 5 workings days in a week and 11 months in a year. Thus, we first randomly generate the number of transactions per unit of labour in the range of 6000–7200 to accommodate unknown differences across branches. In turn, the number of referrals randomly varies between 10% and 40% of the number of transactions, once again, to allow for unknown differences between branch clientele and business mix. This estimate does not include any non-discretionary input of referrals reaching a bank from phone and Internet banking services. According to Paradi, Rouatt and Zhu (2011: 99), '…it is through a branch that customers do a large percentage of their more value added banking activities…'

- Finally, the number of referral sales in the division of relationship bankers depends on the number of referrals received from the division of tellers. Thus, referral sales are allowed to randomly vary between 50% and 90% of the number of referrals to reflect the differences in successful selling activities of relationship bankers (see the mathematical algorithm in Appendix 8.A to Avkiran and McCrystal (2012) for more information on data generation).

8.6.3   Network slacks-based measure of efficiency

We execute weighted NSBM assuming VRS and non-orientation. In applications of DEA, SBM has gradually become the non-radial model of choice. SBM also has the desirable property of being unit-invariant. Our choice of VRS accounts for the often different scale of operations found in business organizations. VRS is more desirable because it processes information on returns-to-scale of operations for each DMU while avoiding inappropriate extrapolation of performance (see Smith, 1997: 244). VRS is particularly suitable in network DEA which attempts to capture the interactions among multiple divisions within each DMU, where the latter may operate on different scales. Essentially, VRS modelling purges efficiency estimates of the impact of scale of operations.

The choice of *non-orientation* and *non-radial* modelling further makes frontier efficiency studies more relevant to the world of business. For example, non-orientation ensures the analysis simultaneously captures slacks on both sides as the linear program minimizes inputs and maximizes outputs. The choice of non-orientation is also a more neutral option because it avoids arguments about the extent managers and employees have control over input and output variables – a situation of particular concern in NDEA where intermediate products are part of the estimation process.

Similarly, use of non-radial modelling accommodates the often non-proportional nature of slacks in organizations where production relationships demand different proportions of changes in inputs and outputs. Here, we pause to expand this point further with the intention of showing that radial changes can be inappropriate unless proportionality is justified based on an intimate knowledge of production processes. Let us assume that the two key inputs in a bank profitability analysis are interest expense and non-interest expense, and the two key outputs are interest income and non-interest income. A bank that observes a policy of paying higher salaries to retain employees who excel in customer service will incur relatively more non-interest expense, but could compensate for this by paying lower interest rates on its deposit accounts (thus, incurring relatively less interest expense). The same bank may also choose to focus its attention on the sale of those services that generate more fee income than loans (thus, earning relatively more non-interest income). Proportional projections assumed in the radial and input- or output-oriented DEA formulations cannot properly reflect the bank's operational decisions. On the other hand, estimating non-proportional projections through a non-radial model is a more realistic representation of a complex business world.

Equation (8.1) is adapted from Tone and Tsutsui (2009) and represents the core equation behind NSBM. Equation (8.1) defines the objective function for DMU level network efficiency estimate. The weighted summation in Equation (8.1) also implies that in order for the overall NSBM estimate to equal 1, all divisions of a branch must also be efficient. This, in turn, suggests that unlike traditional DEA, an overall efficiency estimate of 1 cannot be guaranteed for a given sample.

$$\rho_o^* = \min_{z^j, s^{j-}, s^{j+}} \frac{\sum_{j=1}^{J} w_j \left[1 - (1/N_j) \left(\sum_{n=1}^{N_j} \left(s_n^{j-} / x_o^j \right) \right) \right]}{\sum_{j=1}^{J} w_j \left[1 + (1/M_j) \left(\sum_{m=1}^{M_j} \left(s_m^{j+} / y_o^j \right) \right) \right]} \tag{8.1}$$

subject to

$$x_o^j = X^j z^j + s^{j-} \quad (j = 1, \dots, J) \tag{8.1a}$$

$$y_o^j = Y^j z^j - s^{j+} \quad (j = 1, \dots, J) \tag{8.1b}$$

$$\sum_{k=1}^{K} z_k^j = 1 \ (\forall j), \quad z_k^j \geq 0 \ (\forall k, j), \quad \text{and} \tag{8.1c}$$

$$t^{(j,h)}z^h = t^{(j,h)}z^j, \quad \left(\forall\left(j,h\right)\right), \quad t_k^{(j,h)} = \left(t_1^{(j,h)},...,t_K^{(j,h)}\right)\in R^{T^{(j,h)}\times K} \tag{8.1d}$$

$$s_n^{j-} \geq 0, s_m^{j+} \geq 0\left(\forall j,n,m\right). \tag{8.1e}$$

where
  o = the observed DMU, o = 1, ..., K;
  k = a DMU (K = number of DMUs);
  j = a division (J = number of divisions);
  N = number of inputs;
  M = number of outputs;
  s_n^{j-} = input slack;
  s_m^{j+} = output slack;
  z^j = intensity;
  (j, h) = intermediate product link between division j and division h;
  $t^{(j,h)}$ = intermediate product;
  $T^{(j,h)}$ = number of intermediate products;
  X^j is the input matrix for division j where $X^j = \left(x_1^j,...,x_K^j\right)\in R^{N_j\times K}$;
  Y^j is the output matrix for division j where $Y^j = \left(y_1^j,...,y_K^j\right)\in R^{M_j\times K}$;
  $\sum_{j=1}^{J}w_j = 1, \quad w_j \geq 0\left(\forall j\right)$, where w_j is the relative weight of division j determined exogenously.

Non-negative weights for all the divisions add up to 1, thus allowing separate accounting for the importance of the divisions. Constraint (8.1c) indicates non-negative divisional intensities that add up to 1 across the sample, thus enabling efficiency estimates to reflect VRS. Constraint (8.1d), introduces free linking where linking activities are discretionary, while maintaining continuity between inputs and outputs and enabling an intermediate product (Tone and Tsutsui, 2009).

8.6.4 A brief numerical example

The roadmap for the numerical illustration is as follows:

- compare black box scores with NSBM scores;
- profile divisional inefficiencies; and
- swap divisional weights and re-visit the previous two steps.

The divisional weights indicated in Figure 8.2 place more emphasis on relationship bankers based upon the expectation that this sub-unit is likely to produce higher revenue products/services as they turn referrals to sales. Later in this section, we shift the emphasis to tellers as we swap the weights – on the premise that business generated by relationship bankers is

Table 8.1 Descriptive summary of efficiency estimates (scores) for branches and their divisions.

BBB-s[c]	Mean	Median	Range	CV (%)[a]	Skewness	Proportion of efficient units (%)[b]
	0.6840	0.6627	0.7178	29.74	0.0820	12/80 = 15.00
Panel A (assuming a weight of 0.3 for tellers and 0.7 for relationship bankers)						
NSBM-s[d]	0.5824	0.5595	0.7589	34.55	0.4120	3/80 = 3.75
TEL-s[e]	0.9271	0.9534	0.2379	7.87	−0.6829	19/80 = 23.75
RB-s[e]	0.5077	0.4686	0.8388	44.93	0.7252	7/80 = 8.75
Panel B (assuming a weight of 0.7 for tellers and 0.3 for relationship bankers)						
NSBM-s	0.7266	0.7361	0.5710	19.53	−0.1033	3/80 = 3.75
TEL-s	0.9150	0.9285	0.2527	8.13	−0.4359	17/80 = 21.25
RB-s	0.5130	0.4779	0.8388	44.97	0.6740	7/80 = 8.75

[a] Coefficient of variation equals standard deviation over mean, multiplied by 100.
[b] Ratio of the number of efficient units to the sample size.
[c] BBB-s, branch black box score.
[d] NSBM-s, network slacks-based measure score of branches.
[e] These divisional scores represent the production processes depicted in Figure 8.2 (TEL-s, tellers' score; RB-s, relationship bankers' score).

reliant on referrals from tellers. Of course, if there are no compelling arguments either way, the weights can be left at 0.5.

Initial calculations involve the efficiency of branches using non-oriented VRS SBM. In Table 8.1, this estimate is named the 'branch black box score' where the three exogenous inputs are total capital, labour and customer service training, and the two final outputs are numbers of transactions and referral sales generated by the overall effort of a branch (excluding intermediate products). The network model results based on Figure 8.2 show greater inefficiencies compared to the black box approach as evidenced by a much smaller proportion of efficient units (a similar observation was reported by Fukuyama and Weber, 2010). The rank correlations (Kendall's *tau-b*) – while statistically significant – are substantially lower than 1 at 0.667 and 0.770 under the two sets of divisional weights, also underlining these differences.

In panels A and B of Table 8.1, we can see that inefficiencies are mostly found in the division of relationship bankers, that is, in stage 2 of the production. It is also possible to scrutinize inefficiencies for potential improvements at the input/output level but for brevity such an analysis has not been illustrated in this simulated example (interested readers are referred to Table 5 in Avkiran (2009) for such an example). Finally, the most noticeable impact as we swap the divisional weights is greater NSBM scores as we place more weight on the scores of the more efficient division. Discrimination – as measured by the proportion of efficient units – remains unchanged with the exception of a small change for tellers. More importantly, the rank correlation between the two sets of NSBM branch scores is 0.828 – significant at the 1% level – a figure that may help executive management promote the analysis at branch level by allaying any fears of being grossly disadvantaged by the choice of divisional weights.

8.6.5   Jackknifing versus bootstrapping

Horsky and Nelson (2006: 130) remind us that whenever we can make distributional assumptions about error terms, distributions of parameter estimates are known asymptotically and statistical significance tests such as the *t*-test can be designed. Yet, linear programming behind DEA makes no distributional assumptions. According to Simar and Wilson (1998), because estimators of the non-parametric frontier are obtained from finite samples, efficiency estimates are sensitive to sampling variations. In fact, it is well established in DEA literature that efficiency estimates can be sensitive to sample composition. Sampling variations around the observed efficient frontier may cast doubt on the validity of estimates (Odeck, 2009: 1009).

Initially, employing the sample re-use method of *jackknifing* we show how to test for the potential instability of NSBM estimates for our sample of $N=80$. In a similar manner to that of *leave-one-out* method, a branch is omitted from the original sample and NSBM efficiencies re-estimated for the emerging sample sub-set. The originally omitted branch is then returned before the next branch is removed (thus maintaining the same degrees of freedom throughout re-testing). Network DEA is repeated until all possible samples are exhausted, that is, 80 times. This is followed by computation of pseudovalues, a jackknife estimator and standard error in order to establish confidence intervals (i.e. upper and lower limits).

If the mean of the estimates from the original sample falls within the confidence intervals generated through re-sampling, then we can reasonably conclude that NSBM estimates are unlikely to be artefacts of the sample used. Paraphrasing Crask and Perreault (1977), the essence of jackknifing is to separate the effect of a given branch on the NSBM estimate obtained from the full sample or population. The procedure detailed in Appendix 8.A to this chapter results in upper and lower limits of 0.4342 and 0.5947, respectively. Since the mean NSBM estimate (0.5824) from our model falls within this interval, we conclude that there is no reason to be alarmed about the observed estimates being artefacts of sampling.

In this chapter, we have used jackknifing as an easy-to-implement approach to generate confidence intervals around the sample mean of NDEA estimates. Jackknifing in DEA can be traced to Färe, Grosskopf and Weber (1989). However, Ondrich and Ruggiero (2002) deemed jackknifing as inappropriate for detecting outliers in DEA because computation of standard deviation is not meaningful in distinguishing between outliers caused by measurement error versus outliers that are actually efficient. Nevertheless, we believe that the leave-one-out method (and we do not have to refer to it as 'jackknifing'!) is still useful for the simple purpose of testing the *stability of estimates to re-sampling*. Given that jackknifing can be viewed as an elementary form of bootstrapping, this brings us to a brief discussion on bootstrapping.

Since the 1980s, one of the criticisms levelled at DEA is that the technique is deterministic and does not offer the range of diagnostics that often accompany econometric methods of frontier estimation (see various studies such as Banker, 1996, and Simar and Wilson, 1998, 2000, that respond to this criticism from different perspectives). That is, while traditional DEA reports DMU performance in unique levels of inefficiency, it fails to account for the uncertainty that may be embedded in such estimates. Where it is not possible to separate random noise, DEA incorrectly accounts for the latter as part of inefficiency. Management scientists concerned with this have employed such methods as bootstrapping that can help generate confidence intervals for every estimate and also assist in testing hypotheses on group comparisons.

Nevertheless, despite examples of bootstrapping in DEA methodology literature, the great majority of papers that apply DEA still do not use bootstrapping for inferring uncertainty in efficiency estimates – a point also conceded by Odeck (2009) for the agriculture sector. In the context of this chapter, definition of an NSBM frontier and the data generating process remains elusive. In network DEA, the production process would be difficult to model as a small number of inputs leading to a single output. For example, how can 'true' levels of efficiency be estimated using a Cobb–Douglas type function? These problems add to the difficulty of generating true efficiency estimates and bootstrapping in non-radial as well as network DEA, reminding us that examples of bootstrapping in literature are still limited to the radial models.

Summing up Section 8.6, NDEA is a sophisticated approach to relative performance modelling that takes advantage of divisional operational practices interrelated via shared exogenous inputs and final outputs. Network efficiency estimates generated by NDEA that account for divisional, departmental or sub-structure interactions are more representative of today's complex business organizations than measures that report overall performance without opening the black box. Readers interested in an introduction to some of the widely accepted mathematical formulations that enable NDEA are referred to Färe and Grosskopf (2000) and Tone and Tsutsui (2009), where the latter represents the study behind NDEA code in the software DEA Solver Pro (by SAITECH) used in this chapter.

8.7   Moving forward with DEA

Occasionally, technical details in academic publications may be incorrectly reported due to inadequate language skills or incomplete literature review on part of authors, journal referees who may not be fully qualified or simply due to carelessness. Here, we quote one such error from a 2012 publication in a reputable internationally refereed journal but do not provide the citation as a matter of courtesy: '...it [DEA] permits efficiency to change over time and *allows for the existence of random errors*'. An error is often a misrepresentation of well-established knowledge in the field (e.g. DEA estimates do not discriminate between random error and managerial inefficiency), whereas criticizing a much older article on the basis of most current knowledge should be handled with more tact. After all, in the great majority of cases, pushing the knowledge envelope occurs in incremental steps rather than big steps. Avkiran and Parker (2010) provide a critical look at how the DEA envelope has been expanding since 1978 and make suggestions for innovative steps going forward. One such line of thinking is discussed next.

A limitation in the application versus formulation of DEA rarely attended to in the literature – involves what Adler et al. (2009) recently discussed under the heading of 'productivity dilemma' (also see Abernathy, 1978). Improving efficiency is usually a short-term activity that may reduce long-term organizational adaptability. While most authors of DEA report potential efficiency improvements (i.e. exploitation of production possibilities), they rarely discuss the impact of such recommendations on organizational adaptability and learning (i.e. exploration of new possibilities). Yet, *exploration* is a key ingredient of organizational innovation essential for survival in the dynamic business environment of the global economy. Quoting March (1991: 71), '...systems that engage in exploitation to the exclusion of exploration are likely to find themselves trapped in suboptimal stable equilibria'. Future DEA papers stand to benefit if their conclusions were to include a sub-section under managerial implications that offered reflections on the productivity dilemma, as well as the closely related topic of *rational inefficiency* mentioned in Section 8.3.

Given the space devoted to network DEA in this chapter, we now return to this exciting field. NDEA can provide insight into the specific sources of organizational process inefficiency and enable management to devise targeted remedial action. In other words, NDEA permits a fuller access to the underlying diagnostic information that may otherwise remain outside management's reach in the organizational black box. For example, NDEA could help identify potential merger synergies trapped in organizational sub-units that are often monitored at a managerial accounting level but normally not investigated as interacting components of a larger network or organizational structure. Nevertheless, application of NDEA where an organization's departments comprise the sub-DMUs is not practical unless the exercise is undertaken by someone who has access to internal managerial information. That is, information often garnered through sources in the public domain, such as financial statements or regulatory reports, may not be sufficiently disaggregated. Formulating numerous interactions amongst multiple functions and their corresponding sub-functions would require an internal audit via privileged access.

Fortunately, there are exceptions to the potential data access problem. That is, if the conceptual framework involves a higher-level organization, data collection is less likely to present a problem. For example, a cross-country benchmarking of banking systems where central banks and regulators can be designated as sub-DMUs could constitute a network. In general, data access is less likely to be a source of difficulty for those researchers who successfully promote DEA to the business community. This sentiment brings us to the final topic of the discussion, namely, making DEA more acceptable to bank management.

Paradi, Rouatt and Zhu (2011) introduce an intuitive two-stage approach to making DEA results more credible in the eyes of bank management. Starting from academic literature where various bank behaviour models such as *production, intermediation* and *profitability* are often used, they illustrate how separate multiple efficiency perspectives can be brought together in a scalar value. Thus, the first stage involves input-oriented DEA using three different performance models – an approach that the authors claim makes DEA more acceptable to the branch management who is likely to identify with at least one of the performance models. In the second stage, another DEA run (output-oriented) uses the scores from stage 1 as outputs while inputs are held at unity across the branches of the major Canadian bank investigated. The authors demonstrate that scores from stage 2 are more discriminating and results from this stage are most likely to be used for ranking of branches by senior management.

8.8   Conclusions

We began this chapter by stating that the financial and social costs associated with inefficiencies make a compelling case for identifying and quantifying inefficient production. We also underlined that responding to change can be more effective if management can identify operational inefficiencies within an organization's structure.

We reiterate that bank executives cannot take sustainable profits for granted in a mature and highly regulated banking sector with limited scope for product innovation. Given the impact of financial crises on the global banking sector and its equally damaging impact on tax payers, efficiency analysis will continue being an important part of the toolkit in maintaining sustainable operations. We expect bank efficiency analysis to continue to be of interest to management, investors and regulators alike.

We can regard network DEA as a peer benchmarking method useful in comparing performance of institutions of similar structure and detecting divisional inefficiencies that exist out-of-sight and reduce overall performance. Yet, until some technical problems are satisfactorily resolved, the full benefit of NDEA will not be realized. For example, definitions of an NDEA frontier and an accompanying data generating process have not been fully resolved. The production process is difficult to model because of the complicated linkages in a network structure, and a Cobb–Douglas-type function is not directly useable. These problems spill over to generating true efficiency estimates and bootstrapping in network DEA and non-radial models. Hopefully, researchers interested in this field will attend to these loose ends in a manner that enables an easy incorporation of solutions in applications of DEA. There is always a need for innovative people who can bridge theoretical developments based on mathematics or econometrics and application tools that integrate such developments in a user-friendly interface.

Appendix 8.A: Jackknifing

The jackknifing procedure begins with the generation of NSBM efficiency estimates from a sample where one of the bank branches has been omitted. Next, the omitted branch is returned and another branch taken out before NSBM is repeated on the new sample or sub-set, and so on. For each sub-set, the mean NSBM estimate is then substituted into Equation (8.A.1) to obtain the so-called *pseudovalue*, J_i (Crask and Perreault, 1977):

$$J_i = k\overline{x}^p - (k-1)\overline{x}_i^p, \quad i = 1,\ldots,k \tag{8.A.1}$$

where
k = number of sub-sets (equals sample size N);
$\overline{x}^p$ = mean NSBM estimate for the original sample prior to re-sampling;
$\overline{x}_i^p$ = mean NSBM estimate for sub-set i following re-sampling.
Assuming k different pseudovalues are independent, identically distributed random variables (Crask and Perreault, 1977), then the mean of pseudovalues becomes the *jackknife estimator*, $\hat{J}$:

$$\hat{J} = \frac{\left(\sum_{i=1}^{k} J_i\right)}{k} \tag{8.A.2}$$

Equation (8.A.3) provides the *standard error* of the jackknife estimator:

$$\text{S.E.}_{\hat{j}} = \frac{s}{\sqrt{n}}, \tag{8.A.3}$$

where
n = sub-set size (i.e. $N - 1$);
s = standard deviation of pseudovalues.

Equation (8.A.4) delivers the *confidence interval* around the jackknife estimator:

$$\hat{J} \pm t \text{ critical} \times (\text{S.E.}). \tag{8.A.4}$$

where the critical t value for two-tailed, 95% confidence level and $N-1$ degrees of freedom is used. If the mean NSBM estimate for the sample, $\bar{x}^{\rho}$, falls within the confidence interval thus determined, then we would be less concerned with sampling variations.

References

Abernathy, W.J. (1978) *The Productivity Dilemma Roadblock to Innovation in the Automobile Industry*, Johns Hopkins University Press, Baltimore.

Adler, P.S., Benner, M., Brunner, D.J. et al. (2009) Perspectives on the productivity dilemma. *Journal of Operations Management*, **27** (2), 99–113.

Avkiran, N.K. (2006) *Productivity Analysis in the Service Sector with Data Envelopment Analysis*, N K Avkiran, Ipswich.

Avkiran, N.K. (2007) Stability and integrity tests in data envelopment analysis. *Socio-Economic Planning Sciences*, **41** (3), 224–234.

Avkiran, N.K. (2009) Opening the black-box of efficiency analysis: an illustration with UAE banks. *OMEGA, The International Journal of Management Science*, **37** (4), 930–941.

Avkiran, N.K. and Cai, L. (2012) Predicting bank financial distress prior to crises, New Zealand Finance Colloquium, February.

Avkiran, N.K. and Goto, M. (2011) A tool for scrutinizing bank bailouts based on multi-period peer benchmarking. *Pacific-Basin Finance Journal*, **19** (5), 447–469.

Avkiran, N.K. and McCrystal, A. (2012) Sensitivity analysis of network DEA: NSBM versus NRAM. *Applied Mathematics and Computation*, **218** (22), 11226–11239.

Avkiran, N.K. and Morita, H. (2010) Predicting Japanese bank stock performance with a composite relative efficiency metric: a new investment tool. *Pacific-Basin Finance Journal*, **18** (3), 254–271.

Avkiran, N.K. and Parker, B.R. (2010) Pushing the DEA research envelope. *Socio-Economic Planning Sciences*, **44** (1), 1–7.

Avkiran, N.K. and Rowlands, T. (2008) How to better identify the true managerial performance: state of the art using DEA. *OMEGA, The International Journal of Management Science*, **36** (2), 317–324.

Banker, R.D. (1996) Hypothesis tests using data envelopment analysis. *Journal of Productivity Analysis*, **7** (2–3), 139–159.

Bartlett, C.A. and Ghoshal, S. (1998) Beyond strategic planning to organizational learning: lifeblood of the individualized corporation. *Strategy and Leadership*, **26** (1), 34–39.

Bogetoft, P. and Hougaard, J.L. (2003) Rational inefficiencies. *Journal of Productivity Analysis*, **20** (3), 243–271.

Chapparo, F.P., Salinas-Jimenez, J. and Smith, P. (1999) On the quality of the data envelopment analysis. *Journal of the Operational Research Society*, **50** (6), 636–644.

Chiu, Y.H. and Chen, Y.C. (2009) The analysis of Taiwanese bank efficiency: incorporating both external environment risk an internal risk. *Economic Modelling*, **26** (2), 456–463.

Chiu, Y.H., Jan, C., Shen, D.B. and Wang, P.C. (2008) Efficiency and capital adequacy in Taiwan banking: BCC and super-DEA estimation. *The Service Industries Journal*, **28** (4), 479–496.

Chiu, Y.H., Chen, Y.C. and Bai, X.J. (2011) Efficiency and risk in Taiwan banking: SBM super-DEA estimation. *Applied Economics*, **43** (5), 587–602.

Cobb, C.W. and Douglas, P.H. (1928) A theory of production. *American Economic Review*, **18** (1), 139–165.

Coelli, T.J., Prasada, R.D.S., O'Donnell, C.J. and Battese, G.E. (2005) *An Introduction to Efficiency and Productivity Analysis*, 2nd edn, Springer, New York.

Cooper, W.W., Huang, Z. and Li, S. (1996) Satisficing DEA models under chance constraints. *Annals of Operations Research*, **66** (4), 279–295.

Cooper, W.W., Seiford, L.M. and Tone, K. (2007) *Data envelopment Analysis: A Comprehensive Text with Models, Applications, References and DEA-Solver Software*, 2nd edn, Springer, New York.

Cooper, W.W., Seiford, L.M. and Zhu, J. (2011) *Handbook on Data Envelopment Analysis*, 2nd edn, Kluwer Academic Publishers, Norwell.

Crask, M.R. and Perreault, W.D. (1977) Validation of discriminant analysis in marketing research. *Journal of Marketing Research*, **14** (1), 60–68.

Crouhy, M.G., Jarrow, R.A. and Turnbull, S.M. (2008) The subprime credit crisis of 2007. *The Journal of Derivatives*, **16** (1), 81–110.

Delmas, M. and Tokat, Y. (2005) Deregulation, governance structures, and efficiency: the U.S. electric utility sector. *Strategic Management Journal*, **26** (5), 441–460.

Drake, L., Hall, M.J.B. and Simper, R. (2006) The impact of macroeconomic and regulatory factors on bank efficiency: a non-parametric analysis of Hong Kong's banking system. *Journal of Banking & Finance*, **30** (5), 1443–1466.

Färe, R. and Grosskopf, S. (1996) *Intertemporal Production Frontiers: With Dynamic DEA*, Kluwer Academic Publisher, Boston.

Färe, R. and Grosskopf, S. (2000) Network DEA. *Socio-Economic Planning Sciences*, **34** (1), 35–49.

Färe, R., Grosskopf, S. and Weber, W.L. (1989) Measuring school district performance. *Public Finance Review*, **17** (4), 409–428.

Fethi, M.D. and Pasiouras, F. (2010) Assessing bank efficiency and performance with operational research and artificial intelligence techniques: a survey. *European Journal of Operational Research*, **204** (2), 189–198.

Fries, S. and Taci, A. (2005) Cost efficiency of banks in transition: evidence from 289 banks in 15 post-communist countries. *Journal of Banking and Finance*, **29** (1), 55–81.

Fukuyama, H. and Weber, W. (2010) A slacks-based inefficiency measure for a two-stage system with bad outputs. *OMEGA, The International Journal of Management Science*, **38** (5), 398–409.

Gaganis, C., Liadaki, A., Doumpos, M. and Zopounidis, C. (2009) Estimating and analysing the efficiency and productivity of bank branches: evidence from Greece. *Managerial Finance*, **35** (2), 202–218.

Galagedera, D. and Silvapulle, P. (2003) Experimental evidence on robustness of data envelopment analysis. *Journal of the Operational Research Society*, **54** (6), 654–660.

Gelade, G. and Gilbert, P. (2003) Work climate and organizational effectiveness: the application of data envelopment analysis in organizational research. *Organizational Research Methods*, **6** (4), 482–501.

Horsky, D. and Nelson, P. (2006) Testing the statistical significance of linear programming estimators. *Management Science*, **52** (1), 128–135.

Kao, C. and Hwang, S.N. (2010) Efficiency measurement for network systems: IT impact on firm performance. *Decision Support Systems*, **48** (3), 437–446.

Luo, X. and Donthu, N. (2005) Assessing advertising media spending inefficiencies in generating sales. *Journal of Business Research*, **58** (1), 28–36.

March, J.G. (1991) Exploration and exploitation in organizational learning. *Organization Science*, **2** (1), 71–87.

Odeck, J. (2009) Statistical precision of DEA and Malmquist indices: a bootstrap application to Norwegian grain producers. *OMEGA, The International Journal of Management Science*, **37** (5), 1007–1017.

Ondrich, J. and Ruggiero, J. (2002) Outlier detection in data envelopment analysis: an analysis of jackknifing. *Journal of the Operational Research Society*, **53** (3), 342–346.

Paradi, J.C. and Schaffnit, C. (2004) Commercial branch performance evaluation and results communication in a Canadian bank – a DEA application. *European Journal of Operational Research*, **156** (3), 719–735.

Paradi, J.C. and Zhu, H. (2013) A survey on bank branch efficiency and performance research with data envelopment analysis. *OMEGA, The International Journal of Management Science*, **41** (1), 61–79.

Paradi, J.C., Rouatt, S. and Zhu, H. (2011) Two-stage evaluation of bank branch efficiency using data envelopment analysis. *OMEGA, The International Journal of Management Science*, **39** (1), 99–109.

Paradi, J.C., Yang, Z. and Zhu, H. (2011) Assessing bank and bank branch performance: modelling considerations and approaches, in *Handbook on Data Envelopment Analysis*, 2nd edn (eds W.W. Cooper, L.M. Seiford, and J. Zhu), Kluwer Academic Publishers, New York, pp. 315–361.

Pasiouras, F. (2008) Estimating the technical and scale efficiency of Greek commercial banks: the impact of credit risk, off-balance sheet activities, and international operations. *Research in International Business and Finance*, **22** (3), 301–318.

Pastor, J.M. (2002) Credit risk and efficiency in the European banking system: a three-stage analysis. *Applied Financial Economics*, **12** (12), 895–911.

Pille, P. and Paradi, J.C. (2002) Financial performance analysis of Ontario (Canada) Credit Unions: an application of DEA in the regulatory environment. *European Journal of Operational Research*, **139** (2), 339–350.

Ruggiero, J. (1998) A new approach for technical efficiency estimation in multiple output production. *European Journal of Operational Research*, **111** (2), 369–380.

Seiford, L. and Zhu, J. (1999) Sensitivity and stability of the classifications of returns to scale in data envelopment analysis. *Journal of Productivity Analysis*, **12** (1), 55–75.

Sherman, H.D. and Gold, F. (1985) Bank branch operating efficiency: evaluation with data envelopment analysis. *Journal of Banking and Finance*, **9** (2), 297–315.

Simar, L. and Wilson, P.W. (1998) Sensitivity analysis of efficiency scores: how to bootstrap in nonparametric frontier models. *Management Science*, **44** (1), 49–61.

Simar, L. and Wilson, P.W. (2000) Statistical inference in nonparametric frontier models: the state of the art. *Journal of Productivity Analysis*, **13** (1), 49–78.

Smith, P. (1997) Model misspecification in data envelopment analysis. *Annals of Operations Research*, **73** (1), 233–252.

Tone, K. (2001) A slacks-based measure of efficiency in data envelopment analysis. *European Journal of Operational Research*, **130** (3), 498–509.

Tone, K. and Tsutsui, M. (2009) Network DEA: a slacks-based measure approach. *European Journal of Operational Research*, **197** (1), 243–252.

Tortosa-Ausina, E. (2002) Bank cost efficiency and output specification. *Journal of Productivity Analysis*, **18** (3), 199–222.

Yang, C. and Liu, H.M. (2012) Managerial efficiency in Taiwan bank branches: a network DEA. *Economic Modelling*, **29** (2), 450–461.

9

A dynamic network DEA model with an application to Japanese Shinkin banks

Hirofumi Fukuyama[1] and William L. Weber[2]

[1]*Department of Business Management, Faculty of Commerce, Fukuoka University, Japan*
[2]*Department of Economics and Finance, Southeast Missouri State University, USA*

9.1   Introduction

Bank managers face a dynamic network problem in their attempts to generate and transform deposits into a portfolio of interest-bearing assets. Past successes or failures will enhance or constrain their choices today, which in turn, will affect future production possibilities. Moreover, various departments within the bank might have conflicting goals and the successful manager will have to coordinate production to ensure that each department contributes to the common goal. One type of network production model allows an intermediate output to be produced at one stage or by one division of a firm and then subsequently used as an input at a second stage to generate final outputs, which consist of desirable outputs and undesirable by-products. Managers of stage one might seek to maximize production of the intermediate output while the managers of stage two might seek to minimize its use. For instance, we consider banks that use labor, physical capital, and equity capital in a first stage to produce the intermediate output of deposits. In the second stage, those deposits are used to produce a portfolio of desirable interest-bearing assets and other fee-generating activities. However, the presence of risk and uncertainty usually means that an undesirable output is also produced: some loans becoming nonperforming. In a dynamic framework, the choices made and the efficiency with which resources are used in either stage will tend to impact future production. Nonperforming loans generated in stage 2 will tend to negatively affect the ability of stage 1 managers to

Efficiency and Productivity Growth: Modelling in the Financial Services Industry, First Edition. Edited by Fotios Pasiouras.
© 2013 John Wiley & Sons, Ltd. Published 2013 by John Wiley & Sons, Ltd.

generate deposits in a subsequent period. In addition, regional or macroeconomic upturns and downturns will sometimes mean that the managers of stage 2 might more profitably forego current production and 'save' deposits in an effort to minimize nonperforming loans in the current period and subsequently increase future desirable outputs.

The purposes of this chapter are, first, to develop a dynamic network production model that accounts for the scenario in the first paragraph and can be estimated using data envelopment analysis (DEA) and, second, to apply the theoretical production model to analyze the performance of Japanese Shinkin banks during fiscal years 2002–2009. Shinkin banks are cooperative institutions which collect deposits from members and then use those deposits to finance regional economic activities. We consider a three-year bank production horizon so that bank performance depends not just on production within a single period, but instead allows bank managers to account for regional/macroeconomic conditions and optimize across periods. The dynamic portion of our model builds on the models of Shephard and Färe (1980) and Färe and Grosskopf (1996) and Färe and Grosskopf (2000) and is related to research by Bogetoft et al. (2009) and Tone and Tsutsui (2009). Following Fukuyama and Weber (2010) and Fukuyama and Weber (2012), we control for undesirable outputs in a network production model. In addition, we extend the research of Akther, Fukuyama, and Weber (2013) by allowing the undesirable outputs generated in the second stage of production to have a negative effect on stage 1 production in a subsequent period. In the next section, we provide some background on Japanese Shinkin banks and briefly discuss the limited literature that has examined Shinkin bank performance. Then, in Sections 9.3 and 9.4, we develop and integrate the dynamic two-stage network production model allowing for undesirable outputs to be by-products of final desirable output production. In Section 9.5, we discuss the data and estimates of Shinkin bank performance. Section 9.6 summarizes and concludes.

9.2   Literature review of productivity analysis in credit banks in Japan

Shinkin banks are cooperative financial institutions organized under the Credit Associations Law of 1951. These banks accept deposits from the general public, but limit their lending to members: mortgage and personal loans are made to household members and commercial loans within a prefecture are made to small and medium-sized business members. The Financial Services Agency regulates Shinkin banks and the Shinkin Central Bank (formerly Zenshiren) provides member banks with deposit and lending services and helps facilitate foreign exchange transfers. From March 1998 to March 2011 the number of Shinkin bank head offices declined from 401 to 271 as the number of members grew from 8.6 to 9.3 million. During the same period, the total number of employees shrunk from 152 000 to 116 000 and total deposits grew from 98 to 120 trillion yen. In 1998, time and savings deposits were 79% of total deposits and demand deposits were 20% of total deposits. By 2011, time and savings deposits declined to 66% of total deposits and demand deposits grew to 33% of total deposits. In 1998, Japanese individuals accounted for 74% of total deposits and corporations held 20% of total deposits, with foreigners and the public holding the remainder. By 2011, individuals held 81% of total deposits and corporations held 16%. From 1998 to 2011, total loans and bills discounted by Shinkin banks shrunk from 70.4 to 63.4 trillion yen while investment securities grew from 16.3 to 34.4 trillion yen. Corporations received 70% of total Shinkin bank loans in 1998 and 65% in 2011. Of the corporate loans made in 1998, 23% were to manufacturing, 17% were to construction, 19% to wholesale and retail trade, 10.4% to real estate, and the remainder to corporations in the service industry. By 2011, those shares were 17% to manufacturing, 13% to construction, 15% to wholesale and retail trade, and 19.2% to real estate. Loans to individuals shrunk from 19.8 trillion yen in 1998 to 18.1 trillion yen in 2011 with the share of individual loans for housing increasing from 56% to 82%.[1]

Nishikawa (1973) estimated a simple log-linear cost function for Shinkin banks in 10 Japanese regions in 1968 and found that Shinkin banks operated in the range of constant returns to scale, except in the Tokai region where Shinkin banks had increasing returns to scale. Miyamura (1992) used data on 456 Shinkin banks in 1985 and 451 Shinkin banks in 1990 and estimated a translog cost function for six different outputs including interest income, dividends from trust accounts, noninterest income, other fees and commissions, and loans per branch office. He found that city banks and rural banks faced different production/cost technologies. In 1985, both city and rural banks operated in the range of constant returns to scale for their respective technologies. By 1990, banks in both areas operated in the range of increasing returns to scale and city banks exhibited greater scale economies than rural banks. Miyakoshi (1993) used data from 1989 to 1998 for 114 Shinkin banks in the Kanto area including Tokyo and 123 Shinkin banks in other areas including Hokkaido, Tohoku, Koshinetsu, and Hokuriku. Using a translog cost function with two outputs – loans and securities – he found scope economies for banks in the Kanto area and significant scale economies for banks in both the Kanto area and other areas of Japan. Hirota and Tsutui (1992) also used a translog cost function to test for scope economies in production loans, securities, and deposits. For 452 banks operating in 1987, they found no scope economies between any of the three pairs of outputs, although significant scale economies were found.

Fukuyama (1996) used DEA to estimate technical efficiency for 435 Shinkin banks in 1992. He found that larger banks were more efficient than smaller banks, but the enhanced efficiency was primarily due to better managerial oversight in minimizing input use rather than efficiency gains due to larger banks operating in the range of constant returns to scale. Harimaya (2004) showed similar results using DEA and a stochastic frontier cost function and also found that bank efficiency declined as bank's ratio of cost to deposits increased.

The market structure hypothesis posits that banks in concentrated markets can charge higher rates on loans and pay lower rates on deposits due to their market power, thus increasing their profits. In contrast, the efficient structure hypothesis posits that efficient banks obtain lower costs and higher profits because of their efficiency, leading to a concentrated market. A stochastic frontier analysis by Satake and Tsutsui (2002) found evidence supporting the efficient structure hypothesis only up until the 1980s for Shinkin banks in Kyoto prefecture. Like Harimaya (2004), Satake and Tsutsui (2002) also reported a negative relation between Shinkin bank inefficiency and the ratio of costs to deposits. Fukuyama and Weber (2009) estimated slacks-based inefficiency for between 289 and 298 Shinkin banks in 2002–2005 and found that banks with a higher ratio of equity capital to total assets were less inefficient, suggesting that owners with more money at stake were able to exert more pressure on bank managers to be technically efficient.

Several papers investigated the actual or potential consequences of merger and acquisition activities among Shinkin banks. During the 1990–2002 period, Hosono, Sakai, and Tsuru (2007) compared the five-year period before and after a merger by constructing *pro forma* balance sheets of target and acquired Shinkin banks for 97 mergers and acquisitions. Larger

---

[1] Shinkin Central Bank Research Institute http://www.scbri.jp/e_statistics.htm

Shinkin banks were more likely to acquire smaller and slower growing banks. However, their findings did not support the efficient structure hypothesis, but instead suggested that mergers occurred as banks tried to be designated as 'too big to fail'. In contrast to the work of Hosono, Sakai, and Tsuru (2007), Färe, Fukuyama, and Weber (2010) allowed the potential gains from mergers and acquisitions to be estimated *ex ante*, rather than be inferred from an *ex post* examination of balance sheet and income statement data. For Shinkin banks on Kyushu Island in Japan, the largest potential gain in final outputs for infra-prefecture mergers was for banks in Nagasaki and the smallest potential gain in final outputs occurred for infra-prefecture mergers in Fukuoka and Saga. For inter-prefecture mergers, banks in Miyazaki and Nagasaki had the largest *ex ante* gains, while potential mergers between banks located in Fukuoka and the other six prefectures on Kyushu Island had the smallest potential gains in final loan outputs. Barros, Managi, and Matousek (2009) and Assaf, Barros, and Matousek (2011) used DEA to estimate a Malmquist productivity index. They found that Shinkin banks experienced small average declines in annual productivity attributable to negative technical change during the 2000–2006 period which they attributed to slow growth in Japanese economic activity.

The studies mentioned earlier did not account for the undesirable output of nonperforming loans when measuring bank performance. To provide a more complete representation of the bank technology, Fukuyama and Weber (2008) estimated a parametric directional distance function accounting for nonperforming loans and desirable bank outputs of loans and securities investments. During 2001–2004, they found that regional banks which focused on profits were more efficient, had faster technological progress, and a higher shadow cost of reducing nonperforming loans than cooperative Shinkin banks.

9.3   Dynamic network production

9.3.1   The two-stage technology

In this section, we present a network production technology for banks that use inputs, $x^t \in \mathfrak{R}_+^N$, in one stage to produce the intermediate output of deposits, $z^t \in R_+$, which are then used in a subsequent stage to produce a portfolio of final desirable outputs, $fy^t \in R_+^M$, and undesirable outputs of nonperforming loans, $b^t \in R_+$. We assume there are $j = 1, \dots, J$ banks and production takes place in periods $t = 0, \dots, T$. Nonperforming loans produced in stage 2 during period $t - 1 (b^{t-1})$ act as an undesirable input to stage 1 production in period t. Undesirable inputs shrink the bank's production possibilities set and require greater use of desirable inputs to offset their effects. For instance, banks that generate nonperforming loans are usually constrained in their ability to raise deposits unless they offset those nonperforming loans with an injection of equity capital.

To allow a dynamic aspect to our model, we allow a bank to use deposits in the current period to produce final outputs, or, the bank can save deposits for use in a subsequent period. We denote carryover assets as c^t. Bank managers might want to carryover some assets to a future period when they expect a recession or other events to cause too many loans to become nonperforming, or when expected increases in interest rates would reduce the market value of securities investments. In such an environment, managers might find it more efficient to forego purchasing securities or making loans in period t so that nonperforming loans are also reduced and future production possibilities are expanded. Therefore, total output in the second stage of production is

$$y^t = fy^t + c^t, \qquad (9.1)$$

where total output consists of the sum of final outputs and carryover outputs.

The stage 1 production possibility set in year t is denoted by

$$P1^t = \left\{ \left(b^{t-1}, z^t, x^t\right) \text{ such that } \left(b^{t-1}, x^t\right) \text{ can produce } z^t \right\} \tag{9.2}$$

and the stage 2 production possibility set in the same year is given by

$$P2^t = \left\{ \left(c^{t-1}, z^t, b^t, fy^t + c^t\right) \text{ such that } \left(c^{t-1}, z^t\right) \text{ can produce } \left(b^t, fy^t + c^t\right) \right\} \tag{9.3}$$

In stage 1, banks combine desirable inputs (x^t) with undesirable inputs from the previous period (b^{t-1}) to produce the intermediate output (z^t). In stage 2, the bank combines the intermediate output from the first stage (z^t) and carryover assets from the previous period (c^{t-1}) to produce total outputs (y^t) which equal the sum of final outputs (fy^t) and carryover assets (c^t). Combining Equations (9.2) and (9.3), the network production possibility set[2] is

$$N^t = \left\{ \left(b^{t-1}, x^t, z^t, c^{t-1}, b^t, fy^t + c^t\right) \text{ such that } \right.$$
$$\left. \left(b^{t-1}, x^t, z^t\right) \in P1^t \text{ and } \left(c^{t-1}, z^t, b^t, fy^t + c^t\right) \in P2^t \right\}. \tag{9.4}$$

The directional distance function was introduced by Chambers, Chung, and Färe (1996) and Chambers, Chung, and Färe (1998), and Färe and Grosskopf (2004) provided further theory and applications. This distance function gives the maximum contraction in inputs and undesirable outputs and simultaneous expansion in desirable outputs given a production technology. The directional distance function has been used in numerous empirical applications and has recently been used to measure performance for firms that face a network technology (Akther, Fukuyama, and Weber 2013). Let $\mathbf{g} = \left(g_x, g_b, g_y\right) \in R_+^{N+1+M}$ be a directional vector used to scale inputs and outputs to the production frontier. For the network technology given by Equation (9.4), the directional distance function takes the form

$$\vec{D}^t\left(x^t, b^{t-1}, b^t, c^{t-1}, fy^t + c^t\right)$$
$$= \max\left\{ \beta \text{ subject to } (x^t - \beta g_x, b^{t-1}, b^t - \beta g_b, \hat{z}^t, fy^t + \beta g_y + c^t) \in N^t \right\}. \tag{9.5}$$

Here we note that the intermediate outputs of the first stage (z^t) which become an input to the second stage are not a parameter of $\vec{D}(\cdot)$, but are instead optimally chosen to maximize the size of the network technology. The optimal values are represented by $\hat{z}^t$ and these optimal values provide the link between stage 1 and stage 2 production. The network technology represented by Equation (9.4) is illustrated in Figure 9.1.

Instead of measuring network performance for a single period, we want to allow bank managers to optimize production over several periods by choosing not only the intermediate output of deposits, but also the amount of assets to carryover from one period to the next. In the empirical section of the chapter, we consider a three-year dynamic planning horizon for production, although longer horizons can be used. To streamline notation, let $b = (b^{t-1}, b^t, b^{t+1}, b^{t+2})$, $x = (x^t, x^{t+1}, x^{t+2})$, $z = (z^t, z^{t+1}, z^{t+2})$, $y = (y^t, y^{t+1}, y^{t+2})$, $fy = (fy^t, fy^{t+1}, fy^{t+2})$, and

---

[2] The standard two-stage network model without either bad outputs or carryover variables is presented by Kao and Hwang (2008) and studied further by Chen, Cook, and Zhu (2010).

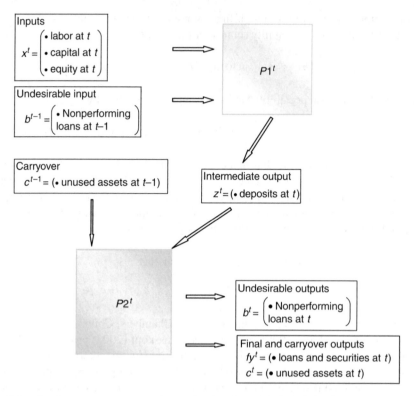

Figure 9.1   Static two-stage network production for a bank. $P1^t$ is the stage 1 production possibility set and $P2^t$ is the stage 2 production possibility set.

$c = (c^{t-1}, c^t, c^{t+1}, c^{t+2})$. The three-year dynamic technology (DN) is illustrated in Figure 9.2 and is denoted by the dynamic network production possibility set

$$\text{DN} = \left\{ (b, x, z, c, fy) \text{ such that } \left(b^{t-1}, x^t, z^t, c^{t-1}, b^t, fy^t + c^t \right) \in N^t, \right.$$

$$\left(b^t, x^{t+1}, z^{t+1}, c^t, b^{t+1}, fy^{t+1} + c^{t+1} \right) \in N^{t+1}, \text{ and} \qquad (9.6)$$

$$\left. \left(b^{t+1}, x^{t+2}, z^{t+2}, c^{t+1}, b^{t+2}, fy^{t+2} + c^{t+2} \right) \in N^{t+2} \right\}.$$

We measure bank performance relative to Equation (9.6) by a three-year dynamic network directional distance function:

$$\vec{D}\left(b, x, c^{t-1}, c^{t+2}, fy; g\right) = \max \left\{ \beta^t + \beta^{t+1} + \beta^{t+2} \quad \text{subject to:} \right.$$

$$\begin{pmatrix} b^{t-1}, \ x^t - \beta^t g_x, \ \hat{z}^t, \ c^{t-1}, \\ fy^t + \beta^t g_y + \hat{c}^t, \ b^t - \beta^t g_b \end{pmatrix} \in N^t,$$

$$\begin{pmatrix} b^t - \beta^t g_b, \ x^{t+1} - \beta^{t+1} g_x, \ \hat{z}^{t+1}, \ \hat{c}^t, \\ fy^{t+1} + \beta^{t+1} g_y + \hat{c}^{t+1}, \ b^{t+1} - \beta^{t+1} g_b \end{pmatrix} \in N^{t+1} \qquad (9.7)$$

$$\left. \begin{pmatrix} b^{t+1} - \beta^{t+1} g_b, \ x^{t+2} - \beta^{t+2} g_x, \ \hat{z}^{t+2}, \ \hat{c}^{t+1}, \\ fy^{t+2} + \beta^{t+2} g_y + c^{t+2}, \ b^{t+2} - \beta^{t+2} g_b \end{pmatrix} \in N^{t+2} \right\}.$$

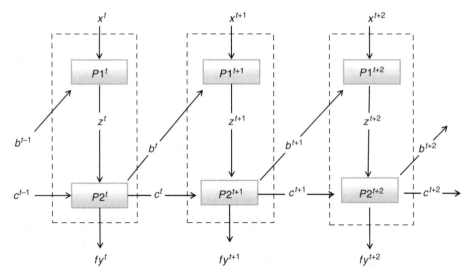

Figure 9.2   Three-period dynamic network production.

In Equation (9.7), banks choose intermediate outputs (deposits) $\hat{z}^t$, $\hat{z}^{t+1}$, and $\hat{z}^{t+2}$ to maximize each period's production possibility set. The choice of these intermediate outputs provides the network link between stage 1 and stage 2 in period t, $t+1$, and $t+2$. The dynamic link between the three periods occurs because banks also choose the amount of assets to carryover from period t to $t+1\left(\hat{c}^t\right)$ and the amount of assets to carryover from period $t+1$ to $t+2\left(\hat{c}^{t+1}\right)$. These carryover assets are chosen to maximize the size of the dynamic production possibility set. The carryover assets from period $t-1$ and from the final period $t+2(c^{t-1}$ and $c^{t+2})$ are taken as given to satisfy tranversality conditions.

9.3.2   Three-year dynamic DEA

In this section, we show how data envelopment analysis (DEA) of Charnes, Cooper, and Rhodes (1978) and Farrell (1957) can be used to represent the best-practice dynamic network technology. In each period, we observe the inputs and outputs for $j=1,...,J$ banks. The method of DEA forms linear combinations of the observed inputs and outputs for the J banks to generate a best-practice technology. The directional distance function is estimated using linear programming methods given the best-practice frontier. An advantage of DEA over stochastic methods of measuring performance is that it does not require the researcher to specify an ad hoc functional form such as Cobb–Douglas, translog, or quadratic, and it does not require the researcher to specify a form for the error structure. However, a limitation of DEA is that all deviation of a firm's outputs and inputs from the frontier is attributed to inefficiency on the part of the bank's managers, even though some of the deviation might be due to luck or measurement error.

We define the intensity variables for stage 1 as $\lambda^t = \left(\lambda_1^t,...,\lambda_J^t\right) \in \Re_+^J$ and the intensity variables for stage 2 as $\Lambda^t = \left(\Lambda_1^t,...,\Lambda_J^t\right) \in \Re_+^J$. The intensity variables form linear combinations of all banks' observed inputs and outputs for the two stages. Extending the network model of Akther, Fukuyama, and Weber (2013) to allow for carryover assets, we define the year t network technology as

$$N^t = \left\{ \left(b^{t-1}, x^t, z^t, c^{t-1}, b^t, fy^t + c^t \right) \text{ such that} \right.$$

$$b^{t-1} = \sum_{j=1}^{J} b_j^{t-1} \lambda_j, \quad x^t \geq \sum_{j=1}^{J} x_j^t \lambda_j, \quad z^t \leq \sum_{j=1}^{J} z_j^t \lambda_j, \quad \lambda_j \geq 0, \quad j = 1, \ldots, J \tag{9.8}$$

$$\left. z^t \geq \sum_{j=1}^{J} z_j^t \Lambda_j, \quad b^t = \sum_{j=1}^{J} b_j^t \Lambda_j, \quad fy^t + c^t \leq \sum_{j=1}^{J} y_j^t \Lambda_j, \quad c^{t-1} \geq \sum_{j=1}^{J} c_j^{t-1} \Lambda_j, \quad \Lambda_j \geq 0, \quad j = 1, \ldots, J \right\}.$$

The right-hand side of each constraint equals a linear combination of the observed values of inputs and outputs with the intensity variables for the $j = 1, \ldots, J$ banks and the left-hand side variables $b^{t-1}, x^t, z^t, b^t, fy^t + c^t$, and c^{t-1} consist of the set of inputs and outputs that satisfy the constraints. The equality constraint $b^{t-1} = \sum_{j=1}^{J} b_j^{t-1} \lambda_j$ is associated with stage 1 of production and corresponds with the assumption that nonperforming loans produced in period $t - 1$ are an undesirable input to stage 1 of the subsequent period. The equality models the notion of weak disposability of inputs. Weak disposability of inputs means that if more undesirable input is used, more desirable inputs must also be used to offset their negative effect if the intermediate output is to remain constant. In stage 2, the equality constraint $b^t = \sum_{j=1}^{J} b_j^t \Lambda_j$ models the notion of weak disposability of outputs. Weak disposability of outputs means that if undesirable outputs (nonperforming loans) are to be reduced, some desirable outputs (loans) must also be foregone.[3] The remaining inequality constraints allow the standard assumption of strong (free) disposability of inputs and outputs. Carryover assets from the previous period (c^{t-1}) expand the stage 2 production possibility set. This effect is seen in the constraints $c^{t-1} \geq \sum_{j=1}^{J} c_j^{t-1} \Lambda_j$, since increases in carryover assets (c^{t-1}) relax the constraint and allow the right-hand side technology to become larger. Finally, the link between stage 1 and stage 2 is seen in the two constraints $z^t \leq \sum_{j=1}^{J} z_j^t \lambda_j$ and $z^t \geq \sum_{j=1}^{J} z_j^t \Lambda_j$. These two constraints can be combined to yield $\sum_{j=1}^{J} \left(\lambda_j - \Lambda_j \right) z_j^t \geq 0$. Thus, the intensity variables for stage 1 $\left(\lambda_j^t \right)$ and stage 2 $\left(\Lambda_j^t \right)$ are chosen to satisfy the constraint and this constraint provides the network link between the two stages of production.

Using the DEA network technology (9.8), the network directional distance function for bank k is estimated as

$$\vec{D}^t (b_k^{t-1}, x_k^t, b_k^t, c_k^{t-1}, fy_k^t + c_k^t) = \max \left\{ \beta \text{ subject to:} \right.$$

$$b_k^{t-1} = \sum_{j=1}^{J} b_j^{t-1} \lambda_j, \quad x_k^t - \beta g_x \geq \sum_{j=1}^{J} x_j^t \lambda_j, \quad \sum_{j=1}^{J} \left(\lambda_j^t - \Lambda_j^t \right) z_j^t \geq 0, \quad b_k^t - \beta g_b = \sum_{j=1}^{J} b_j^t \Lambda_j,$$

$$\left. fy_k^t + \beta g_y + c_k^t \leq \sum_{j=1}^{J} y_j^t \Lambda_j, \quad c_k^{t-1} \geq \sum_{j=1}^{J} c_j^{t-1} \Lambda_j, \quad \lambda_j^t \geq 0, \Lambda_j^t \geq 0, \quad j = 1, \ldots, J \right\}. \tag{9.9}$$

---

[3] Formally, weak disposability of inputs in stage 1 means that if $(b^{t-1}, x^t, z^t) \in P1^t$ then $(\theta b^{t-1}, \theta x^t, z^t) \in P1^t$, for $\theta \geq 1$. Weak disposability of undesirable outputs in stage 2 means that if $(c^{t-1}, b^t, z^t, fy^t + c^t) \in P2^t$ then $(c^{t-1}, \phi b^t, z^t, \phi(fy^t + c^t)) \in P2^t$ for $0 \leq \phi \leq 1$.

We extend the DEA network technology to a dynamic framework by allowing production to take place in three periods: t, $t+1$, and $t+2$. The dynamic network production technology given by Equation (9.6) is defined using DEA as

$$\text{DN} = \Bigg\{ (b, x, z, c, y) \ \text{such that in period} \ t$$

$$b^{t-1} = \sum_{j=1}^{J} b_j^{t-1} \lambda_j^t, \ x^t \geq \sum_{j=1}^{J} x_j^t \lambda_j^t, \ \sum_{j=1}^{J} \left(\lambda_j^t - \Lambda_j^t \right) z_j^t \geq 0, \ \lambda_j^t \geq 0, \quad j = 1,\ldots,J,$$

$$b^t = \sum_{j=1}^{J} b_j^t \Lambda_j^t, \ fy^t + c^t \leq \sum_{j=1}^{J} y_j^t \Lambda_j^t, \ c^{t-1} \geq \sum_{j=1}^{J} c_j^{t-1} \Lambda_j^t, \ \Lambda_j^t \geq 0, \quad j = 1,\ldots,J,$$

(9.10)

in period $t+1$

$$b^t = \sum_{j=1}^{J} b_j^t \lambda_j^{t+1}, \ x^{t+1} \geq \sum_{j=1}^{J} x_j^{t+1} \lambda_j^{t+1}, \ \sum_{j=1}^{J} \left(\lambda_j^{t+1} - \Lambda_j^{t+1} \right) z_j^{t+1} \geq 0, \ \lambda_j^{t+1} \geq 0, \quad j = 1,\ldots,J,$$

$$b^{t+1} = \sum_{j=1}^{J} b_j^{t+1} \Lambda_j^{t+1}, \ fy^{t+1} + c^{t+1} \leq \sum_{j=1}^{J} y_j^{t+1} \Lambda_j^{t+1}, \ c^t \geq \sum_{j=1}^{J} c_j^t \Lambda_j^{t+1}, \ \Lambda_j^{t+1} \geq 0, \quad j = 1,\ldots,J,$$

(9.11)

and in period $t+2$

$$b^{t+1} = \sum_{j=1}^{J} b_j^{t+1} \lambda_j^{t+2}, \ x^{t+2} \geq \sum_{j=1}^{J} x_j^{t+2} \lambda_j^{t+2}, \ \sum_{j=1}^{J} \left(\lambda_j^{t+2} - \Lambda_j^{t+2} \right) z_j^{t+2} \geq 0, \ \lambda_j^{t+2} \geq 0, \quad j = 1,\ldots,J,$$

$$b^{t+2} = \sum_{j=1}^{J} b_j^{t+2} \Lambda_j^{t+2}, \ fy^{t+2} + c^{t+2} \leq \sum_{j=1}^{J} y_j^{t+2} \Lambda_j^{t+2}, \ c^{t+1} \geq \sum_{j=1}^{J} c_j^{t+1} \Lambda_j^{t+2}, \ \Lambda_j^{t+2} \geq 0, \quad j = 1,\ldots,J \Bigg\}.$$

(9.12)

The network links between stage 1 and stage 2 are provided by the constraints
$\sum_{j=1}^{J} \left(\lambda_j^t - \Lambda_j^t \right) z_j^t \geq 0$ in period t, $\sum_{j=1}^{J} \left(\lambda_j^{t+1} - \Lambda_j^{t+1} \right) z_j^{t+1} \geq 0$ in period $t+1$, and
$\sum_{j=1}^{J} \left(\lambda_j^{t+2} - \Lambda_j^{t+2} \right) z_j^{t+2} \geq 0$ in period $t+2$. The dynamic links between period t and $t+1$ are provided by two sets of constraints. First, the undesirable outputs produced in period t at stage 2 become inputs in stage 1 during period $t+1$. This link means that the intensity variables $\Lambda_j^t, j=1,\ldots,J$ in period t and $\lambda_j^{t+1}, j=1,\ldots,J$ period $t+1$ must be chosen to satisfy both $b^t = \sum_{j=1}^{J} b_j^t \Lambda_j^t$ and $b^t = \sum_{j=1}^{J} b_j^t \lambda_j^{t+1}$. Second, carryover assets (c^t) in period t become an input in period $t+1$, so the intensity variables $\Lambda_j^t, \ j=1,\ldots,J$ and $\Lambda_j^{t+1}, \ j=1,\ldots,J$ and the choice of c^t must satisfy $fy^t + c^t \leq \sum_{j=1}^{J} y_j^t \Lambda_j^t$ and $c^t \geq \sum_{j=1}^{J} c_j^t \Lambda_j^{t+1}$. The dynamic links between period $t+1$ and $t+2$ are similar. For the undesirable outputs, the intensity variables $\Lambda_j^{t+1}, \ j=1,\ldots,J$ and $\lambda_j^{t+2}, j=1,\ldots,J$ must satisfy both $b^{t+1} = \sum_{j=1}^{J} b_j^{t+1} \Lambda_j^{t+1}$ and $b^{t+1} = \sum_{j=1}^{J} b_j^{t+1} \lambda_j^{t+2}$. For carryover assets,

the intensity variables $\Lambda_j^{t+1}, j = 1,...,J$ and $\Lambda_j^{t+2}, j = 1,...,J$ and the choice of c^{t+1} must satisfy

$$fy^{t+1} + c^{t+1} \le \sum_{j=1}^{J} y_j^{t+1}\Lambda_j^{t+1} \quad \text{and} \quad c^{t+1} \ge \sum_{j=1}^{J} c_j^{t+1}\Lambda_j^{t+2}.$$

Using Equation (9.12), we define the dynamic three-period network directional distance function for bank k as

$$\vec{D}(b_k, x_k, c_k^{t-1}, c_k^{t+2}, fy_k) = \max \left\{ \beta^t + \beta^{t+1} + \beta^{t+2} \quad \text{subject to:} \right.$$

$$b_k^{t-1} = \sum_{j=1}^{J} b_j^{t-1}\lambda_j^t, \quad x_k^t - \beta^t g_x \ge \sum_{j=1}^{J} x_j^t\lambda_j^t, \quad \sum_{j=1}^{J}\left(\lambda_j^t - \Lambda_j^t\right)z_j^t \ge 0,$$

$$b_k^t - \beta^t g_b = \sum_{j=1}^{J} b_j^t\Lambda_j^t, \quad fy_k^t + \beta^t g_y + \hat{c}^t \le \sum_{j=1}^{J} y_j^t\Lambda_j^t,$$

$$c^{t-1} \ge \sum_{j=1}^{J} c_j^{t-1}\Lambda_j^t, \quad \lambda_j^t \ge 0, \quad j = 1,...,J, \quad \Lambda_j^t \ge 0, \quad j = 1,...,J,$$

$$b_k^t - \beta^t g_b = \sum_{j=1}^{J} b_j^t\lambda_j^{t+1}, \quad x_k^{t+1} - \beta^{t+1} g_x \ge \sum_{j=1}^{J} x_j^{t+1}\lambda_j^{t+1}, \quad \sum_{j=1}^{J}\left(\lambda_j^{t+1} - \Lambda_j^{t+1}\right)z_j^{t+1} \ge 0,$$

$$b_k^{t+1} - \beta^{t+1} g_b = \sum_{j=1}^{J} b_j^{t+1}\Lambda_j^{t+1}, \quad fy_k^{t+1} + \beta^{t+1} g_y + \hat{c}^{t+1} \le \sum_{j=1}^{J} y_j^{t+1}\Lambda_j^{t+1},$$

$$\hat{c}^t \ge \sum_{j=1}^{J} c_j^t\Lambda_j^{t+1}, \quad \lambda_j^{t+1} \ge 0, \quad j = 1,...,J, \quad \Lambda_j^{t+1} \ge 0, \quad j = 1,...,J$$

$$b_k^{t+1} - \beta^{t+1} g_b = \sum_{j=1}^{J} b_j^{t+1}\lambda_j^{t+2}, \quad x_k^{t+2} - \beta^{t+2} g_x \ge \sum_{j=1}^{J} x_j^{t+2}\lambda_j^{t+2}, \quad \sum_{j=1}^{J}\left(\lambda_j^{t+2} - \Lambda_j^{t+2}\right)z_j^{t+2} \ge 0,$$

$$b_k^{t+2} - \beta^{t+2} g_b = \sum_{j=1}^{J} b_j^{t+2}\Lambda_j^{t+2}, \quad fy_k^{t+2} + \beta^{t+2} g_y + c_k^{t+2} \le \sum_{j=1}^{J} y_j^{t+2}\Lambda_j^{t+2},$$

$$\left. \hat{c}^{t+1} \ge \sum_{j=1}^{J} c_j^{t+1}\Lambda_j^{t+2}, \quad \lambda_j^{t+2} \ge 0, \quad j = 1,...,J, \quad \Lambda_j^{t+2} \ge 0, \quad j = 1,...,J \right\}.$$

(9.13)

The distance function given by Equation (9.13) is maximized by choosing the intensity variables λ_j^τ and Λ_j^τ for $\tau = t$, $t+1$, $t+2$ and $j = 1,...,J$; carryover assets $\hat{c}^t$ and $\hat{c}^{t+1}$; and the variables β^t, β^{t+1}, and β^{t+2} subject to the constraints.

9.4   Cooperative Shinkin banks: An empirical illustration

9.4.1   Defining bank inputs and outputs

We apply the dynamic network model to 269 cooperative Japanese Shinkin banks operating during 2002–2009. There exists some disagreement on whether deposits should be treated as an input or an output. Berger and Humphrey (1992) and Berger and Humphrey (1997) reviewed various financial institution efficiency studies and the various methods used to define inputs and outputs. Sealey and Lindley's (1977) asset approach assumes deposits are an input and loans and other interest-bearing assets are outputs. The value-added approach defines outputs as any liability or asset that adds significant value to a bank with inputs equal to labor and the value of fixed assets, including premises and physical capital. The user cost approach of Hancock (1985) defines outputs as any asset or liability that contributes to revenues and inputs equal labor and any asset or liability that adds significantly to costs. Berger, Hanweck, and Humphrey (1987) and Goddard, Molyneux, and Wilson (2001) provided further discussion about the treatment of deposits. Fukuyama and Weber (2009) used the asset approach in measuring Shinkin bank performance and assumed that a bank uses deposits, labor, and physical capital to produce a portfolio of assets, including loans and securities investments. Fukuyama and Weber (2010) proposed a two-stage network model for Japanese credit cooperative Shinkin banks where deposits are an intermediate output of the first stage of production and an input to a second stage where they produce the portfolio of assets. More recently, Akther, Fukuyama, and Weber (2013) extended their network model by including nonperforming loans from a preceding year to estimate the technical inefficiency of a bank in the current year. Wang, Gopal, and Zionts (1997) and Chen and Zhu (2004) treated deposits as an intermediate product in a two-stage network problem. Their initial inputs are fixed assets, employees, and information technology investments, and the final outputs are profits, loans recovered, and marketability. Therefore, the Wang-Gopal-Zionts and Chen-Zhu frameworks differ from Sealey and Lindley's asset approach and the network version of Fukuyama and Weber (2010). Given the disagreement on deposits, a feature of our network model is that we treat deposits as an intermediate output of the first stage of production and as an input to the second stage of production.

9.4.2 NPLs in the efficiency/productivity measurement

We assume that nonperforming loans from a previous period are an undesirable input to the first stage of production in a subsequent period. As such, these nonperforming loans require the use of other inputs to offset their negative effects, or else output will fall. Consider the following example which examines a hypothetical bank balance sheet in period t and period $t+1$. The balance sheet for period t is given in Table 9.1.

The leverage ratio (Equity/Assets) for the bank is $(0.074 = 8/108)$ and the bank is classified as 'well-capitalized' (see Saunders and Cornett, 2008, p. 595). Suppose that during the period, 5 out of 60 loans become nonperforming or bad loans. The new balance sheet at the start of period $t+1$ is given in Table 9.2.

In period $t+1$ the bank's leverage ratio equals 0.029 ($=3/103$) and the bank is categorized as 'significantly undercapitalized'. One of two things (or some combination) must now occur: the bank can shrink deposits and with it, loans and securities, until it meets the leverage ratio requirement of 4% to be categorized as 'adequately capitalized' or, it must raise additional equity capital. One possible way to reorganize the balance sheet and become 'adequately capitalized' is given in Table 9.3.

Thus, the bad loans that occur in period t act as an undesirable input in period $t+1$, and require additional inputs (equity) to offset their effect, or, constrain the amount of the intermediate output (deposits) that can be produced, which in turn causes securities to shrink.

Table 9.1 Hypothetical bank's balance sheet for period t.

Assets		Liabilities + equity	
Cash	10	Deposits	100
Total loans	60	Equity	8
Securities	38		

Table 9.2 Hypothetical bank's balance sheet for period $t+1$.

Assets		Liabilities + equity	
Cash	10	Deposits	100
Performing loans	55		
Total loans	60	Equity	3
Bad loans	−5		
Securities	38		

Table 9.3 Hypothetical bank's reorganized balance sheet for period $t+1$.

Assets		Liabilities + equity	
Cash	10	Deposits	72
Performing loans	55		
Total loans	60	Equity	3
Bad loans	−5		
Securities	10		

Table 9.4 Number of existing and sample Shinkin banks.

No. of banks	2002	2003	2004	2005	2006	2007	2008	2009
Existing	326	306	298	292	287	281	279	272
Sample	269	269	269	269	269	269	269	269

9.4.3   Data

We employ a balanced panel data set consisting of 269 Japanese Shinkin banks during a span of eight fiscal years from 2002 to 2009. The Japanese fiscal year begins on 1 April and ends on 31 March of the subsequent year, thus our data is from the period beginning 1 April 2002 and ending 31 March 2010. During the last year there were 272 Shinkin banks operating, but we use only 269 banks because some data was missing for three of the banks.[4] The data source is Nikkei's Financial Quest (Table 9.4).

The decline in the number of operating banks from 326 in 2002 to 272 in 2009 reflects consolidation in the Shinkin banking industry. We follow Fukuyama and Weber (2010) in defining inputs and outputs. The first-stage inputs are labor (x_1), fixed capital (x_2), and net assets (x_3). Labor (x_1) equals the unconsolidated total number of employees excluding directors holding concurrent posts, temporary employees, and temporary retired workers. Fixed capital equals the asset value of tangible and intangible fixed assets. Net assets for cooperative Shinkin banks correspond to equity capital (assets minus liabilities) for joint stock commercial banks.

The intermediate output of the first stage is deposits (z), which equals the sum of current deposits, ordinary deposits, savings deposits, deposits at notice, time deposits, and other deposits. Deposits produced in the first stage are used as an input in the second stage to

---

[4] We deleted Tsurugi Shinkin Bank, Himifushiki Shinkin Bank, and Hinase Shinkin Bank.

Table 9.5 Descriptive statistics for the pooled sample ($N=2152$).

	Mean	Std. dev.	Minimum	Maximum
y_1 = loans	246.2	321.7	18.6	2409.3
y_2 = securities	118.8	139.7	2.0	1119.1
$c_1 + c_2$ = carryover assets	90.9	111.2	5.4	1023.2
x_1 = labor	412	408	35	2651
x_2 = physical capital	7.2	9.7	0.2	69.3
x_3 = net assets (equity)	23.8	27.7	0.9	204.6
z = deposits	431.0	523.4	33.1	4263.6
b = nonperforming loans	19.5	24.5	0.8	211.9

Labor equals number of employees. Physical capital, net assets (equity), deposits, loans, nonperforming loans, securities investments, and carryover assets are in billions of Japanese yen deflated by the Japanese GDP deflator (base year = 2000).

produce the final outputs of total loans (fy_1) and securities (fy_2), carryover assets (c_1 and c_2), and the undesirable output of nonperforming loans (b). Loans equal the sum of total loans and bills discounted. Nonperforming loans (b) equal the unconsolidated bank account sum of loans to customers in bankruptcy, nonaccrual delinquent loans, loans past due more than three months, and restructured loans. The data does not allow us to distinguish between carryover assets that come from loans (c_1) and carryover assets that come from securities (c_2). However, total carryover assets ($c_1 + c_2$) are derived as $c_1 + c_2 =$ assets − (required reserves + x_2 + $fy_1 + fy_2$). That is, total carryover assets equal the difference between total assets and the sum of required reserves, fixed capital, loans, and securities investments. The required reserve ratio for Shinkin banks depends on bank size and varies for time deposits and other deposits (O'Brien, 2007). For banks with deposits between 50 and 500 billion yen, the required reserve ratio for time deposits (other deposits) is 0.05% (0.1%). For banks with deposits between 500 billion yen and 1.2 trillion yen, the required reserve ratio for time deposits (other deposits) is 0.05% (0.8%). For banks with deposits between 1.2 and 2.5 trillion yen, the required reserve ratio for time deposits (other deposits) is 0.9% (1.3%). For banks with deposits greater than 2.5 trillion yen, the required reserve ratio for time deposits (other deposits) is 1.2% (1.3%). To estimate the model we arbitrarily assume that all carryover assets correspond with securities. Thus, we assume $c_1 = 0$ and $y_1 = fy_1$.

Table 9.5 presents descriptive statistics of the inputs and outputs. All financial data are in billions of Japanese yen deflated by the Japanese GDP deflator. The average Shinkin bank uses 412 workers, approximately 7.2 billion yen in fixed capital, and 23.8 billion yen in net assets (equity capital). These inputs are used to generate an average of 430.9 billion in deposits which are then transformed into 246.2 billion in loans, 118.8 billion in securities, and 91 billion yen in carryover assets. The average Shinkin bank had 19.5 billion yen in nonperforming loans.

9.5 Estimates

To estimate performance we must first choose a directional vector to scale outputs and inputs for each bank to the frontier of the network technology. An infinite number of directional vectors are possible and are dependent on the objective of the researcher, bank manager, and/or policy-maker. For instance, if policy-makers are interested in reducing bad loans holding inputs and desirable outputs constant, an appropriate choice of directional vector might be $\mathbf{g}=(0,g_b,0)$. If instead, the decision-maker is interested in seeing how much desirable outputs could be expanded holding inputs and nonperforming loans constant, then they might instead choose $\mathbf{g}=(0,0,g_y)$. Briec (1997) suggested using the firm's observed values of inputs and outputs as the directional vector. That is, for firm k, the directional vector would be $\mathbf{g}=(x_k,b_k,y_k)$. This directional vector means that the estimated directional distance function multiplied by 100% gives the percent expansion in desirable outputs and simultaneous percent contraction in inputs and undesirable outputs making it easier to compare with Shephard (1970) distance functions. As an alternative, Färe and Grosskopf (2004) showed that when all firms are evaluated using a common directional vector, it is possible, under certain conditions, to aggregate the individual firm's directional distance functions to an industry directional distance function. Such an aggregation is not possible with the Briec (1997) specification. One possible directional vector that would be common to all firms is $\mathbf{g}=(x,b,y)=(1,1,1)$. In this case, the directional distance function gives the simultaneous unit contraction in inputs and undesirable outputs and unit expansion in outputs. Another possibility would be to evaluate each firm for the directional vector $\mathbf{g}=\left(g_x,g_b,g_y\right)=\left(\bar{x},\bar{b},\bar{y}\right)$. This choice of directional vector implies that the estimates of dynamic network inefficiency, $\widehat{D(\cdot)}$, multiplied by 100%, gives the simultaneous percentage contraction in inputs and undesirable outputs, and percentage expansion in desirable outputs relative to the mean values that are feasible given the technology. We choose this mean directional vector using the mean values of inputs and outputs reported in Table 9.5.

To estimate the three-period dynamic network directional distance function requires data from four periods, since nonperforming loans and carryover assets from period $t-1$ are part of the network technology in period t. Our data corresponds with fiscal years 2002–2009. Therefore, the first three-period problem we estimate corresponds to fiscal years 2003–2005. We report estimates of the dynamic network directional distance function for each year in Table 9.6. To illustrate, consider the estimates for the 2003–2005 period. The mean estimate for $\bar{D}(\cdot)=\hat{\beta}^1+\hat{\beta}^2+\hat{\beta}^3=0.137$, which indicates that each of the three inputs and nonperforming loans could be reduced by 13.7% of the mean values for those variables reported in Table 9.5, while loans and securities could be increased by 13.7% of their mean values. For the 2003–2005 period, the year 2005 is the most inefficient year with $\hat{\beta}^3=0.047$ which indicates 4.7% of the inefficiency occurred in that year. From 2003–2005 to 2006–2008, bank inefficiency rises from 13.7% to 15% and then declines slightly to 14.4% in the 2007–2009 period.

A bank is efficient in a particular year of a three-year period if either $\hat{\beta}^1=0$ or $\hat{\beta}^2=0$ or $\hat{\beta}^3=0$. A bank produces on the frontier of the dynamic network technology if $\widehat{D(\cdot)}=\hat{\beta}^1+\hat{\beta}^2+\hat{\beta}^3=0$. Table 9.6 also reports the number of banks that are efficient in at least one subperiod of each three-year period. For 2003–2005, 10 Shinkin banks were efficient for the 2003 subperiod $\left(\hat{\beta}^1=0\right)$, nine banks were efficient for the 2004 subperiod $\left(\hat{\beta}^2=0\right)$, and nine banks were efficient for the 2005 subperiod $\left(\hat{\beta}^3=0\right)$, but only six banks were efficient for all three subperiods $\left(\hat{\beta}^1+\hat{\beta}^2+\hat{\beta}^3=0\right)$. Six banks were efficient during 2004–2006, five banks were efficient during 2005–2007, three banks were efficient during 2006–2008, and four banks were efficient during 2007–2009. Table 9.7 reports the names of the efficient banks in each year. Kochi Shinkin Bank was efficient in each three-year period, 2003–2005 to 2007–2009 and Osaka Higashi Shinkin Bank was efficient in 2004–2006 to 2007–2009. Two banks, Kyoto and Himawari, were efficient in three out of the five periods

Table 9.6   Estimates of inefficiency $\vec{D}(x,b,c,y) = \beta^t + \beta^{t+1} + \beta^{t+2}$.

		Mean	Std. dev.	Minimum	Maximum	No. on frontier
2003–2005[a]	$\hat{\beta}^1$	0.045	0.039	0	0.238	10
	$\hat{\beta}^2$	0.045	0.038	0	0.225	9
	$\hat{\beta}^3$	0.047	0.042	0	0.257	9
	$\hat{\beta}^1 + \hat{\beta}^2 + \hat{\beta}^3$	0.137	0.115	0	0.674	6
2004–2006	$\hat{\beta}^1$	0.045	0.038	0	0.220	10
	$\hat{\beta}^2$	0.048	0.043	0	0.260	10
	$\hat{\beta}^3$	0.051	0.045	0	0.256	7
	$\hat{\beta}^1 + \hat{\beta}^2 + \hat{\beta}^3$	0.144	0.125	0	0.709	6
2005–2007	$\hat{\beta}^1$	0.047	0.043	0	0.257	11
	$\hat{\beta}^2$	0.052	0.047	0	0.264	8
	$\hat{\beta}^3$	0.051	0.046	0	0.264	6
	$\hat{\beta}^1 + \hat{\beta}^2 + \hat{\beta}^3$	0.150	0.134	0	0.756	5
2006–2008	$\hat{\beta}^1$	0.051	0.046	0	0.275	9
	$\hat{\beta}^2$	0.053	0.048	0	0.268	5
	$\hat{\beta}^3$	0.046	0.043	0	0.26	6
	$\hat{\beta}^1 + \hat{\beta}^2 + \hat{\beta}^3$	0.150	0.135	0	0.777	3
2007–2009	$\hat{\beta}^1$	0.051	0.047	0	0.262	8
	$\hat{\beta}^2$	0.047	0.045	0	0.261	7
	$\hat{\beta}^3$	0.046	0.043	0	0.245	7
	$\hat{\beta}^1 + \hat{\beta}^2 + \hat{\beta}^3$	0.144	0.132	0	0.765	4

[a] 269 Shinkin banks are used in each year.

Table 9.7   Efficient Shinkin banks.

	2003–2005	2004–2006	2005–2007	2006–2008	2007–2009
Karatsu Shinkin Bank	x				
Kanonji Shinkin Bank	x	x			
The Kyoto Shinkin Bank	x	x	x		
Yamanashi Shinkin Bank	x				
Sapporo Shinkin Bank					x
Johnan Shinkin Bank		x	x		
Choshi Shinkin Bank	x				
Sawayaka Shinkin Bank					x
Osaka Higashi Shinkin Bank		x	x	x	x
Himawari Shinkin Bank		x	x	x	
Kochi Shinkin Bank	x	x	x	x	x

Table 9.8 Average optimal and actual carryover assets (std. dev.).

	Actual c^t	Optimal $\hat{c}^t$	t-Value (prob>t)	Actual c^{t+1}	Optimal $\hat{c}^{t+1}$	t-Value (prob>t)
2003–2005	83.4	53.9	10.81	87.8	71.5	6.82
	(104.7)	(78.6)	(.01)	(107.9)	(100.2)	(0.01)
2004–2006	87.8	70.2	7.28	86.7	64.1	8.54
	(107.9)	(94.6)	(0.01)	(108.7)	(87.1)	(0.01)
2005–2007	86.7	63.4	8.5	89.8	57.9	9.93
	(108.7)	(88.1)	(0.01)	(106.0)	(75.2)	(0.01)
2006–2008	89.8	56.4	10.53	97.5	52.2	11.33
	(106.0)	(74.6)	(0.01)	(117.8)	(80.1)	(0.01)
2007–2009	97.5	48.7	11.67	97.2	58.5	10.43
	(117.8)	(66.6)	(0.01)	(119.4)	(94.9)	(0.01)

and two banks, Kanonji and Johnan, were efficient in two out of the five periods. Five other banks, Karatsu, Yamanashi, Sapporo, Choshi, and Sawaka, were efficient in one of the three-year periods.

For comparison purposes, we also estimated the Shephard output distance function for each year for two standard models that are found in the literature. In the first standard model, we assumed that banks produce loans and securities investments using labor, physical capital, and equity capital. In both 2003 and 2009, average output efficiency[5] was 74% with 17 banks operating on the frontier. Other years had similar levels of efficiency and frontier banks. In the second standard model, we assumed that banks produce loans and securities investments as desirable outputs and nonperforming loans as an undesirable output, using labor, physical capital, equity capital, and deposits. Output efficiency averages 0.92 in 2003 and falls to 0.88 in 2009 with 43 banks operating on the frontier in both years. Since the network model that we estimate allows a larger production possibility set, there are fewer banks producing on the frontier in the dynamic network model than the standard single-period production model. This finding indicates that bank managers and regulators who use single-period benchmarks are potentially getting a misleading picture of bank performance.

As part of the solution to Equation (9.13), carryover assets are chosen for periods t and $t+1$ ($\hat{c}^t$ and $\hat{c}^{t+1}$), given carryover assets from period $t-1$ (c^{t-1}) and period $t+2$ (c^{t+2}). Recall that although we were able to identify total carryover assets for loans and securities ($c_1 + c_2$), we could not identify the specific amounts associated with loans (c_1) and securities (c_2) and thus assumed that all carryover assets would be put into securities. The optimal values for carryover assets that could go into securities are compared with the actual values and are reported in Table 9.8. As seen in the table, the actual values are always significantly greater than the optimal values. The ratios of actual to optimal carryover assets (not reported), $c^t / \hat{c}^t$ and $c^{t+1} / \hat{c}^{t+1}$, average between 1.43 in 2004–2006 and 2.38 in 2007–2009 for $c^t / \hat{c}^t$ and average between 1.40 in 2004–2006 and 2.24 in 2007–2009 for $c^{t+1} / \hat{c}^{t+1}$. These results indicate that Shinkin banks could reduce inefficiency by reducing carryover assets and simultaneously expanding securities.

---

[5] The Shephard output distance function measures output efficiency as the ratio of actual output to maximum potential output.

Next, we compare actual deposits with optimal deposits by examining the network link between deposits produced as an intermediate output of stage 1 and used as an input in stage 2. To calculate optimal deposits we combine the intensity variables for stages 1 and 2 with actual deposits for the $j = 1, \ldots, J$ Shinkin banks. In the dynamic problem, in period t, optimal deposits in stage 1 must satisfy the constraint that $z^t \leq \sum_{j=1}^{J} \lambda_j^t z_j^t$, and in stage 2, optimal deposits must satisfy the constraint that $z^t \geq \sum_{j=1}^{J} \Lambda_j^t z_j^t$. Let $\hat{\lambda}_j^t$, $j = 1, \ldots, J$ and $\hat{\Lambda}_j^t$, $j = 1, \ldots, J$ represent the optimal period t intensity variables for the dynamic problem. Combining the two constraints shows that the intensity variables must be chosen so that

$$\sum_{j=1}^{J} \hat{\Lambda}_j^t z_j^t \leq z^t \leq \sum_{j=1}^{J} \hat{\lambda}_j^t z_j^t \tag{9.14}$$

Let the minimum value of deposits that satisfies both constraints in Equation (9.14) equal $\hat{z}_{min}^t = \sum_{j=1}^{J} \hat{\Lambda}_j^t z_j^t$ and let the maximum value of deposits that satisfies both constraints in Equation (9.14) equal $\hat{z}_{max}^t = \sum_{j=1}^{J} \hat{\lambda}_j^t z_j^t$. Similar minimum and maximum values for optimal deposits can be calculated for periods $t+1$ and $t+2$ in the three-period dynamic problem. If the two constraints in Equation (9.14) are binding, then $\hat{z}_{min}^t = \hat{z}_{max}^t$. The two constraints were binding for all but two or three banks in each year. Table 9.9 reports the average ratios of optimal to actual deposits. On average, the ratios range from 0.855 to 0.922, but some Shinkin banks would use only 48% of their actual deposits and other Shinkin banks would use as much as 136% of actual deposits if they were to produce the optimal level of deposits consistent with the dynamic network technology in a given three-year period. We used a t-test to test the null hypothesis that the ratio of optimal to actual deposits equals 1. In every year the t-test rejected the null. These results indicate that on average, Shinkin banks produce too many deposits in stage 1 and then use too many deposits relative to the amount needed to efficiently produce the portfolio of loans and securities in stage 2. We also examined the Pearson and Spearman correlation coefficients and there was no significant correlation between the ratios of optimal to actual deposits with total Shinkin bank assets in any of the years. Thus, bank size has no systematic correlation with the tendency of bank managers to overuse or underuse deposits relative to their optimal levels.

9.6   Summary and conclusions

Shinkin banks are small cooperative banks operating in regional markets in Japan. They collect deposits from members and nonmembers and then use those deposits to purchase securities and make loans within their prefecture. In this chapter, we used DEA to examine the performance of Shinkin banks. Our method provided some structure to the production technology that is not often found in bank efficiency/productivity studies. We allowed Shinkin banks to have a network structure where deposits were produced as an intermediate output in one stage of production and then used as an input to produce the portfolio of loans and securities investments in a subsequent stage. We also allowed a dynamic structure to the production technology by allowing Shinkin banks to carryover some assets from one period to the next.

Table 9.9 Ratios of optimal deposits to actual deposits.

	$\frac{\hat{z}^t}{z^t}$			$\frac{\hat{z}^{t+1}}{z^{t+1}}$			$\frac{\hat{z}^{t+2}}{z^{t+2}}$		
	Mean (s)	Minimum	Maximum	Mean (s)	Minimum	Maximum	Mean (s)	Minimum	Maximum
2003–2005	0.869 (0.097)	0.612	1.313	0.868 (0.095)	0.516	1.190	0.893 (0.082)	0.545	1.132
2004–2006	0.863 (0.100)	0.511	1.362	0.870 (0.095)	0.500	1.118	0.895 (0.080)	0.544	1.114
2005–2007	0.868 (0.099)	0.499	1.178	0.862 (0.097)	0.479	1.115	0.903 (0.077)	0.574	1.175
2006–2008	0.859 (0.106)	0.473	1.268	0.856 (0.097)	0.488	1.203	0.921 (0.072)	0.646	1.253
2008–2009	0.855 (0.099)	0.480	1.122	0.874 (0.097)	0.546	1.328	0.922 (0.070)	0.658	1.252

Optimal deposits are $\hat{z}$ and actual deposits are z.

The effect of carryover assets was to allow bank managers to choose the period in which to use deposits and equity capital to produce the portfolio of loans and securities. If poor economic conditions in one period would cause many loans to become nonperforming, bank managers could effectively carryover assets to a subsequent period when economic conditions might have improved enough for them to make the same amount of loans, but with fewer loans becoming nonperforming.

We estimated the dynamic network model giving bank managers three-year horizons using data from 2002 to 2009. We found that if Shinkin banks were to become efficient and produce on the frontier of the dynamic network technology, they could simultaneously reduce inputs and nonperforming loans and expand performing loans and securities investments by an average of 14.4–15.6% of average inputs and outputs. We also found that Shinkin banks could improve performance by producing fewer deposits and by carrying over fewer assets from one period to the next.

References

Akther, S., Fukuyama, H. and Weber, W.L. (2013) Estimating two-stage network slacks-based inefficiency: an application to Bangladesh banking. *OMEGA (International Journal of Management Science)*, **41** (1), 88–96.

Assaf, A.G., Barros, C.P. and Matousek, R. (2011) Productivity and efficiency analysis of Shinkin banks: evidence from bootstrap and Bayesian approaches. *Journal of Banking and Finance*, **35** (2), 331–342.

Barros, C.P., Managi, S. and Matousek, R. (2009) Productivity growth and biased technological change: credit banks in Japan. *Journal of International Financial Markets, Institutions & Money*, **19**, 924–936.

Berger, A.N. and Humphrey, D. (1992) Measurement and efficiency issues in commercial banking, in *Output Measurement in the Service Industry* (ed. Z. Griliches), University of Chicago Press, National Bureau of Economic Research, Chicago, pp. 245–296.

Berger, A.N. and Humphrey, D. (1997) Efficiency of financial institutions: international survey and directions for future research. *European Journal of Operational Research*, **98**, 175–212.

Berger, A.N., Hanweck, G.A. and Humphrey, D.B. (1987) Competitive viability in banking: scale, scope and product mix economies. *Journal of Monetary Economics*, **20**, 501–520.

Bogetoft, P., Färe, R., Grosskopf, S., et al. (2009) Dynamic network DEA: an illustration. *Journal of the Operations Research Society of Japan*, **52** (2), 147–162.

Briec, W. (1997) A graph type extension of Farrell technical efficiency. *Journal of Productivity Analysis*, **8**, 95–110.

Chambers, R.G., Chung, Y. and Färe, R. (1996) Benefit and distance functions. *Journal of Economic Theory*, **70**, 407–419.

Chambers, R.G., Chung, Y. and Färe, R. (1998) Profit, directional distance functions and Nerlovian efficiency. *Journal of Optimization Theory and Applications*, **98** (2), 351–364.

Charnes, A., Cooper, W.W. and Rhodes, E. (1978) Measuring the efficiency of decision-making units. *European Journal of Operational Research*, **2**, 429–444.

Chen, Y. and Zhu, J. (2004) Measuring information technology's indirect impact on firm performance. *Information Technology and Management Journal*, **5** (1–2), 9–22.

Chen, Y., Cook, W.D. and Zhu, J. (2010) Deriving the DEA frontier for two-stage processes. *European Journal of Operational Research*, **202**, 138–142.

Färe, R. and Grosskopf, S. (1996) Productivity and intermediate products: a frontier approach. *Economics Letters*, **50**, 65–70.

Färe, R. and Grosskopf, S. (2000) Network DEA. *Socio-Economic Planning Sciences*, **34**, 35–49.

Färe, R. and Grosskopf, S. (2004) *New Directions: Efficiency and Productivity*, Kluwer Academic Publishers, Boston/Dordrecht/London.

Färe, R., Fukuyama, H. and Weber, W.L. (2010) A Mergers and acquisitions index in data envelopment analysis: an application to Japanese Shinkin banks in Kyushu. *International Journal of Information Systems and Social Change*, **1** (2), 1–18.

Farrell, M.J. (1957) The measurement of production efficiency. *Journal of the Royal Statistical Society Series A (General)*, **120** (Part 3), 253–281.

Fukuyama, H. (1996) Returns to scale and efficiency of credit associations in Japan. *Japan and the World Economy*, **8**, 259–277.

Fukuyama, H. and Weber, W.L. (2008) Estimating inefficiency, technological change and shadow prices of problem loans for regional banks and Shinkin banks in Japan. *The Open Management Journal*, **1**, 1–11.

Fukuyama, H. and Weber, W.L. (2009) A directional slacks-based measure of technical inefficiency. *Socio-Economic Planning Sciences*, **43** (4), 274–287.

Fukuyama, H. and Weber, W.L. (2010) A slacks-based inefficiency measure for a two-stage system with bad outputs. *Omega: The International Journal of Management Science*, **38** (5), 239–410.

Fukuyama, H. and Weber, W.L. (2012) Estimating two-stage network technology inefficiency: an application to cooperative Shinkin banks in Japan. *International Journal of Operations Research and Information Systems*, **3** (2), 1–22.

Goddard, J.A., Molyneux, P. and Wilson, J.O.S. (2001) *European Banking: Efficiency, Technology, and Growth*, John Wiley & Sons, Inc., New York.

Hancock, D. (1985) The financial firm: production with monetary and non-monetary goods. *Journal of Political Economy*, **93**, 859–880.

Harimaya, K. (2004) Measuring the efficiency in Japanese credit cooperatives. *Review of Monetary and Financial Studies*, **21**, 92–111 (in Japanese).

Hirota, S. and Tsutsui, Y. (1992) Ginkogyo niokeru Han-i no Keizai [Scope economies in banking], *Structural Analyses of the Japanese Financial System* (eds A. Horiuchi and N. Yoshino), University of Tokyo Press, Tokyo (in Japanese).

Hosono, K., Sakai, K. and Tsuru, K. (2007) Consolidation of banks in Japan: causes and consequences. Research Institute of Economy, Trade, and Industry Discussion Paper Series 07-E-059.

Kao, C. and Hwang, S.N. (2008) Efficiency decomposition in two-stage data envelopment analysis: an application to non-life insurance companies in Taiwan. *European Journal of Operational Research*, **185** (1), 418–429.

Miyakoshi, T. (1993) Shinyokinko ni okeru Han-i no Keizai [Scope economies of credit associations]. *Keizai Kenkyu*, **44** (3), 233–242 (in Japanese).

Miyamura, K. (1992) Shinyokinko no Hiyo to Kibo no Keizaisei [Costs and scope economies of Shinkin banks]. *Keiei Ronshu*, **38**, 63–83 (in Japanese).

Nishikawa, S. (1973) Ginko: Kyoso to sono Kisei [Banks: competition and its regulation], in *Nihon no Sangyo Soshiki I [Japanese industrial organization I]* (ed. H. Kumagai), Chuo-Koronsha, Tokyo (in Japanese).

O'Brien, Y.-Y.C. (2007) Reserve requirement systems in OECD countries. Finance and economics discussion series, Division of Research and Statistics and Monetary Affairs, Federal Reserve Board, Washington, DC.

Satake, M. and Tsutsui, Y. (2002) Why is Kyoto where Shinkin Banks Reign? An analysis based on the efficient structure hypothesis, in *Regional Finance: A Case of Kyoto* (ed. T. Yuno), Nippon Hyoron Sha, Tokyo (in Japanese).

Saunders, A. and Cornett, M. (2008) *Financial Institutions Management: A Risk Management Approach*, 6th edn, McGraw-Hill Irwin, New York.

Sealey, C. and Lindley, J. (1977) Inputs, outputs, and a theory of production and cost at depository financial institutions. *Journal of Finance*, **32**, 1251–1266.

Shephard, R.W. (1970) *Theory of Cost and Production Functions*, Princeton University Press, Princeton.

Shephard, R.W. and Färe, R. (1980) *Dynamic Theory of Production Correspondences*. Verlag Anton Hain, Meisenheim.

Tone, K. and M. Tsutsui (2009) Network DEA: a slacks-based measure approach. *European Journal of Operational Research*, **197** (1), 243–252.

Wang, C.H., Gopal, R.D. and Zionts, S. (1997) Use of data envelopment analysis in assessing information technology impact on firm performance. *Annals of Operations Research*, **73**, 191–213.

10

Effects of specification choices on efficiency in DEA and SFA

Michael Koetter[1] and Aljar Meesters[2]

[1] *Finance Department, Frankfurt School of Finance and Management, Germany*

[2] *Global Economics and Management Department, Faculty of Economics and Business, University of Groningen, The Netherlands*

10.1 Introduction

Assessing the efficiency of financial intermediaries has a long-standing tradition in the banking and finance literature.[1] Measuring the relative ability of financial institutions to employ resources and transform them into some form of output is of interest to practitioners and policy makers alike. Benchmarking permits the identification of very well and very poor performing banks. Such benchmarking may inform managers and shareholders of banks about best practice behavior. Policy makers, such as prudential regulators, may learn in turn which institutions deserve a more intensive supervision.[2] However, robust inference for managerial and policy purposes requires knowledge about how efficiency scores change in response to different specification choices of the benchmark.

According to Hughes and Mester (2010), efficiency benchmarking of financial institutions can broadly follow two approaches: a structural or a nonstructural approach. The former approach entails a microeconomic foundation of the banking firms' objectives, constraints, and choices whereas the latter approach essentially compares accounting-based performance indicators of

---

[1] Already Berger and Humphrey (1997) reviewed more than 130 bank efficiency studies. Berger (2007) reviews 100 efficiency studies focusing on cross-border comparisons of banks.

[2] For instance, Fiordelisi, Marques-Ibanez, and Molyneux (2011) report that bank efficiency Granger-causes bank risk and that capitalization levels of banks tend to increase if efficiency is improved. Hence, long-term efficiency seems an important performance indicator to monitor.

Efficiency and Productivity Growth: Modelling in the Financial Services Industry, First Edition. Edited by Fotios Pasiouras.
© 2013 John Wiley & Sons, Ltd. Published 2013 by John Wiley & Sons, Ltd.

interest, for instance, profitability or risk proxies. This chapter focuses on the implications of a number of specification choices faced by researchers pursuing the former approach to benchmarking banks using stochastic frontier analysis (SFA) or data envelopment analysis (DEA).

The purpose of this chapter is *not* to identify which method is 'superior'. Instead, we aim to highlight the choices researchers have to take and document some of the implications for a large sample of European banks obtained from Bankscope, the most frequently used source for non-US bank efficiency research.

We begin in Section 10.2 by reviewing literature related to (a) the theoretical foundations of each structural benchmarking exercise, (b) popular benchmarking methodologies employed, and (c) specification options so as to illustrate the choices a research has to make. Section 10.3 describes the basic SFA and DEA models used in this chapter to illustrate how each method is affected by specification choices. Section 10.4 describes the data on European banks. The results are shown in Section 10.5 and discuss the properties of efficiency scores from different models regarding their distribution, rankings, extreme performer identification, and correlation with accounting-based performance measures. We conclude in Section 10.6.

10.2   Bank benchmarking background

10.2.1   Theoretical foundations

Assessing the relative performance of banks requires the specification of some form of benchmark. The nature of this benchmark depends on the theoretical model. Efficiency is defined, most generally, as the deviation of observed bank performance from the benchmark. The seminal example of Farrell (1957) is a production function, where the ratio of observed output and predicted equilibrium output is a measure of the efficiency of the firm.

Only few banking studies consider production benchmarks because banking is a multi-output production process. Conventionally, benchmarks are formulated instead in terms of dual problems.[3] Most bank efficiency studies assume that banks minimize cost C and demand input quantities x_k in complete factor markets at given factor prices w_l to produce outputs y_m. Costs are minimized subject to a technology constraint $T(y, x)$.[4] The solutions of this cost minimization problem are optimum input demand functions and the optimum cost function:

$$C^* = f(y,w). \qquad (10.1)$$

Cost efficiency (CE) is the systematic deviation of observed cost from predicted optimal cost.

Alternatively, the benchmark can be formulated as a profit frontier, and the alternative profit model of Humphrey and Pulley (1997) is the most popular in the literature. Banks demand factors in complete markets as before, but possess some pricing power in output markets. They maximize profit before tax, PBT $= py - wx$, subject to the technology constraint $T(y, x)$ and a so-called pricing opportunity constraint $H(p, y, w)$. Put differently, banks choose optimal input quantities x and output prices p.

---

[3] Exceptions are Nakane and Weintraub (2005), Martin-Oliver and Salas-Fumas (2008), and Koetter and Noth (2013), all of which employ modified versions of the production function estimator of Levinsohn and Petrin (2003) that controls for endogeneity between factor demand and productivity.

[4] Many efficiency studies specify further arguments in the technology constraint, such as equity capital, to account for heterogenous risk preferences of banks (Mester, 1993).

An advantage of the (alternative) profit maximization model is to account for the fact that the inefficiency of banks when maximizing profits are much larger compared to those arising relative to a cost frontier. The assumption of some market power in output markets furthermore circumvents the empirical challenge that price data is usually not available and/or lumpy regarding interest and fee components (Mountain and Thomas, 1999). However, this behavioral assumption might also be ill-suited for the large number of government-owned banks, which sometimes explicitly pursue non-profit-maximizing objectives (Hackethal, Vins, and Koetter, 2012).[5] Assuming cost minimization as the behavioral objective, in turn, might very well be defendable because it is a necessary condition for the survival of any bank in the long run.

In addition, both cost and (alternative) profit models largely neglect the core function of banks, namely to efficiently diversify risk. To this end, Hughes et al. (1996) and Hughes et al. (2000) develop a model of utility-maximizing banks that essentially 'demand' two goods, returns and risk, subject to technology and a profit identity constraint. Koetter (2008) compares the three different concepts of cost, alternative profit, and risk–return efficiency for a sample of German universal banks and finds that accounting for different risk-preferences leads to fundamentally lower inefficiencies. However, utility is modeled as a function of profits and risk, which requires the assumption that banks do maximize profits at given risk as well.

In sum, none of these models is the 'correct' one. It depends on the specific research question, 'Which assumption on economic behavior is most adequate?' Given our goal to illustrate the implications of specification choices on DEA and SFA bank efficiency scores, we consider here a model of cost minimization because it is the most encompassing model and suits our sample well, which accounts for roughly a third of savings banks.

10.2.2   Benchmarking techniques

Broadly speaking, two alternative approaches exist to obtain a cost frontier as in Equation (10.1): parametric and nonparametric methods. Among the most established approaches regarding the former is SFA, introduced by Aigner, Lovell, and Schmidt (1977); Battese and Corra (1977); and Meeusen and Broeck (1977).[6] Occasionally also referred to as the econometric approach (Greene, 1993), the specification of a stochastic frontier deliberately accounts for random noise. Thereby, SFA avoids confining random noise with systematic deviations from the benchmark, that is, inefficiencies.[7]

The main disadvantage of these methods is the need for distributional assumptions on error terms and the functional form of the kernel of the frontier. Nonparametric methods, in turn, do not impose any structure on the data and were introduced by Charnes, Cooper, and Rhodes (1978).[8] A drawback of this approach is that inefficiencies are lumped together with random noise, for example, due to measurement error. According to Mountain and Thomas (1999), banking studies are particularly prone to such errors because measurement of prices based on accounting information is notoriously difficult. Furthermore, growing heterogeneity across banks renders a comparison relative to an identical benchmark particularly sensitive to outliers.

---

[5] Government ownership is pervasive in the banking industry and the evidence suggests that it has a number of undesirable effects on lending and failure resolution (see Brown and Dinç, 2005; Dinç, 2005; Sapienza, 2004).

[6] See Kumbhakar and Lovell (2000) for an introduction to stochastic frontier analysis.

[7] Additional parametric methods include the Distribution Free Approach (DFA) developed by Berger (1993) and the Thick Frontier Approach (TFA), introduced by Berger and Humphrey (1991).

[8] See Ali and Seiford (1993) for a synopsis of the development of this approach.

Recent methodological literature made great strides in tackling the shortcomings of both benchmarking methods, for instance, by developing semi- and nonparametric frontier estimators (Daouia, Florens, and Simar, 2012; Florens and Simar, 2005; Kumbhakar et al., 2007; Park, Sickles, and Simar, 1998, 2007), formulating latent class models to account for heterogeneity (Greene, 2005), or using quantile (Daouia and Simar, 2007) and Bayesian estimation approaches (Kumbhakar and Tsionas, 2005). Within the parametric and nonparametric strands of the bank efficiency literature, numerous studies compare modeling approaches.[9]

But few papers compare identical samples across parametric and nonparametric methods.[10] Generally, these comparisons report that efficiency scores from nonparametric methods are smaller compared to those from parametric methods, possibly due to the neglect of random error, and are better able to cope with smaller samples of more homogenous banks.

We contribute to these studies that shed light on the sensitivity of efficiency scores, first, by providing evidence for a large sample of EU banks and, second, by considering the implications of specification choices that are often only cursory mentioned in bank efficiency studies. We follow the taxonomy suggested in Bauer et al. (1998) and applied by Fiorentino, Karmann, and Koetter (2006) to compare parametric and nonparametric efficiency scores along five consistency conditions. These are: efficiency distributions, rank-order correlations, correspondence of best- and worst-performer identification, and the consistency of efficiency scores with market conditions as well as nonstructural performance indicators.

10.2.3   Specification options

Irrespective of the arguably important choice of the benchmarking methodology, researchers also have to make a number of specification choices regarding the outputs a bank produces, factor price approximation, and how to deal with extreme observations. The effect of these choices on efficiency scores is potentially large and different across parametric and nonparametric methods, but are, to our knowledge, not systematically documented.

10.2.3.1   Bank outputs

The specification of bank outputs remains an open discussion. The two most popular approaches are the intermediation approach (Sealey and Lindley, 1977) and the production approach (Benston, 1965).[11] The former puts the volume of financial funds channeled from savers to investors central. Financial funds borrowed by the bank constitute the main input factor whereas loans and other earning assets represent the major output. The latter approach focuses on the number of transactions a bank conducts and is often applied in studies bank branch efficiency. The number of deposits served is in this context frequently a proxy for transactions executed for the customers of a bank. Therefore, we focus in this chapter on the intermediation approach.

---

[9] See, for example, Bos et al. (2009) for an assessment of specification choices across SFA models and Fethi and Pasiouras (2010) for a comprehensive review of non-parametric bank benchmarking studies.

[10] Comparisons of parametric and non-parametric methods are Ferrier and Lovell (1990) and Bauer et al. (1998) (both US), Resti (1997) and Casu and Girardone (2002) (both IT), Huang and Wang (2002) (TW), Weill (2004), Beccalli and Casu (2006) (both selected EU countries), Fiorentino, Karmann, and Koetter (2006) (DE), and Delis et al. (2009) (GR).

[11] Basu, Inklaar, and Wang (2011) suggest to consider interest margins as output. Contrary to earlier banking studies that also focused on bank revenues, their model ensures consistency of bank output measures with national accounts.

Whereas loans and other earning assets are obvious output candidates in this framework, Stiroh (2004) emphasizes the increasing importance of fee income to substitute for increasingly narrow interest margins in the traditional intermediation business. Some efficiency studies therefore specify off-balance sheet (OBS) activities, such as credit commitments and derivates, as an additional output to loans and other earning assets.

We therefore specify two outputs as a baseline specification, total loans (y_1) and other earning assets (y_2), and assess differences in efficiency scores from SFA and DEA when specifying off-balance sheet activities as a third output y_3.

10.2.3.2   Factor prices

Most applications that use the intermediation approach to specify the cost benchmark in Equation (10.1), approximate three factor prices w. The price of borrowed funds (w_1) is the most relevant one, which we calculate as total interest expenses relative to the sum of deposits and short-term funds. Second, banks hire employees and we measure the wage rate (w_2) as personnel expenditures relative to the number of full-time employees.[12] Third, banks use branches and offices and we approximate the factor price of fixed assets (w_3) by relating total noninterest expenses less personnel expenses to nonearning assets.

The practice of imputing factor prices from observed financial accounts data is problematic, because it violates the underlying theoretical assumption that banks face *exogenous* factor prices when choosing factor quantities (Mountain and Thomas, 1999). As such, we would ideally *observe* prices faced by the bank, which is usually not possible. Alternatively, Koetter (2006) approximates exogenous factor prices by specifying for each bank(-year observation) the average imputed factor price of all other banks in a defined market in Equation (10.1).

We therefore define countries as banks' relevant markets and investigate, if and to what extent SFA and DEA efficiency scores are affected (differently) when using exogenous instead of imputed factor prices.

10.2.3.3   Extreme observations

A seemingly mundane issue in applied bank benchmarking is the treatment of extreme observations. Most applications tend to mention only in passing, how sample 'cleaning' is conducted and usually fail to report systematically the effect on the consistency of efficiency scores. But Bos and Koetter (2011) show that these simple choices can distort efficiency scores substantially.[13]

Especially across methods, different outlier treatments are likely to have different effects since DEA is by definition more sensitive to outliers due to its linear programming nature. Samples with large measurement error and more observations will generally add to more extreme observations that constitute the efficient hull, which should lead to lower mean efficiency scores. But what is the effect of different truncation points in the data and does it matter which variables are considered? And do such choices have any effect on SFA efficiency scores, which account for random noise to begin with?

---

[12] Note, that many applications relate the former to total assets because employment data is frequently missing. This convention is problematic because personnel expenditures per unit of assets is hardly reflecting the equilibrium price in labor markets.

[13] Specifically, they show that the conventional way to deal with negative profits in translog specifications of profit frontiers, namely to exclude them from the sample or to scale the entire distribution, leads to significantly different efficiency (rank) estimates.

We document the differences in SFA and DEA efficiency scores for three different truncation percentiles (1%, 5%, and 10%) at the top and at the bottom of the distribution of (a) all cost function arguments and (b) generated factor prices only. We do so for the EU sample, a large country sample, and a small country sample, respectively.

10.3   Methodologies

10.3.1   Stochastic frontier analysis

We specify Equation (10.1) as a simple baseline stochastic cost frontier that is closely related to the linear regression

$$\ln C^* = g\left(\ln y, \ln w; \beta\right) + \mathbf{v} + \mathbf{u}, \tag{10.2}$$

where $g(\cdot)$ is a function that is linear in parameters and β is the parameter vector of interest. As indicated in Section 10.2.2, much more advanced specifications of SFA and DEA benchmarks are constantly developed. But given our present focus to assess the implications of specification choices on SFA and DEA efficiency scores, we prefer to compare very basic models that are most often used in the applied bank efficiency literature.

The departure of SFA estimators from OLS estimators is that the residuals comprise two components. The first is the stochastic component of the frontier and the vector $\mathbf{v}$ captures random error. Generally, $\mathbf{v}$ is assumed to be independently identically normally distributed with an expected mean of 0, just like under the OLS specification.

The second component of the compound residual is the vector $\mathbf{u}$. The elements of this vector contain the *in*efficiency of each bank, that is, more inefficient banks exhibit larger $\mathbf{u}$. The underlying notion is that costs C^* of a bank can be higher than what is explained by the frontier $g(\cdot)$ due to inefficiency. In order to identify the estimator an assumption for the distribution of the elements of vector $\mathbf{u}$ is necessary. Often assumed distributions for these elements are the exponential, half normal, and truncated normal distribution. Since cost can only increase due to inefficiency, it is important that the distribution of the elements of $\mathbf{u}$ has a nonnegative support. Empirically, the specification of Equation (10.2) requires one to assume a functional form, very frequently the translog, which we employ here as well:

$$\ln C = \alpha_0 + \sum_k \beta_k \ln y_k + \sum_l \beta_l \ln w_l + \frac{1}{2}\sum_k \sum_m \beta_{k,m} \ln y_k \ln y_m$$
$$+ \frac{1}{2}\sum_l \sum_m \beta_{l,m} \ln w_l \ln w_m + \sum_k \sum_l \beta_{k,l} \ln y_k \ln w_l + \mathbf{v} + \mathbf{u}, \tag{10.3}$$

where y and w are the outputs and input prices as defined in Section 10.2.3 and C denotes total operating costs.[14] Note that $\beta_{k,m} = \beta_{m,k}$, $\beta_{l,m} = \beta_{m,l}$, and $\beta_{k,l} = \beta_{l,k}$. Moreover, in order to impose homogeneity of degree one in prices, $\Sigma_l \beta_l = 1$, $\Sigma_m \beta_{k,m} = 0$, and $\Sigma_l \Sigma_m \beta_{l,m} = 0$. These restrictions are imposed by dividing C and all input prices over one input price.

Estimating Equation (10.3) yields estimates for all βs and one can calculate the values of $\mathbf{v} + \mathbf{u}$ for each bank. Jondrow et al. (1982) show how $E[\mathbf{u} | \mathbf{v} + \mathbf{u}]$ can be calculated and Battese

---

[14] Many alternatives exist, for instance the Cobb-Douglas or the Fourier Flexible forms.

and Coelli (1988) show how $E[exp(\mathbf{u})|\mathbf{v}+\mathbf{u}]$ can be calculated where $exp(\mathbf{u})$ is the definition of efficiency used by Farrell (1957).[15]

10.3.2   Data envelopment analysis

The two main limitations of SFA are the need for assumptions regarding the efficiency distribution and the functional form of the frontier, which are not necessary in DEA. In DEA, the production technology constitutes for instance a convex set, which is determined by using piecewise combinations of all efficient banks. A formal program to obtain this set is given in Equation (10.4):

$$
\begin{aligned}
&\min_{\Theta,\lambda} \Theta \\
&st - \mathbf{y}_i + \mathbf{Y}\lambda \geq 0, \\
&\Theta\mathbf{x}_i - \mathbf{X}\lambda \geq 0, \\
&\lambda \geq 0
\end{aligned}
\tag{10.4}
$$

Θ is the component that reflects the efficiency of so-called decision-making unit (DMU) i, which is minimized. Accordingly, the production function is put as far as possible to the outside. $\mathbf{y}_i$ and $\mathbf{x}_i$ are vectors of outputs produced and inputs consumed. $\mathbf{Y}$ and $\mathbf{X}$ are matrices with all the outputs and inputs of all DMUs. λ is a weighting vector, which shows the linear combination of producers corresponding to the lowest Θ. It therefore represents the vector that measures which DMUs outperform DMU i.[16]

Aside from avoiding considerable *a priori* structure as in SFA, the DEA approach is capable of dealing with multiple outputs. However, problems arise due to the assumption of free disposability and the convexity of the production set (Post, Cherchye, and Kuosmanen, 2002). Färe and Grosskopf (1983) show that congestion can violate free disposability and it remains debatable if congestion should be considered as inefficiency or not. One may argue that the DMU is in an area of the production set where it faces congestion and therefore it made a 'wrong' choice that represents inefficiency. Alternatively, a DMU may have to pass such a region to grow and therefore it is an efficient choice. Further, the assumption of convexity can be too restrictive if one permits for economies of scale, or if DMUs can specialize (Farell, 1959).[17] Finally, DEA faces like most nonparametric models the curse of dimensionality and suffers from slower convergence rates than a parametric model.

10.4   Data

The data sampled in this chapter is commercial, savings, and cooperative banks in EU-27 countries for the period 1996–2011. All data are obtained from Bankscope and contain complete information on factor prices and outputs for 5364 banks with a total of 36 432 bank-year observations.

However, the total number of complete observations is much smaller as shown in Table 10.1 and Table 10.2 due to incomplete observations and cleaning procedures described later. These tables show frequency distributions across time, countries, and banking groups.

---

[15] Actually $E[exp(-\mathbf{u})|\mathbf{v}+\mathbf{u}]$ is used since we are interested in cost efficiency.

[16] The constant returns to to scale assumption in Equation (1.4) can be relaxed in factor of a variable returns to scale assumption by adding a convexity constraint (i.e. the sum of the elements of λ should be equal to 1). See Coelli, Rao, and Battese (2005) for an introduction to DEA.

[17] The Free Disposable Hull model of Deprins, Simar, and Tulkens (1984) relaxes the convexity assumption.

Table 10.1   Frequency distribution per banking group 1996–2010.

Banking group	Commercial	Cooperative	Savings	All
Year				
1996	206	799	504	1 509
1997	234	965	586	1 785
1998	235	970	601	1 806
1999	226	1 046	567	1 839
2000	200	1 028	527	1 755
2001	208	990	553	1 751
2002	198	939	513	1 650
2003	190	959	505	1 654
2004	164	915	471	1 550
2005	176	978	481	1 635
2006	160	1 244	475	1 879
2007	161	1 293	513	1 967
2008	176	1 211	494	1 881
2009	136	1 285	472	1 893
2010	91	936	364	1 391
Total	2 761	15 558	7 626	25 945

Complete observations excluding outliers.

Table 10.1 shows that most observations pertain to German cooperative banks. The number of observations over time is relatively stable with a drop in 2010. Table 10.2 shows, in turn, that European banking studies are clearly dominated by the number of German banks.

A notorious problem in banking sector studies pertains to the treatment of consolidated and unconsolidated accounting information. To avoid double counting while keeping the unit of analysis as precise as possible, only data from unconsolidated statements are used. Thus, we assume that even wholly owned subsidiaries have a decision-making unit, or that management from a distance may affect the efficiency level of a subsidiary.

Another problem is missing data that requires the use of proxies. The calculation of input prices is an example. Especially, calculating the price of labor is often problematic since information on the number of employees is frequently missing. Many studies approximate the price of labor by personnel expenses over total assets. The assumption is that larger banks have more employees (e.g., Fries and Taci, 2005; Lensink, Meesters, and Naaborg, 2008). Here, we use the number of employees to calculate the price of labor and note that the number of observations would have increased from 36 432 to 47 314 had we used total assets instead. Hence, there is a trade-off to be made between a reporting bias of number of employees and a proxy that may be rather crude.

The base case data is summarized in Table 10.3. All variables below the 5th and above the 95th percentiles are excluded. Factor prices are exogenous and measured for each bank by the average of all other banks in a given country and year (Koetter, 2006).[18]

---

[18] An approach to increase the number of observations while still using the number of employees is to use the exogenous prices of labor also for banks that are not used to calculate the exogenous prices. We do not to pursue this extrapolation procedure because we compare models with and without exogenous prices.

Table 10.2 Frequency distribution per country 1996–2010.

Country	1996	1997	1998	1999	2000	2001	2002	2003	2004	2005	2006	2007	2008	2009	2010
AT	22	74	74	20	16	23	20	25	31	38	40	61	0	59	5
BE	9	11	8	12	13	11	10	9	6	9	12	0	5	5	5
CY	0	0	0	0	0	0	1	1	2	1	1	1	2	2	2
CZ	0	0	0	0	0	0	0	0	0	0	0	3	1	5	5
DE	1184	1210	1201	1226	1167	1114	998	1015	995	1018	1298	1298	1289	1300	1008
DK	19	29	40	46	42	47	52	43	33	30	48	47	56	53	38
ES	17	19	12	7	9	17	14	12	3	7	17	23	0	0	46
FI	1	2	1	1	2	0	0	0	0	1	0	1	2	1	0
FR	44	49	54	51	43	37	24	21	22	21	21	46	62	45	28
GB	9	9	8	0	0	0	0	10	0	0	0	0	14	0	0
GR	0	0	0	1	1	1	2	3	7	6	3	4	6	0	0
HU	0	0	0	0	0	0	0	0	0	0	2	1	1	0	0
IT	168	339	367	433	425	428	454	443	417	477	418	426	386	412	243
LU	34	35	34	38	29	28	25	24	23	12	0	0	7	0	0
MT	0	0	0	0	0	0	0	0	0	0	1	0	0	0	1
NL	1	4	3	2	3	3	4	1	2	1	0	0	0	0	1
PL	0	0	0	0	0	0	1	2	0	3	1	0	0	0	2
PT	0	2	4	1	2	3	4	3	2	2	1	2	1	0	0
SE	1	2	0	1	3	39	41	42	3	0	7	47	43	1	0
SI	0	0	0	0	0	0	0	0	4	9	9	7	6	9	7
SK	0	0	0	0	0	0	0	0	0	0	0	0	0	1	0
Total	1509	1785	1806	1839	1755	1751	1650	1654	1550	1635	1879	1967	1881	1893	1391

Complete observations excluding outliers.

Table 10.3 Descriptive statistics of bank variables 1996–2010.

Banking type		Commercial		Cooperative		Savings		All		Unit
Observations		2761		15 558		7626		25 945		
Variable		Mean	Std. dev.	Mean	Std. dev.	Mean	Std. dev.	Mean	Std. dev.	
Total loans	y_1	0.741	0.959	0.346	0.573	0.922	0.898	0.557	0.776	mil €
Other earning assets	y_2	0.532	0.734	0.192	0.353	0.545	0.583	0.332	0.511	mil €
Off-balance sheet	y_3	0.216	0.304	0.050	0.117	0.113	0.173	0.086	0.172	mil €
Price of borrowed funds	w_1	3.328	1.283	3.158	1.013	3.266	0.842	3.208	1.001	%
Price of fixed assets	w_2	35.313	19.855	37.276	14.471	34.866	13.127	36.359	14.813	%
Price of labor	w_3	60.390	13.186	52.343	9.164	46.133	9.778	51.374	10.688	tds €
Equity	z	0.096	0.108	0.040	0.071	0.084	0.082	0.059	0.083	bil €
Total assets	TA	1.337	1.463	0.564	0.892	1.535	1.440	0.932	1.233	bil €
Total operating cost	TOC	66.600	72.000	27.400	41.500	75.900	69.400	45.800	59.300	mil €
Profits before tax	PBT	0.011	0.022	0.004	0.008	0.008	0.011	0.006	0.011	mil €

The data is obtained from Bankscope and covers 21 European countries. Based on the full sample, all variables below the 5th and above the 95th percentiles are excluded. Factor prices are exogenous measured for each bank by the average of all other banks in a given country and year (Koetter, 2006). Total loans, other earning assets, and OBS activities are directly obtained from Bankscope. Price of borrowed funds is calculated as total interest expense over deposits and short-term funding. Price of fixed assets is calculated as total noninterest expenses minus personnel expenses over nonearning assets. Price of labor is calculated as personnel expenses over number of employees.

Table 10.3 shows the descriptive statistics per banking group type. Cooperative banks are on average the smallest institutions in EU banking compared to commercial and saving banks.

10.5   Results

We compare DEA and SFA efficiency scores following the consistency conditions pointed out by Bauer et al. (1998) regarding levels, rankings, the identification of best- and worst performers, and the relationship with nonstructural accounting indicators of performance. We begin by comparing the distribution of DEA and SFA efficiency levels and highlight there the implications of specification choices.

10.5.1   Efficiency distributions

Generally, CE levels tend to be lower according to DEA given a larger sensitivity to outliers. To assess the importance of the three discussed choices regarding outlier influence, output choices, and factor price specifications, we present in the following results that show the different implications per choice for either method.

10.5.1.1   Outlier influence

Table 10.4 shows in the upper panel descriptive statistics for CE scores of three samples: All sampled EU countries, a large banking system (Germany, GER), and a small banking system (Belgium, BEL). The lower panel shows corresponding results obtained from DEA. Across columns, we show results for three different truncation thresholds, both applied to the upper and the lower tail of the distribution of all output and factor price variables.[19]

For both SFA as well as DEA scores, the exclusion of extreme observations increases mean efficiency estimates substantially. Mean CE_{SFA} increases from 81% to 99% when excluding the observations at the 10th and 90th percentiles instead of the 1st and 99th percentiles. Hence, even when accounting for random noise, measurement error has a large influence on the level of CE scores. We expect this effect to be even larger for CE_{DEA} that are measured deterministically and the bottom panel of Table 10.4 confirms an increase from a mere mean efficiency of 9–24%, a more than 1.5-fold increase.

These results, thus, confirm the well-documented and large-level gap of CE scores between methods. What is important to note is that random error alone cannot be held accountable for this difference, since also SFA results tend to eliminate any inefficiency on average. An interesting aspect of increasingly restrictive outlier thresholds is that the dispersion of CE_{SFA} also approaches 0 whereas it grows even larger for CE_{DEA}. This result indicates that 'excessive' culling in SFA, for instance, inquest for convergence of maximum likelihood estimates, may lead to completely uninformative results that no longer distinguish meaningful differences across banks regarding their relative positions to the benchmark.

In general, we would expect that aside from extreme observations in the data, sample heterogeneity might explain efficiency differences as well. Whereas SFA specifications can

---

[19] For each variable, we determine observations in the three different definitions of extreme tails for the full sample of 36 432 observations. Observations are then dropped jointly. Alternatively, some studies identify and exclude extreme observations sequentially. The latter practice will generally lead to a greater culling of data.

Table 10.4 Impact of extreme value exclusion on CE scores from SFA and DEA.

Percentile	1			5			10		
Sample	All	GER	BEL	All	GER	BEL	All	GER	BEL
SFA									
Observations	34 256	19 206	216	25 945	17 321	125	17 459	13 134	32
Mean	0.808	0.919	0.874	0.933	0.934	0.840	0.994	0.949	0.998
Std. dev.	0.114	0.054	0.063	0.024	0.041	0.108	0.000	0.026	0.000
Minimum	0.049	0.101	0.562	0.156	0.140	0.383	0.992	0.467	0.998
Maximum	0.972	0.990	0.952	0.974	0.989	0.955	0.995	0.988	0.998
DEA									
Observations	34 256	19 206	216	25 945	17 321	125	17 459	13 134	32
Mean	0.094	0.102	0.326	0.171	0.175	0.393	0.239	0.276	0.777
Std. dev.	0.139	0.111	0.284	0.170	0.134	0.314	0.164	0.139	0.248
Minimum	0.001	0.017	0.023	0.010	0.041	0.044	0.029	0.081	0.244
Maximum	1.000	1.000	1.000	1.000	1.000	1.000	1.000	1.000	1.000

SFA and DEA CE scores for a pooled sample of EU banks, and country-specific samples for Germany (GER) and Belgium (BEL) from 1996 until 2006. All cost function variables are truncated at three different percentile thresholds of the raw data distribution. Based on specification of three outputs and exogenous factor prices. Correlation coefficients are all significantly different from 0 at the 1% level.

be augmented with control variables, either in the kernel or in the error distribution or both, the specification of environmental variables in DEA is nontrivial. Therefore, we show in the columns labeled GER and BEL efficiency results obtained for these subsamples only.

Especially the DEA results in the lower panel vividly illustrate that more homogenous sampling in general and smaller sets of DMU in particular, as for the BEL sample, increase mean CE scores substantially. For example, compared to mean CE_{DEA} of 9% for the entire sample truncated at 1%, the subsample of Belgian banks yields mean efficiency of 33% and even 78% when excluding at the 10% tails. The mere exclusion of too many 'incomparable' banks leads in DEA simply to an efficiency benchmark constituted by just a few 'extreme' banks, from which the mass of observations is fairly far away. SFA scores, in turn, are not necessarily affected in the same way. At the truncation level of 5%, for example, mean CE_{SFA} for Belgian banks is lower (84%) compared to the mean of the entire sample (93%).

Admittedly, our exclusion criterion is fairly rigid since we also include the upper tails of output variables. Extreme observations of factor prices are more likely to reflect measurement error because they are constructed as ratios. But we expect very large outputs to exist, for example, because they reflect the highly skewed size distribution in most banking industries. Excluding observations on very large banks might thus make sense from a technical viewpoint, but it does not necessarily do so from an economic perspective. After all, the entire purpose of benchmarking is to achieve exactly a comparison of decision making units.

Table 10.5, therefore, replicates these results when excluding only extreme factor prices. The decline in sample size is considerably smaller. More importantly, truncating only extreme factor price observations affects CE_{SFA} markedly differently. Whereas the 10% truncation procedure implied before an almost fully efficient banking system with no differences across banks, some differences remain when confining the exclusion only to extreme factor prices.

Mean CE_{DEA} results, in turn, illustrate that determining the optimal hull with linear programming is particularly sensitive to extreme outliers in the output, that is, size, dispersion of a banking industry. Mean efficiency levels are extremely low (3%) for the full sample under the most generous truncation procedure and increase to just 10% when excluding most extreme factor price observations. This result suggests that large samples that are characterized by economically sensible (size) heterogeneity are more difficult to assess with basic DEA models. Consequently, the use of more advanced nonparametric benchmarking methods that permit for environmental controls seem particularly relevant in such settings. Henceforth, we focus on the cross-country sample after excluding the 5th and 95th percentiles of all cost function variables.

10.5.1.2   Cost function arguments

Consider next the choice of different output and factor price specifications in Table 10.6. The top row indicates whether we specified the two standard outputs, total loans and other earning assets, or, in addition, OBS activities. The second row indicates whether factor prices are imputed as usual or calculated as exogenous prices following Koetter (2006).

The left panel shows that neither the mean nor the dispersion of CE_{SFA} scores is affected substantially by these choices for our sample. Interestingly, the same holds for efficiency scores obtained with DEA. Apparently, simple outlier treatment and sampling choices have a much larger impact on efficiency estimates compared to considerations borne out by theory. Regarding the choice of outputs, one explanation might be that OBS activities account overall for only a small share of total bank output (see also Table 10.3) as opposed to the United

Table 10.5 Impact of extreme value exclusion in prices on CE scores from SFA and DEA.

Percentile	1			5			10		
Sample	All	GER	BEL	All	GER	BEL	All	GER	BEL
SFA									
Observations	34 862	19 372	250	29 221	18 296	158	21 984	15 180	47
Mean	0.806	0.914	0.798	0.915	0.930	0.853	0.943	0.935	0.921
Std. dev.	0.114	0.059	0.146	0.035	0.044	0.097	0.020	0.040	0.109
Minimum	0.052	0.080	0.125	0.134	0.138	0.389	0.249	0.172	0.546
Maximum	0.973	0.989	0.957	0.973	0.989	0.956	0.979	0.988	1.000
DEA									
Observations	34 991	19 403	250	29 271	18 310	158	22 012	15 189	47
Mean	0.029	0.070	0.344	0.060	0.118	0.387	0.108	0.120	0.479
Std. dev.	0.058	0.080	0.285	0.073	0.096	0.311	0.071	0.086	0.270
Minimum	0.001	0.013	0.022	0.004	0.025	0.044	0.009	0.029	0.075
Maximum	1.000	1.000	1.000	1.000	1.000	1.000	1.000	1.000	1.000

SFA and DEA CE scores for a pooled sample of EU banks, and country-specific samples for Germany (GER) and Belgium (BEL) from 1996 until 2006. All factor prices are truncated at three different percentile thresholds of the raw data distribution. Based on specification of three outputs and exogenous factor prices. Correlation coefficients are all significantly different from 0 at the 1% level.

Table 10.6   Alternative output and factor price specifications.

Outputs	2	3	3	2	3	3
Exogenous w	No	No	Yes	No	No	Yes
		SFA			DEA	
Mean	0.931	0.927	0.933	0.170	0.171	0.171
Std. dev.	0.021	0.023	0.024	0.168	0.170	0.170
Minimum	0.322	0.283	0.156	0.010	0.011	0.010
Maximum	0.968	0.969	0.974	1.000	1.000	1.000

SFA and DEA CE scores for a sample of 25 945 bank-year observations for a sample of EU banks from 1996 until 2006. All cost function variables are truncated at the 5th and 95th percentile of the raw data distribution. Rank-order correlation coefficients are all significantly different from 0 at the 1%-level. Total loans and other earning assets are specified in all specifications. The third output are OBS activities. Exogenous factor prices are calculated per country and year as in Koetter (2006).

Table 10.7   Cost efficiency according to SFA and DEA.

	SFA				DEA			
	All	Years	Groups	Both	All	Years	Groups	Both
Observations	25 945	25 945	25 944	25 945	25 945	25 945	25 945	25 945
Mean	0.933	0.927	0.978	0.959	0.171	0.216	0.215	0.356
Std. dev.	0.024	0.043	0.053	0.062	0.170	0.185	0.173	0.220
Minimum	0.156	0.078	0.180	0.108	0.010	0.016	0.010	0.016
Maximum	0.974	0.989	0.996	1.000	1.000	1.000	1.000	1.000

SFA and DEA CE scores for a sample of 25 945 bank-year observations for a sample of EU banks from 1996 until 2006. All cost function variables are truncated at the 5th and 95th percentile of the raw data distribution. Based on specification of three outputs and exogenous factor prices.

States, where OBS are important (Stiroh, 2004). Regarding factor prices, an important limitation is that we calculate exogenous input prices per country as opposed to (more numerous) economic agglomeration areas as in Koetter (2006).

In the remainder of this chapter, we continue using the 5%-truncation threshold together with the specification of three outputs and exogenous factor prices.

10.5.1.3   Sample-specific frontiers

Sample heterogeneity and outlier treatment appear important determinants of efficiency score consistency. Next, we show differences across four subsamples for SFA and DEA scores. In each of the two panels corresponding to SFA and DEA scores, Table 10.7 presents results from the entire EU sample (All), year-specific benchmarks (Years), banking group specific benchmarks (Groups), and scores from year and group specific benchmarks (Both). As discussed in Section 10.4, European banks differ substantially in terms of ownership and governance structure. Therefore, we distinguish commercial, savings, and cooperative banks.

Regarding SFA, subsample-specific estimation accounting for either cross-sectional or time-specific effects does not affect mean efficiency scores significantly. DEA scores, in turn, clearly indicate that any reduction in systematic differences immediately translates into higher mean efficiency scores. This result underpins the sensitivity to outliers and the apparently larger importance in DEA to properly account for environmental factors that constitute systematic preferences across benchmarked banks.

The comparison of efficiency levels fails, however, to shed light on the stability of bank rankings resulting from alternative benchmarking methods. Such rankings of best and worst performers, to which we turn next, are probably more important to policy makers than efficiency levels.

10.5.2   Rank correlations

Policy and management might be more interested in the rankings of financial firms rather than point estimates of bank-specific efficiency levels. For example, prudential regulators could use off-site benchmarking scores to identify those institutions that are ranked worst and thus warrant more labor-intensive on-site inspections. Such purposes, however, require that different benchmarking methods result in sufficiently high rank-order stability.

Table 10.8 shows Spearman rank-order correlation coefficients across subsamples and benchmarking methods. Even within parametric methods, CE rankings can differ considerably for different subsamples. CE rankings obtained from year-specific SFA benchmarks correlate only weakly with those for pooled or group-specific samples. Although positive, such low rank-order correlations (19% and 17%) are cumbersome because they indicate that, for instance, scarce prudential resources to conduct on-site visits are difficult to allocate to the most worrisome financial firms if these are identified inconsistently. This result indicates that explicitly accounting for time-specific effects in benchmarking methods is of crucial importance, for instance by modeling efficiency as an autoregressive process (Park, Sickles, and Simar, 2003) or using panel estimators (Greene, 2005; Park, Sickles, and Simar, 2007). The fairly high correlation of 77% between group-specific rankings and those obtained from separate subsamples per period and banking groups indicates, in turn, that accounting for cross-sectional differences across banking groups affects efficiency ranks to a much lesser extent. But the extremely low

Table 10.8   Spearman rank correlations.

		SFA				DEA		
		All	Year	Group	Both	All	Year	Group
SFA	*Year*	0.454						
SFA	*Group*	0.400	0.190					
SFA	*Both*	0.046	0.174	0.774				
DEA	*All*	0.310	0.215	−0.212	−0.304			
DEA	*Year*	0.325	0.245	−0.157	−0.247	0.913		
DEA	*Group*	0.326	0.213	0.178	0.071	0.789	0.840	
DEA	*Both*	0.137	0.268	0.414	0.387	0.309	0.472	0.672

SFA and DEA CE scores for a sample of 25 945 bank-year observations for a sample of EU banks from 1996 until 2006. All cost function variables are truncated at the 5th and 95th percentile of the raw data distribution. Based on specification of three outputs and exogenous factor prices. Rank-order correlation coefficients are all significantly different from 0 at the 1% level.

correlation of just 5% between pooled and separated subsample corroborates that ignoring both time and cross-sectional differences will yield significantly different CE rankings.

These differences are substantially lower in DEA. The lower right part of Table 10.8 shows that rankings from the various subsamples all correlate reasonably well. Note, however, that those rankings obtained from the finest subsamples, by both banking group and year, exhibit also the lowest correlations (31–67%) and year-specific rankings correlate also for DEA the least. Thus, a clear-cut identification of bank rankings on the basis of DEA benchmarking remains challenging.

This conclusion is corroborated when correlating the ranks from SFA and DEA across subsamples in the lower left part of Table 10.8. Spearman's ρ is for most comparisons in-between the levels of the within SFA and DEA comparisons. An important exception is the comparison between SFA rankings accounting for cross-sectional or both types of heterogeneity with pooled and time-specific subsamples in DEA, which correlate negatively.

In sum, high rank-order instability calls for the use of multiple methods to identify those banks that consistently rank best and/or worst if CE scores are to be used for policy and/or managerial purposes. More generally, it seems important to test and report the significance of efficiency scores, for instance by means of bootstrapping methods (Bos et al., 2009).

10.5.3   Extreme performers

Whereas rank-order correlation for all banks in the sample is fairly unstable, it is especially the tails of the efficiency distributions that are most important. We may not be too concerned about the exact rank (or level) of CE in the central mass of the efficiency distribution as long as benchmarking methods identify the extreme performers consistently. Therefore, we show in Table 10.9 the 'overlap' for the top and bottom 25th percentile banks obtained for the four subsamples across SFA and DEA. For example, a value of 0.52 in the lower triangle in Table 10.9 indicates that 52% of the banks that are in the top 25% under one specification are also in the top 25% in the other specification.

Table 10.9   Cost efficiency according to SFA and DEA.

		SFA				DEA			
		All	Year	Group	Both	All	Year	Group	Both
SFA	*All*		0.431	0.672	0.371	0.327	0.321	0.356	0.267
SFA	*Year*	0.520		0.383	0.475	0.293	0.317	0.331	0.419
SFA	*Group*	0.184	0.220		0.607	0.207	0.213	0.365	0.329
SFA	*Both*	0.138	0.215	0.849		0.205	0.208	0.357	0.390
DEA	*All*	0.438	0.386	0.055	0.057		0.779	0.614	0.328
DEA	*Year*	0.462	0.394	0.095	0.094	0.857		0.667	0.421
DEA	*Group*	0.428	0.357	0.217	0.201	0.783	0.846		0.568
DEA	*Both*	0.323	0.352	0.484	0.494	0.436	0.519	0.609	

Percentage of 'overlap' of banks for the top 25% performers (lower left triangle) and 25% top performers (upper right triangle). SFA and DEA CE scores for a sample of 25 945 bank-year observations for a sample of EU banks from 1996 until 2006. All cost function variables are truncated at the 5th and 95th percentile of the raw data distribution. Based on specification of three outputs and exogenous factor prices.

The lower left triangle shows 'overlap' for the top performers according to both benchmarking methods. The pattern within SFA across subsamples mimics that of Table 10.8: especially CE_{SFA} 'overlap' for year-specific subsamples differ considerably from pooled and group-specific estimates. Within DEA, top performers are identified fairly consistently and somewhat better compared to the rank-order correlation of the entire sample. The comparison of top performers across SFA and DEA shows that the negative correlations prevailing in Table 10.8 between group and group-year subsamples of SFA and pooled and year-specific subsamples in DEA vanishes. But overall, 'overlap' remains low and thus indicates that parametric and nonparametric methods identify systematically different banks as top performers.

The upper right triangle compares the worst-performing banks in terms of 'overlap'. Within SFA, 'overlap' is overall higher and indicates that irrespective of subsamples, parametric methods tend to identify similar banks as most inefficient. The same holds for within comparisons of CE_{DEA}. The comparison of worst performers across SFA and DEA exhibits also higher consistency, which bodes well to some extent for the potential practical use of efficiency benchmarking exercises. It must be noted though that the range of 'overlap' between 21% and 42% is too low to render the choice of method an innocent one. We conclude again that it is useful to apply multiple methods and identify those banks that are consistently considered extreme performers.

10.5.4   Accounting-based indicators

In practice, nonstructural performance indicators that are based on accounting data are used more often to assess the performance of financial firms compared to structural benchmarking measures. From a consistency perspective, one might therefore argue that efficiency scores should correlate well with key performance indicators such as return on assets (ROA) or equity (ROE) and cost–income (CI) ratios. However, these indicators are not based on microeconomic theory and might thus contain different information (Hughes and Mester, 2010). CI ratios may indicate inefficient banking operations, but they might just as well indicate competitive markets where marginal costs approach marginal revenues. Therefore, we do not expect perfect correlation between CE scores and accounting-based indicators in Table 10.10.

The rank-order correlations between CE from SFA and DEA for different subsamples and three accounting-based indicators show overall fairly low correlation, which is in line with Bauer et al. (1998) and Koetter (2006). Concerning correlations with CE_{SFA}, especially the

Table 10.10   Correlations of efficiency and accounting-based measures.

	SFA				DEA			
	All	Year	Group	Both	All	Year	Group	Both
ROA	54.12	1.62	32.36	12.80	2.75	6.32	14.61	0.15
ROE	52.14	12.67	5.94	−14.32	33.26	30.74	28.79	−4.19
CI	26.32	32.07	2.27	−4.90	18.88	18.75	16.15	3.43

SFA and DEA CE scores for a sample of 25 945 bank-year observations for a sample of EU banks from 1996 until 2006. All cost function variables are truncated at the 5th and 95th percentile of the raw data distribution. Based on specification of three outputs and exogenous factor prices. ROA denotes return on assets. ROE denotes return on equity. CI denotes the cost-to-income ratio. Rank-order correlation coefficients are all significantly different from 0 at the 1% level.

subsampling per year and/or banking group reduces reasonable pooled correlations with profitability measures (ROE and ROA) substantially. This result suggests that accounting for sample heterogeneity in structural benchmarking, admittedly in a very crude way, leads to very different performance comparisons compared to accounting-based indicators that neglect any conditioning factors. In fact, the negative correlation between ROE and CE_{SFA} when subsampling per year and banking group indicates that these measures convey fundamentally different information. Likewise, the positive correlation between CI and CE_{SFA} for most samples corroborates potentially different interpretations of what accounting-based indicators actually do measure. If higher CI ratios correlate with more efficiency, it might very well be that CI measure competitive conditions rather than managerial skill.

These patterns are by and large confirmed by the comparison of CE_{DEA} and accounting-based indicators. Again, it appears most reasonable not to rely on any single benchmarking method when assessing the relative performance of banks. Instead, we suggest consulting a range of both structural and nonstructural performance measures to identify those financial firms that are consistently identified as best and/or worst in class.

10.6   Conclusion

We assess in this chapter the implications of simple specification choices on CE scores obtained with SFA and DEA. These parametric and nonparametric benchmarking techniques are used pervasively in the banking literature to assess the relative efficiency of banks to convert inputs into outputs. To be informative for policy making and managerial decision making, however, different benchmarking techniques and specification choices should yield consistent results.

For a large cross-country sample of EU banks from 1996 until 2010, we show that both SFA and DEA CE scores are affected by different outlier treatment and subsample selection. Parametric scores are more sensitive to larger sample heterogeneity and measurement error as reflected by significantly increasing mean CE_{DEA} scores when culling extreme data more conservatively and limiting samples to certain countries, time periods, or banking groups. Alternative output and factor price specification choices, in turn, affect efficiency levels only to a very limited extent.

The comparison of SFA and DEA scores across subsamples shows, in turn, that both methods do not identify consistently the same banks as best or worst performers. Rank-order correlations between CE measures from either method are generally low and tend to reduce the most after accounting for time-specific samples. Therefore, accounting for environmental factors more explicitly is important, especially for the more sensitive parametric benchmarking approaches.

Related, we find neither strong evidence that extreme performers are identified consistently across SFA and DEA nor that these structural performance measures correlate highly with frequently employed accounting-based measures. This lack of correlation does not necessarily imply that CE scores are uninformative. It might indicate instead that they are providing additional and/or different information compared to accounting-based ratios that are not rooted in microeconomic theory.

Overall, CE measures do differ depending on specification choices and across parametric and nonparametric methods. This indicates in our view that multiple benchmarking methods should be used to identify those banks that perform extremely well and/or bad irrespective of these choices if benchmarking exercises aim to inform policy and management in banking.

References

Aigner, D., Lovell, C.A.K., and Schmidt, P. (1977) Formulation and estimation of stochastic frontier production function models. *Journal of Econometrics*, **6**, 21–37.

Ali, A.I. and Seiford. L.M. (1993) *The Mathematical Programming Approach to Efficiency Analysis*, Oxford University Press, New York.

Basu, S., Inklaar, R.I., and Wang, J. (2011) The value of risk: measuring the services of U.S. commercial banks. *Economic Inquiry*, **49**, 226–245.

Battese, G.E. and Coelli, T.J. (1988) Prediction of firm-level technical efficiencies with a generalized frontier production function and panel data. *Journal of Econometrics*, **38**, 387–399.

Battese, G. and Corra, G. (1977) Estimation of a production frontier model: with application to the pastoral zone of Eastern Australia. *Journal of Agricultural Economics*, **21**, 169–179.

Bauer, P.W., Berger, A.N., Ferrier, G.D., and Humphrey, D.B. (1998) Consistency conditions for regulatory analysis of financial institutions: a comparison of frontier efficiency methods. *Journal of Economics and Business*, **50**, 85–114.

Beccalli, E. and Casu, B. (2006) Efficiency and stock performance in European banking. *Journal of Business, Finance and Accounting*, **33**, 218–235.

Benston, G. (1965) Branch banking and economies of scale. *Journal of Finance*, **20**, 147–180.

Berger, A.N. (1993) 'Distribution-Free' estimates of efficiency in the U.S. banking industry and tests of the standard distributional assumptions. *Journal of Productivity Analysis*, **4**, 261–292.

Berger, A.N. (2007) International comparisons of banking efficiency. *Financial Markets, Institutions and Instruments*, **16**, 119–144.

Berger, A.N. and Humphrey, D.B. (1991) The dominance of inefficiencies over scale and product mix economies in banking. *Journal of Monetary Economics*, **28**, 117–148.

Berger, A.N. and Humphrey, D.B. (1997) Efficiency of financial institutions: international survey and directions for future research. *European Journal of Operational Research*, **98**, 175–212.

Bos, J.W.B. and Koetter, M. (2011) Handling losses in translog profit models. *Applied Economics*, **43**, 307–312.

Bos, J.W.B., Koetter, M., Kolari, J.W., and Kool, C.J.M. (2009) Effects of heterogeneity on bank efficiency scores. *European Journal of Operational Research*, **195**, 251–261.

Brown, C.O. and Dinç, I.S. (2005) The Politics of bank failures: evidence from emerging markets. *Quarterly Journal of Economics*, **120**, 1413–1444.

Casu, B. and Girardone, C. (2002) A comparative study of the cost efficiency of Italian bank conglomerates. *Managerial Finance*, **28**, 3–23.

Charnes, A., Cooper, W., and Rhodes, E. (1978) Measuring the efficiency of decision making units. *European Journal of Operational Research*, **2**, 429–444.

Coelli, T., Rao, D., and Battese, G. (2005) *An Introduction to Efficiency and Productivity Analysis*, 2nd edn, Springer, New York.

Daouia, A. and Simar, L. (2007) Nonparametric efficiency analysis: a multivariate conditional quantile approach. *Journal of Econometrics*, **140**, 375–400.

Daouia, A., Florens, J.-P., and Simar, L. (2012) Regularization of nonparametric frontier estimators. *Journal of Econometrics*, **168**, 285–299.

Delis, M.D., Koutsomanoli-Fillipaki, K., Staikouras, C.K., and Gerogiannaki, K. (2009) Evaluating cost and profit efficiency: a comparison of parametric and nonparametric methodologies. *Applied Financial Economics*, **19**, 191–202.

Deprins, D., Simar, L., and Tulkens, H. (1984) Measuring labor-efficiency in post offices, in *The Performance of Public Enterprises: Concepts and Measurement* (eds M. Marchand, P. Pestieu, and H. Tulkens), Elsevier, Amsterdam, pp. 243–268.

Dinç, I.S. (2005) Politicians and banks: political influences on government-owned banks in emerging markets. *Journal of Financial Economics*, **77**, 453–479.

Färe, R. and Grosskopf, S. (1983) Measuring congestion in production. *Journal of Economics*, **43**, 257–271.

Farrell, M.J. (1957) The measurement of productive efficiency. *Journal of the Royal Statistical Society Series A*, **120**, 253–281.

Farell, M.J. (1959) The convexity assumption in the theory of competitive markets. *Journal of Political Economy*, **67**, 377–391.

Ferrier, G.D. and Lovell, C.A.K. (1990) Measuring cost efficiency in banking: econometric and linear programming evidence. *Journal of Econometrics*, **46**, 229–245.

Fethi, M.D. and Pasiouras, F. (2010) Assessing bank efficiency and performance with operational research and artificial intelligence techniques: A survey. *European Journal of Operational Research*, **204**, 189–198.

Fiordelisi, F., Marques-Ibanez, D., and Molyneux, P. (2011) Efficiency and risk in European banking. *Journal of Banking and Finance*, **35**, 1315–1326.

Fiorentino, E., Karmann, A., and Koetter, M. (2006) The cost efficiency of German banks: a comparison of SFA and DEA. Bundesbank Discussion Paper Series 2. *Banking and Financial Studies*, **10**, 1–21.

Florens, J.-P. and Simar, L. (2005) Parametric approximations of nonparametric frontiers. *Journal of Econometrics*, **124**, 91–116.

Fries, S. and Taci, A. (2005) Cost efficiency of banks in transition: evidence from 289 banks in 15 post-communist countries. *Journal of Banking and Finance*, **29**, 55–81.

Greene, W.H. (1993) *The Econometric Approach to Efficiency Analysis*, Oxford University Press, New York, pp. 69–119.

Greene, W.H. (2005) Reconsidering heterogeneity in panel data estimators of the stochastic frontier model. *Journal of Econometrics*, **126**, 269–303.

Hackethal, A., Vins, O., and Koetter, M. (2012) Do government owned banks trade market power for slack? *Applied Economics*, **44**, 4275–4290.

Huang, T.-H. and Wang, M.-H. (2002) Comparison of economic efficiency estimation methods: Parametric and non-parametric techniques. *The Manchester School*, **70**, 682–709.

Hughes, J.P. and Mester, L.M. (2010) Efficiency in banking: theory and evidence, in *Oxford Handbook of Banking* (eds A.N. Berger, P. Molyneux, and J. Wilson), Oxford University Press, Oxford, pp. 463–485.

Hughes, J.P., Lang, W., Mester, L.J., and Moon, C.-G. (1996) Efficient banking under interstate branching. *Journal of Money, Credit and Banking*, **28**, 1045–1071.

Hughes, J.P., Lang, W, Mester, L.J., and Moon, C.-G. (2000) Recovering risky technologies using the almost ideal demand system: an application to U.S. Banking. *Journal of Financial Services Research*, **18**, 5–27.

Humphrey, D.B. and Pulley, B.L. (1997) Bank's response to deregulation: profits, technology and efficiency. *Journal of Money, Credit and Banking*, **29**, 73–93.

Jondrow, J., Lovell, C., Materov, I., and Schmidt, P. (1982) On the estimation of technical inefficiency in the stochastic frontier production function model. *Journal of Econometrics*, **19**, 233–238.

Koetter, M. (2006) Measurement matters: input price proxies and bank efficiency in Germany. *Journal of Financial Services Research*, **30**, 199–226.

Koetter, M. (2008) The stability of bank efficiency rankings when risk preferences and objectives are different. *European Journal of Finance*, **14**, 115–135.

Koetter, M. and Noth, F. (2013) IT use, productivity, and market power in banking. *Journal of Financial Stability*, forthcoming.

Kumbhakar, S.C. and Lovell, C.A.K. (2000) *Stochastic Frontier Analysis*, Cambridge University Press, Cambridge.

Kumbhakar, S.C. and Tsionas, E.G. (2005) Measuring technical and allocative inefficiency in the translog cost system: a Bayesian approach. *Journal of Econometrics*, **126**, 355–384.

Kumbhakar, S.C., Park, B.U., Simar, L., and Tsionas, E.G. (2007) Nonparametric stochastic frontiers: a local maximum likelihood approach. *Journal of Econometrics*, **137**, 1–27.

Lensink, R., Meesters, A.J., and Naaborg, I. (2008) Bank efficiency and foreign ownership: do good institutions matter? *Journal of Banking and Finance*, **32**, 834–844.

Levinsohn, J. and Petrin, A. (2003) Estimation production functions using inputs to control for unobservables. *Review of Economic Studies*, **40**, 317–342.

Martin-Oliver, A. and Salas-Fumas, V. (2008) The output and profit contribution of information technology and advertising investments in banks. *Journal of Financial Intermediation*, **17**, 229–255.

Meeusen, W. and Broeck, J.V.D. (1977) Efficiency estimation for Cobb Douglas production functions with composed error. *International Economic Review*, **18**, 435–444.

Mester, L.M. (1993) Efficiency in the savings and loan industry. *Journal of Banking and Finance*, **17**, 267–286.

Mountain, D.C. and Thomas, H. (1999) Factor price misspecification in bank cost function estimation. *Journal of International Financial Markets, Institutions and Money*, **9**, 163–182.

Nakane, M.I. and Weintraub, D. (2005) Bank privatization and productivity: evidence for Brazil. *Journal of Banking and Finance*, **29**, 2259–2289.

Park, B.U., Sickles, R.C., and Simar, L. (1998) Stochastic panel frontiers: a semiparametric approach. *Journal of Econometrics*, **84**, 273–301.

Park, B.U., Sickles, R.C., and Simar, L. (2003) Semiparametric-efficient estimation of AR(1) panel data models. *Journal of Econometrics*, **117**, 279–309.

Park, B.U., Sickles, R.C., and Simar, L. (2007) Semiparametric efficient estimation of dynamic panel data models. *Journal of Econometrics*, **136**, 281–301.

Post, T., Cherchye, L., and Kuosmanen, T. (2002) Nonparametric efficiency estimation in stochastic environments. *Operations Research*, **50**, 645–655.

Resti, A. (1997) Evaluating the cost efficiency of the Italian banking system: what can be learned from the joint application of parametric and non-parametric techniques. *Journal of Banking and Finance*, **21**, 221–250.

Sapienza, P. (2004) The effects of government ownership on bank lending. *Journal of Financial Economics*, **890**, 357–384.

Sealey, C.W. and Lindley, J.T. (1977) Inputs, outputs, and a theory of production and cost and depository financial institutions. *Journal of Finance*, **32**, 1251–1265.

Stiroh, K.J. (2004) Diversification in banking: is noninterest income the answer? *Journal of Money, Credit and Banking*, **36**, 853–882.

Weill, L. (2004) Measuring cost efficiency in European banking: A comparison of frontier techniques. *Journal of Productivity Analysis*, **21**, 133–152.

11

Efficiency and performance evaluation of European cooperative banks

Michael Doumpos and Constantin Zopounidis

Financial Engineering Laboratory, Department of Production Engineering and Management, Technical University of Crete, Greece

11.1 Introduction

Cooperative banks emerged during the 19th and 20th centuries to address important market imperfections and meet the needs of a broad public basis for access to loans and financing (Fonteyne, 2007). Since then, cooperative banks have undergone significant changes. Today they provide savings products and loans to consumers and small and medium-sized enterprises, but they have also evolved to include investment products in their portfolio. Despite that, cooperative banks are still well-distinguished in terms of their operating model from other types of banks and they have attracted considerable interest on their own from researchers, practitioners, and policy makers.

That should not be surprising given the size of the cooperative banking sector, at least in the case of Europe. According to statistics from the European Association of Cooperative Banks,[1] there are about 4000 local cooperative banks in Europe, with 50 million members, serving more than 176 million customers, and with an average market share of 20%. Thus, it is clearly evident that cooperative banks constitute a major part of the European banking sector and they add significantly to Europe's development, competitiveness, and employment policies.

Even though the economic turmoil is not over yet, preliminary studies indicate that cooperative banks responded well to the credit crisis. According to Groeneveld (2011), the direct losses

---

[1] http://www.eurocoopbanks.coop

Efficiency and Productivity Growth: Modelling in the Financial Services Industry, First Edition. Edited by Fotios Pasiouras.
© 2013 John Wiley & Sons, Ltd. Published 2013 by John Wiley & Sons, Ltd.

and write-offs of European cooperative banks as a consequence of the credit crisis account for 8% of the total, which is much lower compared to the market share of cooperative banks. Boonstra (2010) emphasizes that no European cooperative bank has failed nor has been nationalized during the credit crisis and attributes most of the losses that cooperative banks faced to their international activities, which are similar to commercial banks. However, Boonstra also notes that cooperative banks may be more severely affected by an economic downturn (as opposed to the manageable losses due to the credit crisis), due to their strong ties to local economies.

Except for such studies, which explore the impact of the crisis on cooperative banks, a significant part of the literature has focused on the differences between cooperative banks and other banking institutions. Some country-specific studies have been presented by Altunbas, Evans, and Molyneux (2001) for Germany; Bos and Kool (2006) for Netherlands; and Hasan and Lozano-Vivas (2002) for Spain, whereas Girardone, Nankervis, and Velentza (2009) and Kontolaimou and Tsekouras (2010) presented results for samples involving multiple European countries. These studies as well as others (some additional references are given in Goddard et al., 2007) consider the comparison of cooperative to noncooperative banks from different efficiency perspectives (e.g., profit, cost, and productivity). However, the existing empirical results fail to provide clear-cut evidences on whether commercial banks outperform cooperative banks.

In contrast to the vast majority of existing studies, this chapter is not concerned with the comparison of cooperative to commercial banks. Instead, we focus on analyzing the efficiency and performance of European cooperative banks (from five major European countries) using the most up-to-date data covering the period before the credit crisis and up to 2010. On the methodological side, the analysis is performed in two stages. First, data envelopment analysis (DEA) is employed for analyzing the efficiency of the banks. The efficiency analysis is performed under two popular settings in bank efficiency measurement, namely, the profit and the intermediation approaches. However, performing comparisons with respect to different countries on the basis of efficiency scores is troublesome.[2] In addition, the efficiency analysis results do not provide a direct way of comparing and ranking all banks, which is very useful for benchmarking and monitoring purposes. Thus, in a second stage, a multicriteria evalua-tion procedure is employed to analyze the performance of all banks in a common setting based on widely used financial ratios. Multicriteria techniques have been successfully employed for bank performance evaluation (see Doumpos and Zopounidis, 2010 and the references therein), but not in the context of cooperative banks. The results of the empirical analysis in this chapter provide several interesting findings on the effect of the crisis on the efficiency and performance of the banks in each country, the differences between the profit and the intermediation approaches for analyzing cooperative banks, their connections of effi-ciency analysis results with a multicriteria evaluation procedure, and the indicators that best describe the performance of the cooperative banks in Europe.

The rest of the chapter is organized as follows: Section 11.2 provides an overview of the methodological approaches used for analyzing the efficiency and performance of coopera-tive banks in this study. Section 11.3 presents the setting of the empirical analysis (data and variables), as well as the results from the application of DEA and the multicriteria evalua-tion procedure. Finally, Section 11.4 concludes the chapter and discusses some future research directions.

---

[2] Lozano-Vivas, Pastor, and Pastor (2002) discuss this issue in the context of bank efficiency analysis and present a DEA-based approach that enables cross-country comparisons through the consideration of environmental variables.

11.2 Methodology

11.2.1 Data envelopment analysis models

Data envelopment analysis (DEA) is a popular methodology for the estimation of the efficiency of decision-making units (DMUs, e.g., banks) based on the inputs that each unit uses and the outputs it produces (Cooper, Seiford, and Tone, 2007).

In particular, assume that there is data on K inputs and M outputs for N DMUs. For the ith DMU these are represented by the vectors $\mathbf{x}_i$ and $\mathbf{y}_i$, respectively. The $K \times N$ input matrix $\mathbf{X}$ and the $M \times N$ output matrix $\mathbf{Y}$ represent the data for all DMUs. Then, the efficiency of the ith DMU is measured by the ratio

$$\theta_i = \frac{\mathbf{u}_i \mathbf{y}_i}{\mathbf{y}_i \mathbf{x}_i} \in [0,1],$$

where $\mathbf{u}_i, \mathbf{v}_i \geq 0$ are weight vectors corresponding to the outputs and inputs of the ith DMU. DEA provides an assessment of the relative efficiency of a DMU compared to a set of other DMUs. In this relative evaluation setting, each DMU is free to specify its own combination of input–output weights that maximize its performance relative to its 'competitors'. Under constant returns to scale (CRS) and assuming an input orientation, the optimal efficiency for the ith DMU can be estimated through the linear programming formulation introduced by Charnes, Cooper, and Rhodes (1978), which is expressed in primal and dual form as follows (CCR model):

$$
\begin{array}{lll}
\text{Primal:} & & \text{Dual:} \\[4pt]
\max & \mathbf{u}_i \mathbf{y}_i & \min \quad \theta_i^c \\
\text{s.t.} & \mathbf{v}_i \mathbf{X} - \mathbf{u}_i \mathbf{Y} \geq 0 \quad \text{s.t.} & \theta_i^c \mathbf{x}_i - \mathbf{X}\lambda \geq 0. \\
& \mathbf{v}_i \mathbf{x}_i = 1 & \mathbf{Y}\lambda \geq \mathbf{y}_i \\
& \mathbf{u}_i, \mathbf{v}_i \geq 0 & \lambda \geq 0, \theta_i^c \in \mathbf{R}
\end{array}
\tag{11.1}
$$

The estimate θ_i^c obtained from the CCR model provides a global technical efficiency measure without taking into consideration any scale effects. In that sense, it is assumed that all DMUs are operating at an optimal scale (Coelli et al., 2005). To take into account cases where this assumption is not true, variable returns to scale (VRS) can be introduced by simply adding the convexity constraint $\lambda_1 + \lambda_2 + \ldots + \lambda_N = 1$ to the dual CCR model. This constraint ensures that a DMU is benchmarked only against other units of similar size. The resulting model is widely known as the BCC model (Banker, Charnes, and Cooper, 1984).

The combination of the results obtained from the CCR and BCC models provides a decomposition of the global efficiency as follows:

$$\theta_i^c = \theta_i^v \theta_i^s,$$

where $0 \leq \theta_i^v \leq 1$ is the pure efficiency score obtained under VRS from the BCC model and $0 \leq \theta_i^s \leq 1$ is the scale efficiency factor. Thus, the inefficiency of a DMU can be attributed to inefficient operation (e.g., too small θ_i^v), disadvantageous exogenous conditions (corresponding to scale inefficiency), or both.

11.2.2   Multicriteria evaluation

DEA models are useful for evaluating the relative efficiency of DMUs and discriminating between efficient and inefficient DMUs. However, the use of DEA models for evaluating and ranking all DMUs in a common basis is troublesome. An overview of different DEA-based ranking models and an empirical comparative analysis can be found in Sarkis (2000), whereas Bouyssou (1999) provides a critical discussion of the theoretical difficulties that arise when such models are used in a multicriteria evaluation context.

Multicriteria decision making (MCDM), on the other hand, provides a wide range of techniques, which are well-suited to evaluation problems where a complete ranking of a discrete set of alternatives is needed. In the context of this study, the simulation-based Stochastic Multicriteria Acceptability Analysis (SMAA) framework is employed (Lahdelma, Hokkanen, and Salminen, 1998). SMAA provides a general context for multicriteria evaluation problems under uncertainty, but it is also applicable in deterministic problems. The basic underlying idea of SMAA is that the uncertainties involved in multicriteria evaluation problems can be taken into consideration through simulation approaches. Such simulations enable the decision-maker to obtain a holistic view of the evaluation results under different scenarios with regard to the parameters of the decision model and/or the evaluation data. SMAA can be used with any multicriteria evaluation model and for different types of decision problems (e.g., choice, ranking, classification, or description). An overview of the SMAA modeling framework, its extensions, and applications can be found in Tervonen and Figueira (2008).

Simulation approaches for multicriteria performance evaluation problems are particularly useful when specific preferential information on the relative importance of the evaluation criteria and their aggregation is not available for a given decision-maker or a group of decision-makers. In such cases, it is helpful to perform a comprehensive evaluation of the alternatives' performance under different scenarios with respect to the parameters of the evaluation model. Thus, the evaluation takes into account different settings and hypotheses with respect to the judgment policy of a 'hypothetical' decision-maker.

In contrast to DEA-based efficiency analysis, MCDM evaluation models do not distinguish inputs and outputs. Instead, each alternative (i.e., a DMU in the context of DEA) is described over a set of N evaluation criteria, which enable the comparison of all alternatives on a common basis. In this study, the aggregation of the criteria is performed through an additive value function model

$$V(\mathbf{x}) = \sum_{j=1}^{N} w_j v_j (x_{ij}), \tag{11.2}$$

where $\mathbf{x}_i = (x_{i1}, \ldots, x_{iN})$ is the vector with the data for bank i on the evaluation criteria, $w_1, \ldots, w_N$ are nonnegative trade-off constants for the criteria that sum up to 1, and $v_j(\cdot)$ is the marginal value function criterion j normalized in [0, 1]. On the basis of such an additive model, the alternatives under consideration can be ranked from the best to the worst according to their global value score (in descending order, i.e., the best alternatives are those with the highest global value).

The marginal value functions provide a decomposition of the overall performance of an alternative on the set of criteria and can have any monotone form (e.g., nondecreasing for maximization criteria). In order to avoid posing restrictions on the form of the marginal value functions, we employ a piecewise linear modeling approach. In particular, the scale of each criterion j (assumed to be in maximization form) is divided into s_j subintervals defined by breakpoints $b_0^j < b_1^j < \cdots < b_{s_j-1}^j < b_{s_j}^j$, where b_0^j and $b_{s_j}^j$ are the least and most preferred levels of the criterion. Then, for any $b_{\ell-1}^j \le x_{ij} \le b_\ell^j$ (for some $\ell \in \{1,\dots,s_j\}$), the corresponding marginal value (partial score) of bank i on criterion j can be obtained by linear interpolation as follows:

$$v_j\left(x_{ij}\right) = v_j\left(b_{\ell-1}^j\right) + \left[v_j\left(b_\ell^j\right) - v_j\left(b_{\ell-1}^j\right)\right]\frac{x_{ij} - b_{\ell-1}^j}{b_\ell^j - b_{\ell-1}^j}.$$

In a typical MCDM setting, the parameters (i.e., the criteria trade-offs and the associated marginal value functions) of the evaluation model (11.2) are specified by the decision-maker. As noted earlier, when a decision-maker is not available (as in this study), a simulation-scenario analysis approach can be helpful. Thus, through the SMAA framework, a Monte Carlo simulation approach is employed in order to perform the evaluation of the banks under different scenarios with respect to the criteria aggregation model. In particular, each scenario r involves the construction of a random additive value function $V_r(\mathbf{x}) = w_{1r}v_{1r}(x_1) + \dots + w_{Nr}v_{Nr}(x_N)$ through the following two-step process:

1. For each criterion j, a random marginal value function is first constructed by generating $s_j - 1$ uniformly distributed random numbers in $(0, 1)$, which are sorted and then assigned to $v_{jr}\left(b_1^j\right), v_{jr}\left(b_2^j\right),\dots,v_{jr}\left(b_{s_j-1}^j\right)$. For normalization, $v_{jr}\left(b_0^j\right)$ and $v_{jr}\left(b_{s_j}^j\right)$ are set equal to 0 and 1, respectively. In all simulations, four subintervals are used for the criteria (i.e., $s_j = 4$, for all j) defined on the basis of the 25th, 50th, and 75th percentiles of the data.

2. Random trade-off constants $w_1,\dots,w_N \ge \varepsilon$ are generated such that $w_1 + \dots + w_n = 1$. The constant ε is set equal to 0.01 in order to exclude unrealistic scenarios, where a criterion becomes almost irrelevant for the evaluation.

The resulting additive value model $V_r(\mathbf{x})$ is used to evaluate and rank the banks according to their global values. The results of all simulation runs can be aggregated to obtain an overall evaluation score for each bank. In this study, this overall score for each bank i is simply the average score of the bank across all evaluation scenarios:

$$V(\mathbf{x}_i) = \frac{1}{R}\sum_{r=1}^R V_r(\mathbf{x}_i),$$

where R is the number of scenarios explored through the simulation process.

11.3 Empirical results

11.3.1 Data and variables

The sample used in the analysis consists of 4386 bank-year observations during the period 2005–2010, obtained through the Bankscope database. The cooperative banks in the sample originate from Germany, France, Italy, Spain, and Austria. In all these countries, cooperative banks are well-developed and constitute an important part of the banking sector. Details on the number of banks in the sample are given in Table 11.1.

Different approaches have been suggested in the existing literature on bank efficiency evaluation with regard to the specification of the input and output variables. In particular,

Table 11.1 Number of banks in the sample.

	2005	2006	2007	2008	2009	2010	Total
Germany	242	246	247	248	245	243	1471
France	66	63	66	75	74	73	417
Italy	234	239	243	245	247	243	1451
Spain	79	79	69	79	79	77	462
Austria	108	114	111	89	84	79	585
Total	729	741	736	736	729	715	4386

Table 11.2 Input and output variables.

	Profit approach	Intermediation approach
Inputs	Loan loss provisions (LLP)	Deposits and short-term funding (DSTF)
	Personnel expenses (PE)	Fixed assets (FA)
	Other operating expenses (OE)	Loan loss provisions (LLP)
Outputs	Net interest income (NII)	Loans (L)
	Noninterest operating income (OI)	Other earning assets (OEA)

Table 11.3 Financial ratios for performance evaluation.

Net interest margin (NIM)
Cost to income ratio (CIR)
Loan loss provisions/net interest income (LLP/NII)
Equity/total assets (E/A)
Noninterest expenses/total assets (IE/A)
Return on assets (ROA)
Net loans/total assets (L/A)
Liquid assets/deposits and short-term funding (LA/DSTF)

production, intermediation, and profit approaches have been employed in Pasiouras (2008). Berger and Humphrey (1997) argue that a production approach is more suitable for evaluating the efficiency of bank branches, whereas the intermediation approach is better suited when the analysis involves entire banking institutions. On the other hand, Berger and Mester (2003) as well as Drake, Hall, and Simper (2006) note that the profit approach captures the diversity of strategic decisions taken by financial firms in a dynamic context. In this study, both the intermediation and the profit approaches are employed. The input and output variables used in each setting are listed in Table 11.2.

In addition to the variables used in measuring the efficiency of the banks, a set of eight financial ratios is also employed for the evaluation of their performance through the multicriteria methodology (Table 11.3). The ratios are selected on the basis of data availability and their relevance to bank performance evaluation according to the existing literature. The selected ratios cover all major aspects of a bank's performance, including profitability, capital structure, liquidity, and solvency.

Table 11.4 and Table 11.5 present some summary statistics (averages) for the input and output variables as well as for the selected financial ratios, over the period of the analysis and

Table 11.4   Averages of input/output variables and financial ratios by year.

		2005	2006	2007	2008	2009	2010
Inputs and	FA	48.20	55.46	57.99	60.82	58.50	61.28
Outputs[a]	DSTF	3724.12	4362.87	4593.33	5300.42	5160.37	5697.89
	LLP	10.57	12.71	14.68	30.62	38.84	29.93
	PE	57.81	65.67	67.18	74.00	69.21	74.36
	OE	44.26	50.74	48.86	62.32	53.15	56.33
	L	2917.73	3551.25	3884.34	4466.18	4364.40	4800.14
	OEA	3348.51	4357.63	4512.64	5674.39	4778.00	4936.79
	NII	94.01	104.22	97.63	121.56	129.79	133.95
	OI	64.76	83.23	77.06	55.38	61.21	72.61
Ratios	NIM	2.62	2.66	2.66	2.60	2.40	2.26
	CIR	67.92	63.68	64.26	66.27	66.07	67.29
	LLP/NII	15.01	17.60	14.73	18.85	19.67	19.11
	E/A	8.86	9.01	8.95	8.52	8.64	8.61
	IE/A	2.37	2.28	2.17	2.08	2.03	1.99
	ROA	0.58	0.61	0.62	0.44	0.33	0.32
	L/A	63.81	64.10	64.62	64.57	64.23	65.37
	LA/DSTF	21.06	20.91	21.24	20.92	18.68	16.26

[a]In thousand euros.

Table 11.5   Averages of input/output variables and financial ratios by country.

		Germany	France	Italy	Spain	Austria
Inputs and	FA	3320.79	70080.28	2463.35	2748.44	2646.11
Outputs[a]	DSTF	26.00	300.23	36.99	37.07	27.20
	LLP	2074.37	34091.80	1334.32	1947.34	1644.47
	PE	75.50	1600.80	44.71	55.59	82.72
	OE	22.84	476.27	31.31	21.50	18.40
	L	18.33	372.29	21.45	15.27	17.65
	OEA	1392.56	26658.93	1762.97	2184.62	1342.97
	NII	1827.14	38114.73	506.48	421.24	1144.62
	OI	43.39	732.95	55.56	57.31	35.74
Ratios	NIM	2.43	1.83	2.92	2.57	2.32
	CIR	69.08	60.89	65.62	59.67	67.13
	LLP/NII	19.89	16.50	14.83	16.80	19.27
	E/A	6.15	10.48	11.10	9.81	7.51
	IE/A	2.23	1.88	2.31	1.73	2.11
	ROA	0.30	0.66	0.65	0.62	0.32
	L/A	57.53	69.79	71.46	70.94	55.49
	LA/DSTF	17.43	20.91	19.27	20.63	26.10

[a]In thousand euros.

across the different countries in the sample. The effect of the recent crisis is clearly evident in the sharp increase of loan loss provisions in 2008 and 2009 (increase by more than 150% in 2009 compared to 2007) and the decline in ROA (by more than 45% in 2010 compared to 2007). As far as the differences between the countries are concerned, the cooperative banks from France are on average much larger than those from other countries. Furthermore, banks from France, Italy, and Spain are on average more profitable (i.e., higher ROA), have lower loan loss provisions to net interest income, and are more leveraged (high loan/assets ratio). German and Austrian banks depict common characteristics, with the only exception being the higher liquidity of Austrian banks (LA/DSTF ratio).

11.3.2   Efficiency analysis results

In order to analyze the efficiency of the cooperative banks in the sample, the DEA input-oriented models (CCR and BCC) are employed. Five panel data sets are used, one for each country covering all years in the examined time period. In this way, the differences between the country characteristics are controlled, and the bias from the consideration of all countries in a common sample is eliminated. Table 11.6 and Table 11.7 summarize the efficiency results across all countries and years, for both the profit and the intermediation approach.

The comparison between the profit and the intermediation approaches indicates considerable differences. Overall, the CCR efficiency under the profit-based approach is about 23% lower than the intermediation approach, whereas the overall average BCC and scale efficiency scores are lower by about 18% and 7%, respectively. Furthermore, different efficiency patterns among countries are also observed. In particular, Austria is the only country where the efficiency is higher under the profit approach (by 4.6–16% for CCR efficiency and 2.4–9.4% for BCC

Table 11.6   Average efficiency scores by country and year (profit approach).

Type	Country	2005	2006	2007	2008	2009	2010	Average
CCR	Germany	0.553	0.608	0.572	0.558	0.597	0.620	*0.585*
	France	0.676	0.691	0.687	0.612	0.650	0.660	*0.661*
	Italy	0.358	0.374	0.358	0.326	0.324	0.312	*0.342*
	Spain	0.613	0.625	0.636	0.591	0.598	0.577	*0.606*
	Austria	0.676	0.687	0.702	0.702	0.680	0.664	*0.686*
	Average	*0.526*	*0.554*	*0.537*	*0.507*	*0.520*	*0.520*	*0.527*
BCC	Germany	0.606	0.657	0.626	0.613	0.641	0.659	*0.634*
	France	0.781	0.791	0.784	0.714	0.765	0.775	*0.767*
	Italy	0.518	0.532	0.511	0.479	0.465	0.449	*0.492*
	Spain	0.687	0.720	0.757	0.719	0.696	0.650	*0.704*
	Austria	0.724	0.741	0.757	0.747	0.732	0.715	*0.737*
	Average	*0.620*	*0.648*	*0.634*	*0.606*	*0.610*	*0.605*	*0.621*
Scale	Germany	0.921	0.930	0.919	0.916	0.936	0.945	*0.928*
	France	0.871	0.877	0.875	0.853	0.853	0.856	*0.863*
	Italy	0.715	0.726	0.734	0.727	0.738	0.735	*0.729*
	Spain	0.896	0.875	0.847	0.830	0.860	0.889	*0.866*
	Austria	0.941	0.936	0.935	0.944	0.935	0.936	*0.938*
	Average	*0.851*	*0.855*	*0.850*	*0.841*	*0.852*	*0.858*	*0.851*

Table 11.7 Average efficiency scores by country and year (intermediation approach).

Type	Country	2005	2006	2007	2008	2009	2010	Average
CCR	Germany	0.717	0.698	0.714	0.696	0.694	0.701	0.703
	France	0.762	0.801	0.801	0.782	0.786	0.780	0.785
	Italy	0.679	0.671	0.669	0.677	0.675	0.641	0.669
	Spain	0.732	0.746	0.740	0.703	0.664	0.713	0.716
	Austria	0.595	0.601	0.605	0.610	0.650	0.627	0.613
	Average	0.693	0.688	0.693	0.689	0.688	0.682	0.689
BCC	Germany	0.803	0.785	0.792	0.768	0.756	0.762	0.778
	France	0.804	0.826	0.832	0.811	0.810	0.804	0.814
	Italy	0.744	0.737	0.738	0.752	0.752	0.722	0.741
	Spain	0.765	0.785	0.763	0.742	0.726	0.769	0.758
	Austria	0.684	0.685	0.692	0.686	0.703	0.698	0.691
	Average	0.762	0.758	0.760	0.754	0.751	0.746	0.755
Scale	Germany	0.894	0.889	0.901	0.906	0.916	0.917	0.904
	France	0.951	0.969	0.965	0.967	0.971	0.972	0.966
	Italy	0.917	0.915	0.913	0.908	0.904	0.897	0.909
	Spain	0.959	0.952	0.971	0.951	0.920	0.931	0.947
	Austria	0.871	0.881	0.879	0.891	0.922	0.899	0.889
	Average	0.910	0.910	0.914	0.916	0.918	0.916	0.914

Table 11.8 Correlations between the profit and the intermediation efficiency scores.

	CCR	BCC	Scale
Germany	0.343	0.398	0.276
France	0.237	0.274	0.298
Italy	0.358	0.408	0.370
Spain	0.349	0.435	0.222
Austria	0.350	0.549	0.275

efficiency), although the differences have reduced in 2009 and 2010. This indicates that Austrian cooperative banks are more efficient in managing their expenses and generating profits than when evaluated in terms of their ability to produce loans by exploiting their assets. On the other hand, the efficiency of Italian cooperative banks under the profit scenario is considerably lower compared to the intermediation approach (by about 50% in CCR and 33% in BCC).

Table 11.8 provides detailed results on the correlations between the efficiency scores obtained under the two considered variable settings. The results show that in most cases the correlations are rather moderate, yet significant at the 1% level.

In terms of the dynamics of the efficiency results, the efficiency scores under the intermediation approach do not exhibit significant variations over time. Thus, the operation of the banks evaluated in terms of their ability to produce loans has not been considerably affected by the recent crisis. On the other hand, the situation is different from the profit perspective. In particular, the average CCR and BCC efficiency scores under the profit-based approach declined in 2008 by about 5.6% and 4.4%, respectively, compared to 2007. In 2009 and 2010, minor improvements are observed in Germany and France. In particular, the improvement in 2010 compared to 2008 for German banks exceeds 11% under the CCR model and 7% under the BCC model. The corresponding increases for banks in France are 7.9% (CCR) and 8.5% (BCC). In all other countries, both CCR and BCC efficiencies (under the profit approach) continued to decline during 2009–2010.

It is also worth noting that under both the profit and the intermediation approaches, scale efficiency scores remain at almost constant levels, in most cases higher than 85% (with the exception of Italy under the profit approach). Thus, the overall (CCR) efficiency of the cooperative banks is mainly described by their internal operation as captured through the results of the BCC model. This result agrees with the finding of Bos and Kool (2006) on a sample of cooperatives in the Netherlands, who concluded that about 90% of the inefficiencies in their sample originated from managerial inefficiencies.

Finally, Figure 11.1 and Figure 11.2 illustrate the relationship between the size of the banks and their global CCR efficiency scores under the profit and intermediation approaches. The banks in each country are classified as small, medium, and large depending on their assets. Small banks are those with assets in the first asset quartile (i.e., bottom 25%) of all banks in the same country, medium banks have assets in the interquartile range, whereas large

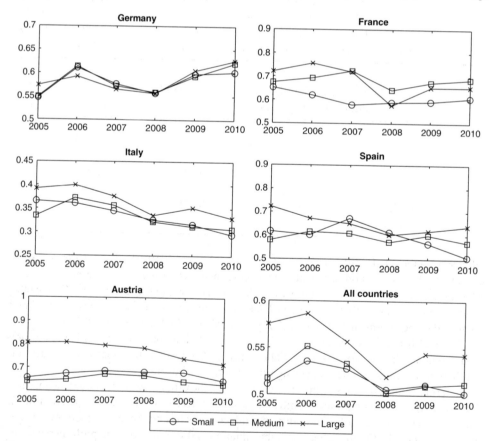

Figure 11.1   CCR efficiency scores by country and asset size (profit-based approach).

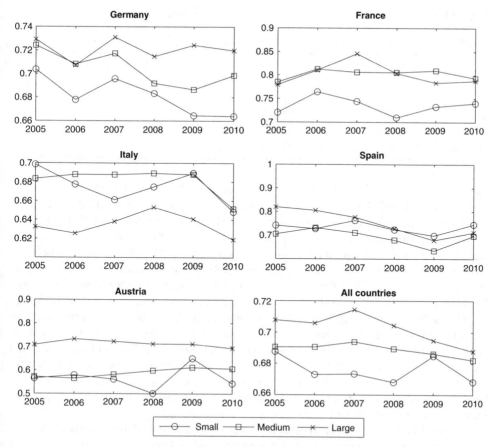

Figure 11.2   CCR efficiency scores by country and asset size (intermediation approach).

banks are those with assets in the third quartile (top 25%). The results show that under both the profit and the intermediation approaches, large banks have (overall) higher efficiency scores, but generally there is no clear-cut conclusion consistent across all countries.

11.3.3   Multicriteria evaluation results

While the DEA results provide useful information about the relative efficiency of the banks, they do not enable performing direct comparisons among all banks from different countries in a common setting. The multicriteria evaluation process is useful in such cases. The application of the SMAA-2 framework in this study was implemented under 10 000 different scenarios generated through a Monte Carlo simulation process. As described in Section 11.2.2, each scenario corresponds to a different additive evaluation model based on different assumptions with respect to the priorities (trade-offs) assigned to the selected financial ratios and the form of the ratios' marginal value functions.

The application of the multicriteria evaluation process is based on the full panel data sample consisting of all bank-year observations from all countries. The average global scores (values) of the banks across all years and countries are summarized in Table 11.9. Overall,

Table 11.9 Average multicriteria evaluation results by country and year.

Country	2005	2006	2007	2008	2009	2010	*Average*
Germany	0.437	0.436	0.432	0.429	0.459	0.483	*0.446*
France	0.550	0.556	0.524	0.469	0.482	0.513	*0.514*
Italy	0.546	0.589	0.600	0.561	0.486	0.431	*0.535*
Spain	0.569	0.590	0.600	0.561	0.530	0.473	*0.553*
Austria	0.506	0.528	0.537	0.503	0.478	0.470	*0.507*
Average	*0.507*	*0.526*	*0.527*	*0.500*	*0.480*	*0.466*	

banks from Spain and Italy performed best (on average), while German banks had the lowest performance. Nevertheless, when examining the variations over the time period of the analysis, it was clearly evident that German banks were the only ones that managed to respond to the crisis in a satisfactory way. In particular, over the period 2005–2008, the performance of German banks remained almost unchanged, whereas in 2009–2010, they improved their performance by more than 12% overall (2010 vs 2008). The performance of the banks in all other countries declined in 2008 by 6.4% (Austria) up to 10.3% (France) compared to 2007. French banks rebounded in 2009–2010 achieving an improvement of 2.6% in 2009 compared to 2008, followed by an additional improvement of 6.5% in 2010. The performance of Austrian banks continued to decline in 2009–2010 but at a reduced rate (−4.9% in 2009 vs 2008, and −1.7% in 2010 vs 2009). On the other hand, Italian and Spanish banks continued their decline, with the overall decrease in 2010 compared to 2008 exceeding 23% in the case of Italy and 15% in the case of Spain. Overall, the multicriteria evaluation results seem to better fit (as opposed to the DEA efficiency results) the tough conditions prevailing in the European banking sector due to the ongoing economic turmoil.

The results of the multicriteria evaluation process can also be used to get an insight into the relationship between the criteria weighting scenarios explored through the simulation process and the performance of the banks. This provides an indication of the strengths and weaknesses of the banks. To perform this analysis, for each criterion j the simulation scenarios in which the criterion is assigned the highest and lowest priority (i.e., the highest/lowest weight among all criteria) are identified. Let the corresponding scenarios be denoted by H_j^+ (highest weight) and H_j^- (lowest weight). Then, we calculate the percentage change between the average performance of the banks under scenarios H_j^+ compared to the average performance under scenarios H_j^-. Table 11.10 presents the obtained results for all countries. According to the results, the performance of German banks is much improved when loan/assets is considered as the most important ratio. The improvement is 15.62% compared to the scenarios where loan/assets is considered as the least important ratio. On the other hand, the performance of German banks deteriorates when equity/assets is given top priority. A similar result is also obtained for the ROA ratio. Thus, the low loan/assets ratio is a strength of German banks, whereas the low equity/assets and ROA ratios are their most important financial weaknesses. Following the same line of reasoning, the strengths and weaknesses of the banks in the other countries are also identified. In Table 11.10, these are marked with (+) for strengths and (−) for weaknesses. Overall, it is worth noting that the loan/assets is the only ratio that has a significant impact (positive or negative) in all countries. The earning/assets and ROA ratios are also strong determinants of the performance of the banks in all countries except Spain.

Figure 11.3 provides some additional results on the performance of the banks in terms of their size. In contrast to the indications derived on the basis of the DEA results, the multicriteria

Table 11.10 Performance changes (in %) with respect to the criteria's priorities.

	Germany	France	Italy	Spain	Austria
NIM	2.49	−14.75(−)	5.24(+)	−2.80	−4.70
CIR	−2.12	6.08(+)	−1.59	4.61	−2.09
LLP/NII	0.80	1.04	2.87	−1.51	−0.41
E/A	−10.89(−)	8.06(+)	9.82(+)	2.80	−5.62(−)
IE/A	1.68	9.69(+)	−4.81	9.64(+)	3.20
ROA	−7.30(−)	7.86(+)	5.22(+)	3.53	−6.45(−)
L/A	15.62(+)	−9.68(−)	−10.30(−)	−10.88(−)	9.07(+)
LA/DSTF	0.86	−6.45(−)	−5.02(−)	−4.37	7.68(+)

The strengths of the banks are indicated by (+) and the weaknesses by (−).

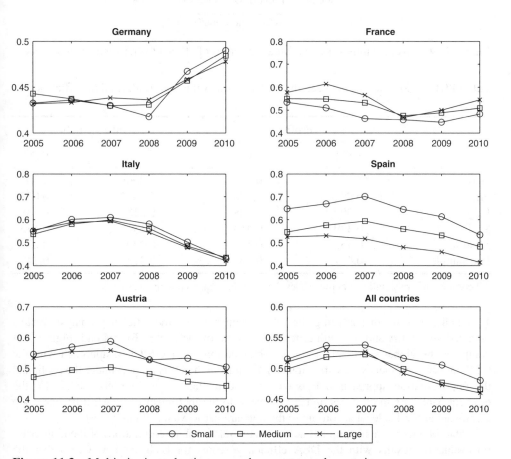

Figure 11.3 Multicriteria evaluation scores by country and asset size.

evaluation of the banks suggests that generally small banks seem to have performed slightly better than larger banks. Again, however, this conclusion depends on the country under consideration. For instance, large banks in France have consistently outperformed smaller ones, whereas the differences in Germany are limited and mixed.

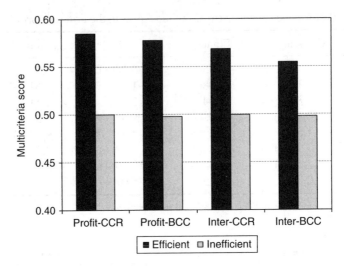

Figure 11.4 Average multicriteria evaluation scores of efficient versus inefficient banks.

Table 11.11 Correlations between the multicriteria evaluation results and the efficiency scores.

	Profit		Intermediation	
	CCR	BCC	CCR	BCC
Germany	0.473	0.477	−0.006*	−0.009*
France	0.498	0.476	−0.003*	−0.053*
Italy	0.392	0.308	0.218	0.138
Spain	0.379	0.413	0.131	0.143
Austria	0.455	0.516	0.229	0.295
Full sample	0.401	0.305	0.312	0.249

All correlations are significant at the 1% level except those marked with an asterisk.

Finally, it is worth analyzing the relationship between the efficiency analysis results obtained with DEA and the multicriteria evaluation of the banks. To this end, Figure 11.4 compares the global scores of the banks obtained through the MCDM approach with the CCR and BCC efficiency classifications obtained from the DEA models under the profit and the intermediation approaches. It is clearly evident that the multicriteria scores for the efficient banks are in all cases higher compared to the inefficient ones. All differences are significant at the 1% level according to the nonparametric Mann–Whitney test.

Table 11.11 presents additional details on the relationship (correlation) of the multicriteria evaluation results with the DEA efficiency scores for each country. Given that the efficiency results are obtained from different samples (i.e., one for each country), we also report the correlations between the multicriteria results and the efficiency estimates obtained from the full sample of all countries in last row of the table. It is clearly evident that the multicriteria evaluation scores of the banks are positively correlated with the efficiency scores obtained from the profit-based DEA models. The correlations for all countries are significant at the 1% level. On the other hand, the correlations when the intermediation approach is employed are much weaker. In fact, for Germany and France the correlations are negative but insignificant at the 10% level. For the rest of the countries, the correlations remain statistically significant at the 1% level, but they are much lower compared to the ones observed with the results of the profit-based approach. Similar results are obtained even if the multicriteria results are compared to efficiency estimates obtained from the full sample, ignoring the differences between the countries.

11.4   Conclusions

In this chapter, an integrated analysis of a large sample of cooperative banks from five major European countries was presented, using the most recent data available covering the period 2005–2010. The analysis covered both the efficiency of the banks as well as the evaluation of their overall performance.

In the first stage of the analysis, DEA was employed for efficiency measurement, under both a profit and an intermediation approach. The results of the two approaches were found to be rather weakly related. Overall, the profit efficiency scores were more affected by the recent crisis, whereas the results derived through the intermediation approach were much more stable.

However, the efficiency results of DEA do not allow direct comparisons between different countries, and they do not provide direct indications on the overall financial performance of the banks. To address these issues, a multicriteria evaluation process was also employed. The results show that cooperative banks from Italy and Spain performed well during the period 2005–2007, but (as expected) they were the ones most affected by the crisis. German banks, on the other hand, were the ones least affected by the crisis and furthermore they managed to improve their performance in 2009–2010. The results of the multicriteria process also highlight the importance of the loan/assets, equity/assets, and ROA ratios as strong descriptors of the performance of the banks.

Future research can focus on a number of issues, which may include, among others, (a) the comparative analysis of cooperative banks as opposed to commercial banks in the face of the new conditions formed due to the ongoing economic turmoil, (b) the development of early warning systems for cooperative banks, (c) the analysis of the specific country characteristics that affect the context in which cooperative banks operate in each country, (d) the re-examination (within the framework of the recent crisis) of the role of cooperative banks in the operation and stability of the European banking sector, and (e) the analysis of the effectiveness and the impact of the transformations imposed on the capital requirements' regulatory framework on the viability and performance of cooperative banks.

References

Altunbas, Y., Evans, L., and Molyneux, P. (2001) Bank ownership and efficiency. *Journal of Money, Credit and Banking*, **33** (4), 926–954.

Banker, R.D., Charnes, A., and Cooper, W.W. (1984) Some models for the estimation of technical and scale inefficiencies in data envelopment analysis. *Management Science*, **30**, 1078–1092.

Berger, A.N. and Humphrey, D.B. (1997) Efficiency of financial institutions: international survey and directions for future research. *European Journal of Operational Research*, **98**, 175–212.

Berger, A.N. and Mester, L.J. (2003) Explaining the dramatic changes in performance of US banks: technological change, deregulation, and dynamic changes in competition. *Journal of Financial Intermediation*, **12**, 57–95.

Boonstra, W.W. (2010) Banking in times of crisis: the case of Rabobank, in *The Quest for Stability: The View of Financial Institutions* (eds M. Balling, J. Berk, and M.O. Strauss-Kahn), SUERF – The European Money and Finance Forum, Vienna, pp. 31–56.

Bos, J.W.B. and Kool, C.J.M. (2006) Bank efficiency: the role of bank strategy and local market conditions. *Journal of Banking & Finance*, **30**, 1953–1974.

Bouyssou, D. (1999) Using DEA as a tool for MCDM: some remarks. *Journal of the Operational Research Society*, **50**, 974–978.

Charnes, A., Cooper, W.W., and Rhodes, E. (1978) Measuring the efficiency of decision making units. *European Journal of Operational Research*, **2**, 429–444.

Coelli, T.J., Rao, P.D.S., O'Donnell, C.J., and Battese, G.E. (2005) *An Introduction to Efficiency and Productivity Analysis*, 2nd edn, Springer, New York.

Cooper, W.W., Seiford, L.M., and Tone, K. (2007) *Data Envelopment Analysis – A Comprehensive Text with Models, Applications, References and DEA-Solver Software*, 2nd edn, Springer, New York.

Doumpos, M. and Zopounidis, C. (2010) A multicriteria decision support system for bank rating. *Decision Support Systems*, **50** (1), 55–63.

Drake, L., Hall, M.J.B., and Simper, R. (2006) The impact of macroeconomic and regulatory factors on bank efficiency: a non-parametric analysis of Hong Kong's banking system. *Journal of Banking & Finance*, **30**, 1443–1466.

Fonteyne, W. (2007) Cooperative banks in Europe – policy issues. Working paper 07/159. International Monetary Fund, Washington, DC.

Girardone, C., Nankervis, J.C., and Velentza, E. (2009) Efficiency, ownership and financial structure in European banking: a cross-country comparison. *Managerial Finance*, **35** (3), 227–245.

Goddard, J., Molyneux, P., Wilson, J.O.S., and Tavakoli, M. (2007) European banking: an overview. *Journal of Banking & Finance*, **31** (7), 1911–1935.

Groeneveld, J.M. (2011) Morality and integrity in cooperative banking. *Ethical Perspectives*, **18** (4), 515–540.

Hasan, I. and Lozano-Vivas, A. (2002) Organisational form and expense preference: Spanish experience. *Bulletin of Economic Research*, **54**, 135–150.

Kontolaimou, A. and Tsekouras, K. (2010) Are cooperatives the weakest link in European banking? A non-parametric metafrontier approach. *Journal of Banking & Finance*, **34** (8), 1946–1957.

Lahdelma, R., Hokkanen, J., and Salminen, P. (1998) SMAA-stochastic multiobjective acceptability analysis. *European Journal of Operational Research*, **106**, 137–143.

Lozano-Vivas, A., Pastor, J.C., and Pastor, J.M. (2002) An efficiency comparison of European banking systems operating under different environmental conditions. *Journal of Productivity Analysis*, **18**, 59–77.

Pasiouras, F. (2008) Estimating the technical and scale efficiency of Greek commercial banks: the impact of credit risk, off-balance sheet activities, and international operations. *Research in International Business and Finance*, **22**, 301–318.

Sarkis, J. (2000) A comparative analysis of DEA as a discrete alternative multiple criteria decision tool. *European Journal of Operational Research*, **123**, 543–557.

Tervonen, T. and Figueira, J. (2008) A survey on stochastic multicriteria acceptability analysis methods. *Journal of Multi-Criteria Decision Analysis*, **15**, 1–14.

12

A quantile regression approach to bank efficiency measurement*

Anastasia Koutsomanoli-Filippaki,[1] Emmanuel Mamatzakis[2] and Fotios Pasiouras[3]

[1] *Bank of Greece, Greece*
[2] *Department of Business and Management, University of Sussex, UK*
[3] *University of Surrey, UK* and *Technical University of Crete, Greece*

12.1 Introduction

The estimation of bank efficiency, whether at the branch or at the institution level, is a topic that has attracted considerable attention in the literature (see Berger and Humphrey, 1997; Berger, 2007; Fethi and Pasiouras, 2010). Over the years, a number of topics have been explored including the relationship of efficiency with ownership (Miller and Parkhe, 2002), regulations (Pasiouras, 2008), institutional development (Lensink, Meesters, and Naaborg, 2008), off-balance-sheet activities (Siems and Clark, 1997), risk (Berger and DeYoung, 1997), stock returns (Chu and Lim, 1998), mergers (Avkiran, 1999), and bank failure (Wheelock and Wilson, 2000), to name a few.

Despite the plethora of efficiency studies in the banking literature, there is no consensus on the preferred approach for the empirical estimation of the frontier (production, cost, profit, etc.) of fully efficient firms. For example, Berger and Humphrey (1997) mention that 69 of the banking studies in their survey used nonparametric methods and 61 used parametric methods. In general, data envelopment analysis (DEA) is the most widely used nonparametric technique, and stochastic frontier analysis (SFA) is the most frequently employed parametric technique. Each approach has its advantages and disadvantages, and as a result it has supporters

---

*The views expressed in this chapter are those of the author(s) and do not necessarily represent those of the Bank of Greece.

Efficiency and Productivity Growth: Modelling in the Financial Services Industry, First Edition. Edited by Fotios Pasiouras.
© 2013 John Wiley & Sons, Ltd. Published 2013 by John Wiley & Sons, Ltd.

and equally dedicated opponents. In general, the econometric approaches have the advantage of allowing for noise in the measurement of the efficiency, but their disadvantages are the imposition of a particular production function form and the requirement of an assumption about the distribution of efficiency. In contrast, the main advantages of DEA are that (a) it avoids the need for a priori specification of function forms and (b) it does not require any assumption to be made about the distribution of inefficiency. On the other hand, the short-comings of DEA are that (a) it assumes data to be free of measurement error, (b) it is sensitive to outliers, and (c) having few observations and many inputs and/or outputs will result in many firms appearing on the frontier.

Lately, a third approach was proposed in the literature, namely, quantile regression analysis. This technique has been frequently employed in the econometrics literature; however, there are only a few studies in the context of efficiency estimation, examining among others the efficiency of hotels (Bernini, Freo, and Gardini, 2004), nursing facilities (Knox, Blankmeyer, and Stutzman, 2007), dairy farms (Chidmi, Solís, and Cabrera, 2011), and check processing operations (Wheelock and Wilson, 2008). In the banking sector, this technique was applied only very recently, with a handful number of studies examining US (Wheelock and Wilson, 2009), German (Behr, 2010), and European banks (Koutsomanoli-Filippaki and Mamatzakis, 2011).

Quantile regression can be particularly useful in the context of efficiency analysis. First, this approach is well-suited for efficiency estimations when there is considerable heterogeneity in the firm-level data (Behr, 2010). In other words, the estimation of conditional quantiles is more robust against outliers, and it also provides the means to obtain different slope parameters describing the production of efficient firms rather than average firms. Furthermore, as discussed in Liu, Laporte, and Ferguson (2008), quantile regression requires an assumption about the functional form of the production frontier (unlike DEA); however, it does not require the imposition of a particular form on the distribution of the inefficiency terms (unlike SFA). Additionally, quantile regression avoids the criticism against DEA of not allowing for random error.

This chapter aims to provide an overview of this promising alternative approach, along with an empirical application in a large international dataset, including 1520 commercial banks operating in 73 countries, between 2000 and 2006. Apparently, with such a wide coverage, our sample is quite heterogeneous both in terms of the countries' development as well as in terms of the banks' characteristics. Given the increasing number of cross-country studies, our approach provides an ideal setting for the application of quantile regression that can be particularly useful in samples with large bank heterogeneity. Section 12.2 discusses the methodological framework of quantile regression. Section 12.3 presents the empirical results. The concluding remarks are discussed in Section 12.4.

12.2   Methodology and data

12.2.1   Methodology

Quantile regression is a statistical technique intended to estimate, and perform inference about, conditional quantile functions. This analysis is particularly useful when the conditional distribution does not have a standard shape, such as an asymmetric, fat-tailed, or truncated distribution. Consequently, quantile regression was recently employed in various strands of the finance and banking literature, including banking risk and regulations (Klomp and de Haan, 2012), the herding behavior in stock markets (Chiang, Li, and Tan, 2010), capital structure (Fattouh, Scaramozzino, and Harris, 2005), bankruptcy prediction (Li and Miu, 2010), ownership and profitability (Li, Sun, and Zou, 2009), the relationship between stock price index and exchange rate (Tsai, 2012), and credit risk (Schechtman and Gaglianone, 2012).[1] In the context of our study, quantile analysis provides an ideal tool to examine bank efficiency heterogeneity, departing from conditional mean models. In other words, the quantile regression approach allows efficient or almost efficient banks to employ production relations that may differ strongly from those of average or less-efficient banks, and in a sense it provides the means for the proper comparison with truly 'benchmark' banks that fall within the chosen quantile (Behr, 2010).

In detail, a quantile regression involves the estimation of conditional quantile functions, that is, models in which quantiles of the conditional distribution of the dependent variable are expressed as functions of observed covariates (Koenker and Hallock, 2000). Using standard formulation, the linear regression model takes the form

$$y_{it} = \mathbf{x}_{it}\beta_{\phi} + \varepsilon_{i\phi},\tag{12.1}$$

where $\varphi \in (0,1)$, $\mathbf{x}_{it}$ is a $K \times 1$ vector of regressors, $\mathbf{x}_{it}\,\beta_{\varphi}$ denotes the φth sample quantile of y (conditional on vector $\mathbf{x}_{it}$), and $\varepsilon_{i\varphi}$ is a random error whose conditional quantile distribution equals 0.

The objective function for efficient estimation of β corresponding to the φth quantile of the dependent variable (y) can be expressed by the following minimization problem:

$$\min_{\beta}\frac{1}{n}\left\{ \sum_{i:y_i \geq x_i\beta} \varphi\left|y_i - x_i\beta\right| + \sum_{i:y_i \leq x_i\beta}(1-\varphi)\left|y_i - x_i\beta\right| \right\}\tag{12.2}$$

which is solved via linear programming. Note that the median estimator, that is, quantile regression estimator for $\varphi = 0.5$, is similar to the least-squares estimator for Gaussian linear models, except that it minimizes the sum of absolute residuals rather than the sum of squared residuals.

For the estimation of efficiency, we opt for a parametric methodology and employ the distribution-free approach (DFA), developed by Berger (1993) who follows Schmidt and Sickless (1984). This approach is a particularly attractive technique due to its flexibility, as it does not impose a priori any specific shape on the distribution of efficiency (DeYoung, 1997). Instead, the DFA methodology assumes that the inefficiency of each bank remains constant across the sample period and that the random error averages out over time.

By averaging the residuals to estimate bank-specific efficiency, DFA estimates how well a bank tends to do relative to its competitors over a range of conditions over time, rather than its relative efficiency at any one point in time (DeYoung, 1997). Berger and Humphrey (1997) argue that the DFA approach gives a better indication of a bank's longer-term performance by averaging over a number of conditions than any of the other methods. Therefore, under DFA, a panel data is required, and only panel estimates of efficiency over the entire time interval are available.[2]

---

[1] For a general discussion of quantile regression see Koenker and Hallock (2001).

[2] However, the rationality of the DFA assumptions depends on the length of the period studied. Choosing a very short period may leave large amounts of random error in the averaged residuals, in which case random error would be attributed to inefficiency. On the other hand, if a very long period is chosen, the firm's average efficiency might not be constant over the time period because of the changes in environmental conditions, making it less meaningful (DeYoung, 1997). Following the empirical literature, the seven-year period of our sample reasonably balances these concerns.

For the estimation of the DFA, we opt for the widely used translog cost function, which gives us the following specification:

$$\ln C_i = \alpha_0 + \sum_i a_i \ln \mathbf{P}_i + \sum_i \beta_i \ln \mathbf{Y}_i + \frac{1}{2}\sum_i \sum_j a_{ij} \ln \mathbf{P}_i \ln \mathbf{P}_j + \frac{1}{2}\sum_i \sum_j \beta_{ij} \ln \mathbf{Y}_i \ln \mathbf{Y}_j$$

$$+ \sum_i \sum_j \delta_{ij} \ln \mathbf{P}_i \ln \mathbf{Y}_j + \sum_i \varphi_i \ln \mathbf{N}_i + \frac{1}{2}\sum_i \sum_j \varphi_{ij} \ln \mathbf{N}_i \ln \mathbf{N}_j + \sum_i \sum_j \xi_{ij} \ln \mathbf{P}_i \ln \mathbf{N}_j \quad (12.3)$$

$$+ \sum_i \sum_j \zeta_{ij} \ln \mathbf{Y}_i \ln \mathbf{N}_j + kD_i + \ln v_i + \ln u_i,$$

where all variables are expressed in natural logs.[3] C_{it} denotes the observed total cost for bank i, $\mathbf{P}_i$ is a vector of input prices, $\mathbf{Y}_j$ is a vector of bank outputs, and $\mathbf{N}$ is a vector of fixed netputs.[4] Moreover, because structural conditions in banking and general macroeconomic conditions may generate differences in banking efficiency from country to country, we also include country effects in the estimation of the cost frontier. Note that u_i is the bank-specific efficiency factor and v_i is the random error term. All elements of Equation (12.3) are allowed to vary across time with the exception of u_i, which remains constant for each bank by assumption. In the estimation, the $\ln v_i$ and $\ln u_i$ terms are treated as composite error terms, that is, $\ln \hat{\varepsilon}_i = \ln \hat{v}_i + \ln \hat{u}_i$. Once estimated, the residuals, $\ln \hat{\varepsilon}_i$, are averaged across T years for each bank i. The averaged residuals are estimates of the X-efficiency terms, $\ln u_i$, because the random error terms, $\ln v_i$, tend to cancel each other out in the averaging. Thus, the bank's i efficiency is defined as

$$\text{EFF}_i = \frac{\exp\left[\hat{f}(p_i y_i)\right]\exp\left[(\ln \hat{u}_{\min})\right]}{\exp\left[\hat{f}(p_i y_i)\right]\exp\left[(\ln \hat{\mathbf{u}}_i)\right]} = \exp\left[(\ln \hat{u}_{\min} - \ln \hat{\mathbf{u}}_i)\right] \quad (12.4)$$

where $\ln \hat{\mathbf{u}}_i$ is the residual vector after having averaged over time, and $\ln \hat{u}_{\min}$ is the most efficient bank in the sample.

12.2.2   Data and specification of the frontier

We start the construction of our dataset by considering all the commercial banks in the Bankscope database. Once we exclude the banks for which we do not have complete data for the period of our study, we end up with a sample of 1520 commercial banks operating in 73 countries, between 2000 and 2006. This sample includes domestic and foreign banks as well as listed and unlisted banks. It is worth emphasizing that there exists a certain degree of heterogeneity across banks as they operate in quite different environments in terms of regulations, institutional infrastructure, market characteristics, and overall development. Despite the bank heterogeneity, it is not uncommon for the recent bank efficiency literature to use such large international datasets (e.g., Lensink, Meesters, and Naaborg, 2008; Barth et al., 2010), making our setting ideal for testing the usefulness of a quantile approach. As discussed in the next section, to reveal potential differences across different levels of development,

---

[3] Standard homogeneity and symmetry restrictions are imposed: $\sum_i a_i = 1$, $\sum_i a_{ij} = 0$, $\sum_i \delta_{ij} = 0$, $\sum_i \xi_{ij} = 0$, $\alpha_{im} = \alpha_{mi}$ and $\alpha_{jk} = \alpha_{kj}$, $\forall i, j, k, m$.
[4] Fixed netputs are quasi-fixed quantities of either inputs or outputs that affect variable costs.

we combine information from the International Monetary Fund (IMF) and the European Bank for Reconstruction and Development (EBRD), and we classify banks into four groups according to the level of development of the country that they operate, namely, major advanced countries, advanced countries, transition countries, and developing countries. Moreover, to examine the impact of the aforementioned country-specific characteristics, we use information from various sources such as the Worldwide Governance Indicators database, the World Bank database on Bank Regulation and Supervision, and the World Bank database on Financial Development and Structure, and we perform second-stage regressions. These results are discussed in detail in Section 12.3.2.

There is a debate in the literature on the selection of inputs and outputs, and in particular on the appropriate treatment of deposits (Berger and Humphrey, 1997). Following Dietsch and Lozano-Vivas (2000), Maudos et al. (2002), Pasiouras, Tanna, and Zopounidis (2009), and others, we adopt the value-added approach which suggests using deposits as outputs since they imply the creation of value added. Therefore, we use the following three outputs: loans (Y_1), other earning assets (Y_2), and total deposits (Y_3). Furthermore, consistent with numerous studies on bank efficiency, we select the following three input prices: cost of borrowed funds (P_1), calculated as the ratio of interest expenses to total deposits; cost of labor (P_2), calculated by dividing the personnel expenses by total assets; and cost of physical capital (P_3), calculated by dividing the expenditures on plant and equipment (i.e., overhead expenses net of personnel expenses) by fixed assets. Thus, our approach recognizes that deposits have both input and output characteristics, the first captured through the inclusion of the interest expense paid on deposits in the input prices vector and the second captured through the stock of deposits in the output vector.

Furthermore, we normalize the dependent variable and the three outputs by equity. Berger et al. (2000) point out that the normalization by equity capital controls for heteroskedasticity, reduces scale biases in estimation, provides the grounds for a more economic interpretation, and controls for financial leverage.

Additionally, to account for technological differences across different levels of a country's overall development, we use dummy variables to distinguish between major advanced, advanced, transition, and developing countries (Lozano-Vivas and Pasiouras, 2010).

12.3   Empirical results

12.3.1   Cost efficiency estimates

We calculate cost efficiency scores for each bank in our sample using the DFA and compare these scores across quantiles and across different levels of development. In order to cover as wide a range of quantiles as possible, we run regressions for quantiles 0.05, 0.25, 0.75, and 0.95.

Figure 12.1 presents the average efficiency scores across quantiles. There are three interesting observations to be made. First, there is a remarkable variation across quantiles. In other words, the average efficiency score for the whole sample ranges from 0.3704 for quantile 0.95–0.9113 for quantile 0.05. Second, cost efficiency estimates across quantiles, and particularly in the tail of the distribution, differ substantially from the conditional mean (OLS) point estimate of efficiency, which is approximated by quantile 0.5 and equals 0.7573. Thus, quantile regression analysis provides a more comprehensive picture of the underlying range of disparities in cost efficiency than the classical estimation. Third, the average efficiency is monotonically decreasing as it follows a negative trend at higher order of quantiles. In other

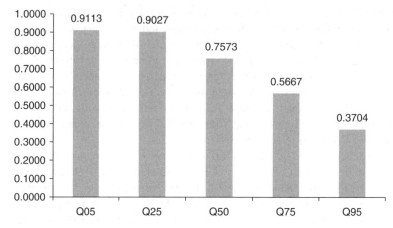

Figure 12.1   Average cost efficiency scores by quantile.

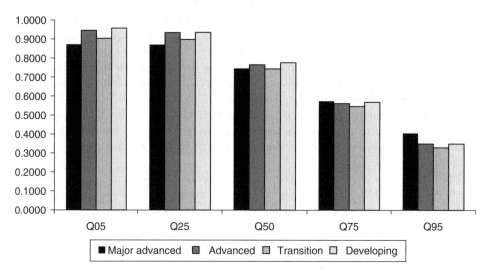

Figure 12.2   Average cost efficiency scores by country development level and quantiles.

words, cost efficiency is estimated at around 0.9113 for quantile 0.05, decreases to 0.9027 for quantile 0.25, dropping further to 0.7573 and 0.5667 for quantiles 0.50 and 0.75, respectively, while it reaches its minimum value at 0.3704 when the cost function is calculated at the 0.95 quantile. In general, these results confirm the ones of Koutsomanoli-Filippaki and Mamatzakis (2011) for European banks; however, the minimum cost efficiency in our case is considerably lower than the one recorded in their study.

Figure 12.2 presents a disaggregation of the estimated cost efficiencies by the level of a country's overall development (i.e., major advanced, advanced, etc.). First, this disaggregation confirms the aforementioned negative trend at higher order of quantiles, irrespective of the level of overall country development. Second, there appears to be some variability in the underlying relationship between the level of a country's overall development and bank efficiency. In other words, we observe that banks operating in major advanced countries appear to be less cost efficient when looking at the 0.05 and 0.25 quantiles, and more cost efficient when looking at the 0.75 and 0.95 quantiles. Additionally, the scores appear to be of similar magnitude when looking at the 0.50 quantile. This would imply that resolving into the classical OLS mean regression analysis would result in the loss of valuable information regarding the bank performance across the world. Finally, the average cost efficiency of banks from major advanced countries is always higher than that of banks from all other categories at the highest quantile, but the former becomes lower than that of the latter at very low quantiles, that is, Q5 and Q25. The differences in cost efficiency between advanced, transition, and developing countries remain quite stable across quantiles, while they are not large in magnitude. It is worth noticing that countries in transition have a record of the lowest bank performance in our sample. This evidence suggests that reform efforts during transition could come at the expense of lowering bank cost efficiency, though once the country becomes advanced, the benefits of these reforms translate into higher scores in cost efficiency.

Table 12.1 presents Spearman's rank correlation coefficients between the cost efficiency scores obtained across different quantiles. As expected, there are similarities and differences depending on whether we compare the rankings that are obtained from estimations at neighboring or distant quantiles.

For example, estimations at the 0.05 and 0.25 quantiles rank the banks in approximately the same way. Additionally, there are moderate correlations between estimations at the 0.50 quantile and the 0.75 quantile, as well as between the 0.75 quantile and the 0.95 quantile. However, there are remarkable differences between estimations at the 0.05 and 0.95 quantiles, as well as between the 0.25 and 0.95 quantiles, as it becomes evident by the negative coefficients. The correlations by level of development reveal the existence of differences across the group of countries. For example, the estimations for major advanced and transition countries are similar to the ones for the whole sample. Nonetheless, in the case of developing countries, we observe that not only the correlation coefficients tend to be higher but there is also a moderate positive correlation between estimations at 0.05 and 0.95 quantiles. In the case of advanced countries, the correlations are similar to the ones of developing countries, although lower in magnitude.

12.3.2   Determinants of cost efficiency

To shed more light on our analysis, we also perform second-stage regressions, where cost efficiency scores derived at different quantiles are regressed on a set of environmental variables. Following recent studies by, among others, Pasiouras (2008), Pasiouras, Tanna, and Zopounidis (2009), and Lozano-Vivas and Pasiouras (2010), we account for regulatory conditions using four indices that control for capital requirements (CAPRQ), private monitoring (PRMONIT), supervisory power (SPOWER), and activity restrictions (ACTRS). To capture the macroeconomic conditions, we use the inflation rate (INFL) and real GDP growth (GDPGR). To account for industry conditions, we use the ratio of bank claims to the private sector over GDP (CLAIMS) and concentration in the banking sector (CONC). Finally, to control for the institutional development (INSTDEV), we use the average of six indicators measuring voice and accountability, political stability, government effectiveness, regulatory quality, rule of law, and control of corruption (see, e.g., Lensink, Meesters, and Naaborg, 2008). Further information about these variables is provided in Appendix 12.A.

The results in Table 12.2 reveal various interesting findings. First, the 0.75 and 0.95 quantiles appear to be of significance for the direction of the impact of various environmental variables

Table 12.1 Spearman's rank correlation coefficients.

All sample	Q05	Q25	Q50	Q75	Q95
Q05	1.000				
Q25	0.908***	1.000			
Q50	0.692***	0.794***	1.000		
Q75	0.230***	0.356***	0.644***	1.000	
Q95	−0.217***	−0.088***	0.203***	0.580***	1.000
Major advanced countries	Q05	Q25	Q50	Q75	Q95
Q05	1.000				
Q25	0.920***	1.000			
Q50	0.661***	0.794***	1.000		
Q75	−0.013	0.147***	0.477***	1.000	
Q95	−0.567***	−0.417***	−0.045	0.574***	1.000
Advanced countries	Q05	Q25	Q50	Q75	Q95
Q05	1.000				
Q25	0.414***	1.000			
Q50	0.659***	0.348***	1.000		
Q75	0.581***	0.266***	0.800***	1.000	
Q95	0.464***	0.634***	0.620***	0.545***	1.000
Transition countries	Q05	Q25	Q50	Q75	Q95
Q05	1.000				
Q25	0.709***	1.000			
Q50	0.439***	0.744***	1.000		
Q75	0.172**	0.665***	0.699***	1.000	
Q95	−0.118*	0.514***	0.620***	0.907***	1.000
Developing countries	Q05	Q25	Q50	Q75	Q95
Q05	1.000				
Q25	0.948***	1.000			
Q50	0.862***	0.911***	1.000		
Q75	0.717***	0.791***	0.915***	1.000	
Q95	0.587***	0.697***	0.773***	0.848***	1.000

*** Statistically significant at the 1% level.
** Statistically significant at the 5% level.
* Statistically significant at the 10% level.

on cost efficiency. Moreover, the positive impact of CAPRQ on cost efficiency that is reported at the 0.05 and 0.25 quantiles is reversed at the 0.75 and 0.95 quantiles. This change in the sign of CAPRQ has important implications as it shows that capital requirements have a positive influence on the more efficient banks and a negative impact on the less efficient banks. Thus, either the most efficient banks are capable of turning the regulatory burden imposed by higher capital requirements to their benefit or supervisors distinguish between efficient and inefficient banks. The latter would be in line with the results of DeYoung,

Table 12.2   Second-stage regressions.

	Q05	Q25	Q50	Q75	Q95
CAPRQ	0.0006**	0.0004*	0.0000	−0.0004*	−0.0008**
	(0.026)	(0.084)	(0.889)	(0.060)	(0.022)
OFFPR	−0.0000	−0.0001	−0.0002	−0.0001	−0.0000
	(0.868)	(0.626)	(0.105)	(0.512)	(0.965)
PRMONIT	−0.0003	−0.0002	−0.0003	−0.0002	0.0001
	(0.565)	(0.639)	(0.375)	(0.572)	(0.915)
ACTRS	−0.0004	−0.0007	−0.0007	−0.0012	−0.0010
	(0.694)	(0.482)	(0.257)	(0.134)	(0.438)
INFL	0.0001*	0.0001	0.0001	−0.0000	−0.0000
	(0.087)	(0.136)	(0.518)	(0.891)	(0.580)
GDPGR	−0.0005***	−0.0004***	−0.0002**	0.0001	0.0004***
	(0.000)	(0.000)	(0.025)	(0.176)	(0.006)
CLAIMS	−0.0256***	−0.0204***	−0.0051***	0.0120***	0.0316***
	(0.000)	(0.000)	(0.001)	(0.000)	(0.000)
CONC	−0.0379***	−0.0296***	−0.0108***	0.0080***	0.0232***
	(0.000)	(0.000)	(0.000)	(0.004)	(0.000)
INSTDEV	0.0130***	0.0089***	0.0012	−0.0055***	−0.0107***
	(0.000)	(0.000)	(0.480)	(0.008)	(0.002)
Constant	0.9378***	0.9267***	0.7716***	0.5662***	0.3514***
	(0.000)	(0.000)	(0.000)	(0.000)	(0.000)

p-values in parentheses; results obtained from fixed effects estimations with the dependent variable being the cost efficiency at different quantiles; variables are defined in Appendix 12.A.
*** Statistically significant at the 1% level.
** Statistically significant at the 5% level.
* Statistically significant at the 10% level.

Hughes, and Moon (2001) who conclude in their US study that '…regulators impose greater discipline and higher distress costs on inefficient banks than on efficient banks' (p. 275). The impact of the institutional development also differs across quantiles, being positive for the more efficient banks and negative for the less efficient banks.

In contrast, GDPGR, CLAIMS, and CONC exercise a negative influence on the efficiency of banks in the case of the 0.05 and 0.50 quantiles and a positive impact in the case of the 0.75 and 0.95 quantiles. The negative impact of GDP growth is in line with the findings of Maudos et al. (2002) who argue that under expansive demand conditions, banks feel less pressured to control their costs and are therefore less cost efficient. However, our results illustrate that there is a turning point after which banks become cautious and take advantage of the growth in the economy so that they can operate more efficiently. A similar picture that emerges in the case of concentration could explain why the results in the literature, as for the impact of concentration on efficiency, are mixed.[5] Overall, our findings indicate that an OLS analysis, which is close to the median quantile (0.5), would be misleading, as it would report

---

[5] For example, Fries and Taci (2005) find that there is no significant association between concentration and cost efficiency, Grigorian and Manole (2006) report a positive association, and Maudos et al. (2002) find a negative association.

an insignificant coefficient for CAPRQ and INSTDEV, and it would also ignore that the impact of environmental factors can vary across different levels of efficiency.

12.4   Conclusions

This chapter presents an application of quantile regression analysis in estimating the cost efficiency of 1520 commercial banks operating in 73 countries during 2000–2006. This approach allows us to estimate banks' cost function for various quantiles of the conditional distribution and to examine the tail behaviors of that distribution. In further analysis, we also examine whether and how the impact of environmental factors differs across the various quantiles of efficiency. The employed methodological framework is of particular importance in light of the heterogeneity in bank efficiency across various countries.

The results can be summarized as follows. First, there is a remarkable variation of efficiency across quantiles. Second, the efficiency estimates across quantiles, and particularly in the tail of the distribution, differ substantially from the conditional mean (OLS) point estimate of efficiency (i.e., quantile 0.5). Third, the average efficiency is monotonically decreasing. We confirm this negative trend at higher order of quantiles for all levels of overall country development (i.e., major advanced, advanced, transition, and developing countries). Fourth, there appears to be some variability regarding the underlying relationship between the level of a country's overall development and bank cost efficiency. Fifth, the results of Spearman's rank correlation coefficients show that there exists variability in the ranking of banks depending on whether we compare estimations from neighboring or distant quantiles. In this case, the results differ among different levels of overall country development. Sixth, the estimations of the second-stage regressions illustrate that there is a turning point as for the direction of the impact of various variables on cost efficiency. Furthermore, our findings indicate that an OLS analysis, which is close to the median quantile (0.5), would be misleading. Overall, we conclude that quantile regressions, by permitting the estimation of various quantile functions of the underlying conditional distribution, provide a more comprehensive picture of the underlying relationships.

Appendix 12.A: Information on variables

Panel A: Variables in the frontier function		Source
TC	Total cost	Bankscope
Q1	Loans	Bankscope
Q2	Other earning assets	Bankscope
Q3	Deposits	Bankscope
W1	Interest expenses/deposits	Bankscope
W2	Personnel expenses/total assets	Bankscope
W3	(Overheads – personnel expenses)/fixed assets	Bankscope
EQ	Equity	Bankscope
MADV	Dummy for major advanced economies	IMF
ADV	Dummy for advanced economies	IMF
TRANS	Dummy for transition economies	EBRD

Panel B: Variables in second-stage regressions

CAPRQ	Index of capital requirements, with higher values indicating higher stringency. This variable is determined by adding 1 if the answer is 'yes' to questions 1–6 and 0 otherwise, while the opposite occurs in the case of questions 7 and 8 (i.e., yes=0, no=1). (1) Is the minimum required capital asset ratio risk-weighted in line with Basel guidelines? (2) Does the ratio vary with market risk? (3–5) Before minimum capital adequacy is determined, which of the following are deducted from the book value of capital: (a) Market value of loan losses not realized in accounting books? (b) Unrealized losses in securities portfolios? (c) Unrealized foreign exchange losses? (6) Are the sources of funds to be used as capital verified by the regulatory/supervisory authorities? (7) Can the initial or subsequent injections of capital be done with assets other than cash or government securities? (8) Can initial disbursement of capital be done with borrowed funds?	WB database on Bank Regulation and Supervision
OFFPR	Index of supervisory power, with higher values indicating greater power. This variable is determined by adding 1 if the answer is yes and 0 otherwise, for each one of the following 14 questions: (1) Does the supervisory agency have the right to meet with external auditors to discuss their report without the approval of the bank? (2) Are auditors required by law to communicate directly to the supervisory agency any presumed involvement of bank directors or senior managers in illicit activities, fraud, or insider abuse? (3) Can supervisors take legal action against external auditors for negligence? (4) Can the supervisory authorities force a bank to change its internal organizational structure? (5) Are off-balance-sheet items disclosed to supervisors? (6) Can the supervisory agency orderthe bank's directors or management to constitute provisions to cover actual or potential losses? (7) Can the supervisory agency suspend the directors' decision to distribute dividends? (8) Can the supervisory agency suspend the directors' decision to distribute bonuses? (9) Can the supervisory agency suspend the directors' decision to distribute management fees? (10) Can the supervisory agency supersede bank shareholder rights and declare the bank insolvent? (11) Does banking law allow the supervisory agency or any other government agency (other than court) to suspend some or all ownership rights of a problem bank? (12) Regarding bank restructuring and reorganization, can the supervisory agency or any other government agency (other than court) supersede shareholder rights? (13) Regarding bank restructuring and reorganization, can the supervisory	WB database on Bank Regulation and Supervision

(*continued overleaf*)

(*continued*)

	agency or any other government agency (other than court) remove and replace management? (14) Regarding bank restructuring and reorganization, can the supervisory agency or any other government agency (other than court) remove and replace directors?	
PRMONIT	Index of private monitoring, with greater values indicating higher incentives for private monitoring. This variable is determined by adding 1 if the answer is 'yes' to questions 1–6 and 0 otherwise, while the opposite occurs in the case of questions 7 and 8 (i.e., yes=0, no=1). (1) Is subordinated debt allowable (or required) as part of capital? (2) Are financial institutions required to produce consolidated accounts covering all bank and any nonbank financial subsidiaries? (3) Are off-balance sheet items disclosed to public? (4) Must banks disclose their risk management procedures to public? (5) Are directors legally liable for erroneous/misleading information? (6) Do regulations require credit ratings for commercial banks? (7) Does accrued, though unpaid, interest/principal enter the income statement while loan is nonperforming? (8) Is there an explicit deposit insurance protection system?	WB database on Bank Regulation and Supervision
ACTRS	Index of restrictions on bank activities. The score for this variable is determined on the basis of the level of regulatory restrictiveness for bank participation in: (1) securities activities, (2) insurance activities, (3) real estate activities, and (4) bank ownership of nonfinancial firms. These activities can be unrestricted, permitted, restricted, or prohibited that are assigned the values of 1, 2, 3, and 4, respectively. We use an overall index by calculating the average value over the four categories.	WB database on Bank Regulation and Supervision
INFL	Annual rate of inflation (%)	GMID
GDPGR	Real GDP growth (%)	GMID
CLAIMS	Bank claims on the private sector/GDP	GMID
CONC	Assets of the three largest banks as a share of assets of all commercial banks in the same country	WB database on Financial Development and Structure
INSTDEV	Index of overall institutional development. It can take values from −2.5 to 2.5 with higher values indicating a better outcome. It is estimated as the average of six indicators measuring voice and accountability, political stability, government effectiveness, regulatory quality, rule of law, and control of corruption.	Worldwide Governance Indicators database

IMF, International Monetary Fund; EBRD, European Bank for Reconstruction and Development; WB, World Bank; GMID, Global Market Information Database.

References

Avkiran, N.K. (1999) The evidence on efficiency gains: The role of mergers and the benefits to the public. *Journal of Banking and Finance*, **23**, 991–1013.

Barth, J.R., Lin, C., Ma, Y. et al. (2010) Do bank regulation, supervision and monitoring enhance or impede bank efficiency? http://dx.doi.org/10.2139/ssrn.1579352 (accessed on 29 November 2012).

Behr, A. (2010) Quantile regression for robust bank efficiency score estimation. *European Journal of Operational Research*, **200**, 568–581.

Berger, A.N. (1993) Distribution-free estimates of efficiency in the US banking industry and tests of the standard distribution assumptions. *Journal of Productivity Analysis*, **4**, 261–292.

Berger, A.N. (2007) International comparisons of banking efficiency. *Financial Markets, Institutions & Instruments*, **16**, 119–144.

Berger, A.N. and DeYoung, R. (1997) Problem loans and cost efficiency in commercial banks. *Journal of Banking and Finance*, **21**, 849–870.

Berger, A.N. and Humphrey, D. (1997) Efficiency of financial institutions: international survey and direction of future research. *European Journal of Operational Research*, **98**, 175–212.

Berger, A.N., Cummins, J.D., Weiss, M.A., and Zi, H. (2000) Conglomeration versus strategic focus: Evidence from the insurance industry. *Journal of Financial Intermediation*, **9**, 323–362.

Bernini, C., Freo, M., and Gardini, A. (2004) Quantile estimation of frontier production function. *Empirical Economics*, **29**, 373–381.

Chiang, T.C., Li, J., and Tan, L. (2010) Empirical investigation of herding behavior in Chinese stock markets: evidence from quantile regression analysis. *Global Finance Journal*, **21**, 111–124.

Chidmi, B., Solís, D., and Cabrera, V.E. (2011) Analyzing the sources of technical efficiency among heterogeneous dairy farms: a quantile regression approach. *Journal of Development and Agricultural Economics*, **3**, 318–324.

Chu, S.F. and Lim, G.H. (1998) Share performance and profit efficiency of banks in an oligopolistic market: evidence from Singapore. *Journal of Multinational Financial Management*, **8**, 155–168.

DeYoung, R. (1997) A diagnostic test for the distribution-free efficiency estimator: an example using U.S. commercial bank data. *European Journal of Operational Research*, **98**, 243–249.

DeYoung, R.E., Hughes, J.P., and Moon, C.-G. (2001) Efficient risk-taking and regulatory covenant enforcement in a deregulated banking industry. *Journal of Economics and Business*, **53**, 255–282.

Dietsch, M. and Lozano-Vivas, A. (2000) How the environment determines banking efficiency: a comparison between French and Spanish industries. *Journal of Banking and Finance*, **24**, 985–1004.

Fattouh, B., Scaramozzino, P., and Harris, L. (2005) Capital structure in South Korea: a quantile regression approach. *Journal of Development Economics*, **76**, 231–250.

Fethi, M.D. and Pasiouras, F. (2010) Assessing bank efficiency and performance with operational research and artificial intelligence techniques: a survey. *European Journal of Operational Research*, **204**, 189–198.

Fries, S. and Taci, A. (2005) Cost efficiency of banks in transition: evidence from 289 banks in 15 postcommunist countries. *Journal of Banking and Finance*, **29**, 55–81.

Grigorian, D.A. and Manole, V. (2006) Determinants of commercial bank performance in transition: an application of data envelopment analysis. *Comparative Economic Studies*, **48**, 497–522.

Klomp, J. and de Haan, J. (2012) Banking risk and regulations: does one size fit all? *Journal of Banking & Finance*, **36**, 3197–3212.

Knox, K.J., Blankmeyer, E.C., and Stutzman, J.R. (2007) Technical efficiency in Texas nursing facilities: a stochastic production frontier approach. *Journal of Economics and Finance*, **9**, 75–86.

Koenker, R. and Hallock, K.F. (2000) *Quantile Regression: An Introduction*. http://www.econ.uiuc. edu/~roger/research/intro/intro.html (accessed on 29 November 2012).

Koenker, R. and Hallock, K.F. (2001) Quantile regression. *Journal of Economic Perspectives*, **15**, 143–156.

Koutsomanoli-Filippaki, A.I. and Mamatzakis, E.C. (2011) Efficiency under quantile regression: what is the relationship with risk in the EU banking industry? *Review of Financial Economics*, **20**, 84–95.

Lensink, R., Meesters, A., and Naaborg, I. (2008) Bank efficiency and foreign ownership: do good institutions matters? *Journal of Banking and Finance*, **32**, 834–844.

Li, M.-Y.L. and Miu, P. (2010) A hybrid bankruptcy prediction model with dynamic loadings on accounting-ratio-based and market-based information: a binary quantile regression approach. *Journal of Empirical Finance*, **17**, 818–833.

Li, T., Sun, L., and Zou, L. (2009) State ownership and corporate performance: a quantile regression analysis of Chinese listed companies. *China Economic Review*, **20**, 703–716.

Liu, C., Laporte, A., and Ferguson, B.S. (2008) The quantile regression approach to efficiency measurement: insights from Monte Carlo simulations. *Health Economics*, **17**, 1073–1087.

Lozano-Vivas, A. and Pasiouras, F. (2010). The impact of non-traditional activities on the estimation of bank efficiency: international evidence. *Journal of Banking and Finance*, **34**, 1436–1449.

Maudos, J., Pastor, J.M., Perez, F., and Quesada, J. (2002) Cost and profit efficiency in European banks. *Journal of International Financial Markets, Institutions and Money*, **12**, 33–58.

Miller, S.R. and Parkhe, A. (2002) Is there a liability of foreignness in global banking? An empirical test of banks' X-efficiency. *Strategic Management Journal*, **23**, 55–75.

Pasiouras, F. (2008) International evidence on the impact of regulations and supervision on banks' technical efficiency: an application of two-stage data envelopment analysis. *Review of Quantitative Finance and Accounting*, **30**, 187–223.

Pasiouras, F., Tanna, S., and Zopounidis, C. (2009) The impact of banking regulations on banks' cost and profit efficiency: cross-country evidence. *International Review of Financial Analysis*, **18**, 294–302.

Schechtman, R. and Gaglianone, W.P. (2012) Macro stress testing of credit risk focused on the tails. *Journal of Financial Stability*, **8**, 174–192.

Schmidt, P. and Sickles, R. (1984) Production frontiers and panel data. *Journal of Business and Economic Statistics*, **2**, 367–374.

Siems, T.F. and Clark, J.A. (1997) *Rethinking Bank Efficiency and Regulation: How Off-Balance-Sheet Activities Make a Difference*. Financial Industry Studies, Federal Reserve Bank of Dallas, Dallas, pp. 1–12.

Tsai, I.-C. (2012) The relationship between stock price index and exchange rate in Asian markets: a quantile regression approach. *Journal of International Financial Markets, Institutions and Money*, **22**, 609–621.

Wheelock, D.C. and Wilson, P.W. (2000) Why do banks disappear? The determinants of U.S. bank failures and acquisitions. *Review of Economics and Statistics*, **82**, 127–138.

Wheelock, D.C. and Wilson, P.W. (2008) Non-parametric, unconditional quantile estimation for efficiency analysis with an application to Federal Reserve check processing operations. *Journal of Econometrics*, **145**, 209–225.

Wheelock, D.C. and Wilson, P.W. (2009) Robust nonparametric quantile estimation of efficiency and productivity change in U.S. commercial banking, 1985–2004. *Journal of Business and Economic Statistics*, **27**, 354–368.

Index

Note: Page numbers in *italics* refers to Figures; those in **bold** to Tables.

Efficiency and Productivity Growth: Modelling in the Financial Services Industry, First Edition. Edited by Fotios Pasiouras.
© 2013 John Wiley & Sons, Ltd. Published 2013 by John Wiley & Sons, Ltd.

Statistics in Practice

Human and Biological Sciences

Berger – Selection Bias and Covariate Imbalances in Randomized Clinical Trials
Berger and Wong – An Introduction to Optimal Designs for Social and Biomedical Research
Brown and Prescott – Applied Mixed Models in Medicine, Second Edition
Carpenter and Kenward – Multiple Imputation and its Application
Carstensen – Comparing Clinical Measurement Methods
Chevret (Ed) – Statistical Methods for Dose-Finding Experiments
Ellenberg, Fleming and DeMets – Data Monitoring Committees in Clinical Trials: A Practical
    Perspective
Hauschke, Steinijans & Pigeot – Bioequivalence Studies in Drug Development: Methods and
    Applications
Källén – Understanding Biostatistics
Lawson, Browne and Vidal Rodeiro – Disease Mapping with Win-BUGS and MLwiN
Lesaffre, Feine, Leroux & Declerck – Statistical and Methodological
Aspects of Oral Health Research
Lui – Statistical Estimation of Epidemiological Risk
Marubini and Valsecchi – Analysing Survival Data from Clinical Trials and Observation Studies
Millar – Maximum Likelihood Estimation and Inference: With Examples in R, SAS and ADMB
Molenberghs and Kenward – Missing Data in Clinical Studies
O'Hagan, Buck, Daneshkhah, Eiser, Garthwaite, Jenkinson, Oakley & Rakow – Uncertain
    Judgements: Eliciting Expert's Probabilities
Parmigiani – Modeling in Medical Decision Making: A Bayesian Approach
Pintilie – Competing Risks: A Practical Perspective
Senn – Cross-over Trials in Clinical Research, Second Edition
Senn – Statistical Issues in Drug Development, Second Edition
Spiegelhalter, Abrams and Myles – Bayesian Approaches to Clinical Trials and Health-Care
    Evaluation
Walters – Quality of Life Outcomes in Clinical Trials and Health-Care Evaluation
Welton, Sutton, Cooper and Ades – Evidence Synthesis for Decision Making in Healthcare
Whitehead – Design and Analysis of Sequential Clinical Trials, Revised Second Edition
Whitehead – Meta-Analysis of Controlled Clinical Trials
Willan and Briggs – Statistical Analysis of Cost Effectiveness Data
Winkel and Zhang – Statistical Development of Quality in Medicine

Earth and Environmental Sciences

Buck, Cavanagh and Litton – Bayesian Approach to Interpreting Archaeological Data
Chandler and Scott – Statistical Methods for Trend Detection and Analysis in the
    Environmental Statistics

Glasbey and Horgan – Image Analysis in the Biological Sciences
Haas – Improving Natural Resource Management: Ecological and Political Models
Helsel – Nondetects and Data Analysis: Statistics for Censored Environmental Data
Illian, Penttinen, Stoyan, H and Stoyan D-Statistical Analysis and Modelling of Spatial Point Patterns
Mateu and Müller (Eds) – Spatio-temporal design: Advances in efficient data acquisition
McBride – Using Statistical Methods for Water Quality Management Webster and Oliver – Geostatistics for Environmental Scientists, Second Edition
Wymer (Ed) – Statistical Framework for Recreational Water Quality Criteria and Monitoring

Industry, Commerce and Finance

Aitken – Statistics and the Evaluation of Evidence for Forensic Scientists, Second Edition
Balding – Weight-of-evidence for Forensic DNA Profiles
Brandimarte – Numerical Methods in Finance and Economics: A MATLAB-Based Introduction, Second Edition
Brandimarte and Zotteri – Introduction to Distribution Logistics
Chan – Simulation Techniques in Financial Risk Management
Coleman, Greenfield, Stewardson and Montgomery (Eds) – Statistical Practice in Business and Industry
Frisen (Ed) – Financial Surveillance
Fung and Hu – Statistical DNA Forensics
Gusti Ngurah Agung – Time Series Data Analysis Using EViews
Kenett (Eds) – Operational Risk Management: A Practical Approach to Intelligent Data Analysis
Kenett (Eds) – Modern Analysis of Customer Surveys: With Applications using R
Kruger and Xie – Statistical Monitoring of Complex Multivariate Processes: With Applications in Industrial Process Control
Jank and Shmueli (Ed.) – Statistical Methods in e-Commerce Research Lehtonen and Pahkinen – Practical Methods for Design and Analysis of Complex Surveys, Second Edition
Ohser and Mücklich – Statistical Analysis of Microstructures in Materials Science
Pasiouras (Ed.) – Efficiency and Productivity Growth: Modelling in the Financial Services Industry
Pourret, Naim & Marcot (Eds) – Bayesian Networks: A Practical Guide to Applications
Taroni, Aitken, Garbolino and Biedermann – Bayesian Networks and Probabilistic Inference in Forensic Science
Taroni, Bozza, Biedermann, Garbolino and Aitken – Data Analysis in Forensic Science